Driver & Vehicle
Standards
Agency

The OFFICIAL DVSA THEORY TEST for Drivers of Large Vehicles

London: TSO

Published with the permission of the Driver and Vehicle Standards Agency on behalf of the Controller of Her Majesty's Stationery Office.

First published 1996
Fourteenth edition 2017
Fourth impression 2018

ISBN 978 0 11 553505 5

A CIP catalogue record for this book is available from the British Library.

Other titles in the Driving Skills series

The Official DVSA Guide to Driving – the essential skills
The Official DVSA Guide to Better Driving
The Official DVSA Theory Test for Car Drivers
The Official DVSA Theory Test for Car Drivers (DVD-ROM)
The Official DVSA Guide to Learning to Drive
The Official DVSA Guide to Hazard Perception (DVD-ROM)

The Official DVSA Theory Test Kit iPhone/Android App
The Official DVSA Highway Code iPhone App
The Official DVSA Hazard Perception Practice iOS App

The Official DVSA Guide to Riding – the essential skills
The Official DVSA Theory Test for Motorcyclists
The Official DVSA Theory Test for Motorcyclists (DVD-ROM)
The Official DVSA Guide to Learning to Ride
Better Biking – the official DVSA training aid (DVD)

The Official DVSA Guide to Driving Buses and Coaches
The Official DVSA Guide to Driving Goods Vehicles
The Official DVSA Theory Test for Drivers of Large Vehicles (DVD-ROM)
Driver CPC – the official DVSA guide for professional bus and coach drivers
Driver CPC – the official DVSA guide for professional goods vehicle drivers

The Official DVSA Guide to Tractor and Specialist Vehicle Driving Tests
The Official DVSA Theory Test for Approved Driving Instructors (DVD-ROM)

All metric and imperial conversions in this book are approximate.

100% recycled
This book is printed on 100% recycled paper

We're turning over a new leaf.

FSC www.fsc.org
RECYCLED
Paper made from recycled material
FSC® C002151

Find us online

GOV.UK – Simpler, clearer, faster

GOV.UK is the best place to find government services and information for

- car drivers
- motorcyclists
- driving licences
- driving and riding tests
- towing a caravan or trailer
- medical rules
- driving and riding for a living
- online services.

Visit **www.gov.uk** and try it out.

You can also find contact details for DVSA and other motoring agencies like DVLA at **www.gov.uk**

You'll notice that links to **GOV.UK**, the UK's new central government site, don't always take you to a specific page. This is because this new kind of site constantly adapts to what people really search for and so such static links would quickly go out of date. Try it out. Simply search what you need from your preferred search site or from **www.gov.uk** and you should find what you're looking for. You can give feedback to the Government Digital Service from the website.

Message from Mark Winn, the Chief Driving Examiner

Learning to drive a large goods vehicle or passenger-carrying vehicle is a great opportunity, which will bring you added responsibility. Whether you'll be carrying goods or passengers, a comprehensive knowledge of driving and road safety is essential, and passing the theory test is an important step towards becoming a professional driver.

The knowledge and understanding that you've acquired should then be put into practice on the road. When you drive a large vehicle, you should demonstrate your ability to drive safely and set a good example to others. Your attitude and approach should be courteous and considerate. The safety of the goods or passengers you carry, as well as other road users, could depend on it.

You'll never know all the answers and, throughout your driving career, there'll always be more to learn. By applying and developing what you've learned during your training, driving with a responsible attitude and in a fuel-efficient manner, you'll be well on your way to becoming a good professional driver.

I wish you a lifetime of safe driving.

Mark Winn

Mark Winn
Chief Driving Examiner

Contents

Introduction

About the theory test

In this section, you'll learn about

- how to use this book
- getting started
- the theory test
- after the theory test
- using the questions and answers sections
- using this book to learn and revise.

How to use this book

To prove that you have the right knowledge, understanding and attitude to be a safe and responsible driver of large vehicles, you'll need to pass the large vehicle theory test.

It includes

- a multiple choice test, to assess your knowledge of driving theory in general, as well as with specific reference to driving larger vehicles
- a hazard perception test, to assess your hazard recognition skills.

This book contains hundreds of questions that are very similar to the questions you'll be asked in the test and cover the same topics. It's easy to read, and explains why the answers are correct. References to the source material also appear with each question.

Everyone learns in different ways, so this book has features to help you understand driving theory whatever kind of learner you are, including

- bite-size chunks of information, which are easier to understand at your own pace
- lots of photographs and images to illustrate what you're learning
- things to discuss and practise with your instructor, to put your learning about each topic into practice
- meeting the standards, to help you understand how each topic relates to the National Driving Standard for lorry and bus drivers.

This book is designed to help you learn about the theory of driving large vehicles and to practise for the test. To prepare thoroughly, you should also study the source materials from which the questions are taken. These are

The Official Highway Code
Know Your Traffic Signs
The Official DVSA Guide to Driving Buses and Coaches
The Official DVSA Guide to Driving Goods Vehicles
The Official DVSA Guide to Driving – the essential skills

If you intend to drive lorries or buses/coaches professionally, you'll find one or both of the following useful in explaining the Driver Certificate of Professional Competence (Driver CPC) qualification and the various test modules

Driver CPC – the official DVSA guide for professional bus and coach drivers
Driver CPC – the official DVSA guide for professional goods vehicle drivers

There's always more you can learn, so keep your knowledge up to date throughout your driving career.

Getting started

Applying for your licence

To obtain your provisional entitlement to drive goods vehicles or buses and coaches, you should apply to the Driver and Vehicle Licensing Agency (DVLA) or, in Northern Ireland, the Driver and Vehicle Agency (DVA). Application forms (D2) and information leaflets (D100) are available from traffic area offices or DVLA. For more information, visit **www.gov.uk**

To drive a large vehicle, you must

- have a full car driving licence
- hold a provisional entitlement for the category of vehicle you want to drive
- meet the detailed and specific eyesight and medical requirements
- be over 18 years old, although other age restrictions may apply depending on vehicle size and category.

In Northern Ireland you must also complete a criminal records check if you want to drive a passenger-carrying vehicle (PCV).

Some details on various licence restrictions can be found in *The Official DVSA Guide to Driving Goods Vehicles* or *The Official DVSA Guide to Driving Buses and Coaches*.

You can get full details from the DVLA enquiry line on 0300 790 6801. (In Northern Ireland the DVA enquiry line is 0300 200 7861.)

Medical requirements

As a driver of large vehicles, you'll usually be responsible for safely transporting goods or passengers, so it's vital that you meet detailed and specific medical standards. The medical requirements apply to both large goods vehicle (LGV) and PCV licences.

You can't hold an LGV or a PCV licence unless your eyesight meets the high standard required.

You must be fit and free from any condition that affects your ability to retain control of a large vehicle. If you're disabled, you may drive a vehicle that has been specially adapted for you.

You must have a medical, carried out by a doctor. The doctor has to complete a medical report form, D4 (DLM1 in Northern Ireland), and then this has to be sent in with your licence application. Information is available at **www.gov.uk** (**nidirect.gov.uk/motoring** in Northern Ireland).

On the road

When you receive your licence with the appropriate category of provisional entitlement added, you should check that all the details are correct. You can then drive that category of large vehicle on the road as long as you

- only drive under the supervision of a person who holds a current licence for the category of vehicle being driven (minimum licence requirements apply; check these with your instructor)
- display L plates (or D plates if you prefer when driving in Wales) at the front and back of the vehicle
- display LGV or PCV plates to the front and back of the vehicle when driving in Northern Ireland.

DVSA recommends that you check the National Register of LGV Instructors (not applicable in Northern Ireland) before choosing an instructor. To be placed on this register, instructors have to pass rigorous tests of their driving and instructional ability. You can obtain details from 01952 520210 or visit **lgvinstructorregister.com** for more information. A national PCV instructor register is available at **careersthatmove.co.uk/PCV**

Driver Certificate of Professional Competence

A qualification for professional bus, coach and lorry drivers – the Driver Certificate of Professional Competence (Driver CPC) – has been introduced across the EU. Bus, coach and lorry drivers wishing to drive professionally now have to hold the Driver CPC qualification.

As part of the initial qualification, you have to pass a theory test based on case studies, in addition to the multiple choice theory test and hazard perception test. There's also an additional practical demonstration test. Further information can be found at **www.gov.uk**

About the theory test

The theory test is an on-screen test and is a module made up of two parts – the multiple choice part and the hazard perception part.

The parts can be taken in any order, and at different sittings, but you must pass both parts to enable you to move on to the next stage of your driving development, which involves taking your practical test. Should you be required to take the Driver CPC case studies test, you may sit this test either before or after completing both parts of the theory test as described above.

Can I take the practical test first?

No. You have to pass both parts of your theory test before you can book a practical test.

Does everyone have to take the theory test?

Normally, if you upgrade your licence within the same category you won't have to sit the test. For example, if you upgrade from a rigid-bodied to an articulated lorry you won't have to take another theory test. However, if you have sub-category C1 and D1 entitlement from when you passed your car test, and you want to upgrade to C or D, you'll have to obtain the correct provisional entitlement and pass both parts of the theory test before booking a practical test.

If you have a full category C licence and wish to take a practical test for category D, you'll have to obtain the correct provisional entitlement, and pass the appropriate theory test, before you can book a practical test, and vice versa.

From the start, you must be aware of the differences between driving smaller vehicles and driving large buses or lorries.

You must have a sound knowledge of The Highway Code, including the meaning of traffic signs and road markings. You must be particularly aware of those signs that show a restriction for lorries or buses.

Your knowledge of the above information is tested in the first part of the theory test, as a series of multiple choice questions. More information about this part of the test is given on pages 19–20 and the revision questions are given in the main part of the book, beginning on page 34.

Each revision question has references to the learning materials; for example

DGV s4, DBC s5, HC r153, p42, KYTS p20

DGV s indicates the section in *The Official DVSA Guide to Driving Goods Vehicles.*

DBC s indicates the section in *The Official DVSA Guide to Driving Buses and Coaches.*

HC r/HC p indicates the rule or page in *The Official Highway Code.*

KYTS p indicates the page in *Know Your Traffic Signs.*

The second part of the theory test is called the hazard perception test; more information about this is given on page 21.

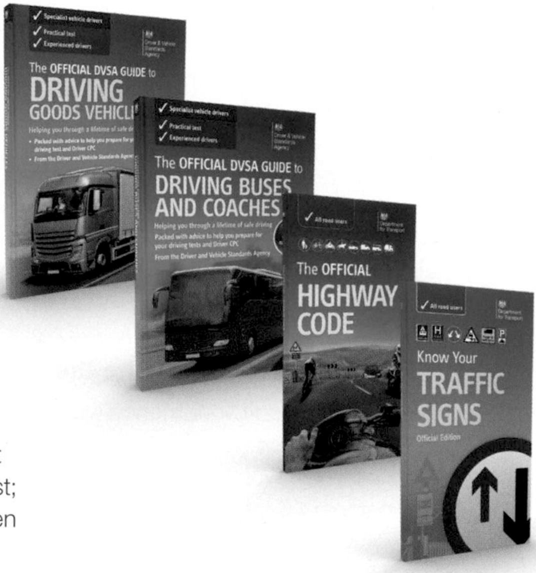

Preparing for your tests

Although you must pass your theory test before you can take your practical test, it's best to start studying for your theory test as soon as possible – but don't actually take it until you have some practical experience of driving larger vehicles.

To prepare for the multiple choice part of the theory test, DVSA strongly recommends that you study the books from which the theory test questions are taken, as well as the revision questions you'll find in this book.

The Official Highway Code This is essential reading for all road users. It contains the very latest rules of the road and up-to-date legislation, as well as advice on road safety and best practice. It's available as a book, eBook and iPhone app.

Know Your Traffic Signs This contains most of the signs and road markings that you're likely to come across.

Specialist publications *The Official DVSA Guide to Driving Goods Vehicles* and *The Official DVSA Guide to Driving Buses and Coaches* cover all aspects of driving these vehicles – in particular the regulations and the way in which they differ from driving smaller vehicles. They also contain the appropriate practical test syllabus. Two other books, *Driver CPC – the official DVSA guide for professional bus and coach drivers* and *Driver CPC – the official DVSA guide for professional goods vehicle drivers*, explain more about gaining the Driver CPC qualification and the new elements of the test procedure, including practice case studies.

The Official DVSA Guide to Driving – the essential skills This contains much general advice about driving that isn't necessarily repeated in the specialist books detailed above.

These books will help you to answer the questions correctly and will also help you when studying for your practical test. The information in them will be relevant throughout your driving life, so make sure you always have an up-to-date copy.

Other study aids

The Official DVSA Theory Test for Drivers of Large Vehicles (DVD-ROM)
This is an alternative way of preparing for the multiple choice part of the theory test. You can randomly test yourself against the complete range of revision theory test questions, create custom tests and take timed mock exams.

The Official DVSA Guide to Hazard Perception (DVD-ROM) We strongly recommend that you use this, preferably with your instructor, to prepare for the hazard perception part of the test. The DVD-ROM is packed with useful tips, quizzes and expert advice. It also includes more than 100 interactive hazard perception clips, which you can use to test yourself and see if you're ready to take the real test.

The Official DVSA Hazard Perception Practice iOS App A simple and convenient way to prepare for your hazard perception test on the go. The app is compatible with both iPhone and iPad, and contains 14 official interactive DVSA practice clips.

Why do the questions change?

To make sure that all candidates are being tested fairly, questions and video clips are under continuous review. Some questions may be changed as a result of customer feedback. They may also be altered because of changes to legislation, and DVSA publications are updated so that the revision questions reflect these changes.

Can I take a mock test?

You can take a mock test for the multiple choice part of the theory test online at **safedrivingforlife.info/practicetheorytest**

You can buy official DVSA learning materials online at **safedrivingforlife.info/shop** or by calling our expert publications team on **0333 200 2401**. The team can give you advice about learning materials and how to prepare for the tests and beyond. They can also help you select a suitable learning material if you have a special need; for example, if you have a learning disability or English isn't your first language.

DVSA publications are also available from book shops and online retailers. DVSA apps can be downloaded from the iOS App Store and eBooks are available from your device's eBook store.

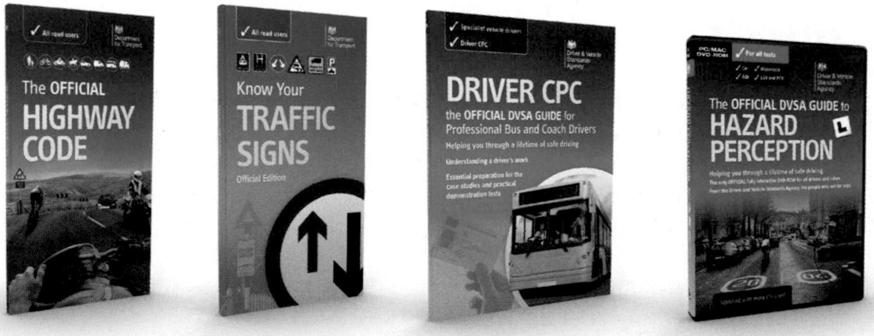

The theory test

Booking your theory test

Visit **www.gov.uk** to book your theory test online (for Northern Ireland, use **nidirect.gov.uk/motoring**).

If you have any special needs for the theory test, call 0300 200 1122 (0845 600 6700 for Northern Ireland). If you're a Welsh speaker, call 0300 200 1133.

If you have hearing or speech difficulties and use a minicom machine, call 0300 200 1166.

You'll need your

- DVLA or DVA driving licence number
- credit or debit card details (if you do this over the phone, the card holder must book the test). We accept Mastercard, Visa, Delta and Visa Electron.

You'll be given a booking number and you'll receive an appointment email on the same day if you book online.

If you book over the phone and don't provide an email address, you'll receive an appointment letter within 10 days.

Where can I take the test?

There are over 150 theory test centres throughout England, Scotland and Wales, and six in Northern Ireland. Most people have a test centre within 20 miles of their home, but this will depend on the density of population in your area. To find your nearest test centre, please visit **www.gov.uk**

What if I don't receive an acknowledgement?

If you don't receive an acknowledgement within the time specified, please visit **www.gov.uk** or telephone the booking office to check that an appointment has been made. We can't take responsibility for postal delays. If you miss your test appointment, you'll lose your fee.

When are test centres open?

Test centres are usually open on weekdays, some evenings and some Saturdays.

How do I cancel or postpone my test?

You can cancel or postpone your test online by visiting **www.gov.uk** or by telephone. You should contact the booking office at least **three clear working days** before your test date, otherwise you'll lose your fee.

Short-notice cancellation and rebooking (not refund) is permitted in the following circumstances:

- if you're ill or injured and have a supporting medical certificate
- if you've been affected by a bereavement
- if you're sitting school examinations.

Booking by post If you prefer to book by post, you'll need to fill in an application form. The form can be downloaded from **www.gov.uk**, or your instructor may have one.

You should normally receive confirmation of your appointment within 10 days of posting your application form. This will be by email if you've provided an email address or by post if not.

If you need support for special needs, please turn to page 18.

Taking your theory test

Arriving at the test centre You must make sure that when you arrive at the test centre you have all the relevant documents with you. If you don't have them, you won't be able to take your test and you'll lose your fee.

You'll need

- your signed photocard licence, or
- your signed old-style paper driving licence and valid passport (your passport doesn't have to be British).

No other form of identification is acceptable in England, Wales or Scotland.

Other forms of identification may be acceptable in Northern Ireland; please check **nidirect.gov.uk/motoring** or your appointment letter.

All documents must be original. We can't accept photocopies.

The test centre staff will check your documents and make sure that you take the right category of test.

Remember, if you don't bring your documents your test will be cancelled and you'll lose your fee.

Make sure you arrive in plenty of time so that you aren't rushed. If you arrive after the session has started, you may not be allowed to take the test.

Watch the 'Driver CPC for lorries' and 'Driver CPC for buses' videos on DVSA's YouTube channel. These explain how to prepare for the LGV or PCV theory test, what to expect on the day and what you need to do to pass.

youtube.com/dvsagovuk

Languages other than English

In Wales, and at theory test centres on the Welsh borders, you can take your theory test with Welsh text on-screen. A voiceover can also be provided in Welsh.

All Great Britain driving test candidates must take their theory test in either English or Welsh. Unless you're taking the test in Northern Ireland

- no voiceovers will be provided in any other language
- translators cannot attend the test with you to translate it from English into any other language.

Provision for special needs

Every effort is made to ensure that the theory test can be taken by all candidates.

It's important that you state your needs when you book your test so that the necessary arrangements can be made.

Reading difficulties There's an English-language voiceover on a headset to help you if you have reading difficulties or dyslexia.

You can ask for up to twice the normal time to take the multiple choice part of the test.

You'll be asked to provide a letter from a suitable independent person who knows about your reading ability, such as a teacher or employer. Please check with the Special Needs section (call on the normal booking number; see page 15) if you're unsure who to ask.

We can't guarantee to return any original documents, so please send copies only.

Hearing difficulties If you're deaf or have other hearing difficulties, the multiple choice part and the introduction to the hazard perception part of the test can be delivered in British Sign Language (BSL) by an on-screen signer.

A BSL interpreter, signer or lip speaker can be provided if requested at the time of booking. If you have any other requirements, please call the Special Needs section on the normal booking number (see page 15).

Physical disabilities If you have a physical disability that would make it difficult for you to use a mouse button in the theory test, we may be able to make special arrangements for you to use a different method if you let us know when you book your test.

Multiple choice questions

The multiple choice part of the theory test consists of 100 multiple choice questions. You select your answers for this part of the test by using a mouse.

Before you start, you'll be given the chance to work through a practice session for up to 15 minutes to get used to the system. Staff at the test centre will be available to help you if you have any difficulties.

The questions will cover a variety of topics relating to road safety, the environment and documents, as well as other factors specific to driving large vehicles, such as passenger care and load safety. Only one question will appear on the screen at a time, and you'll be asked to mark one correct answer.

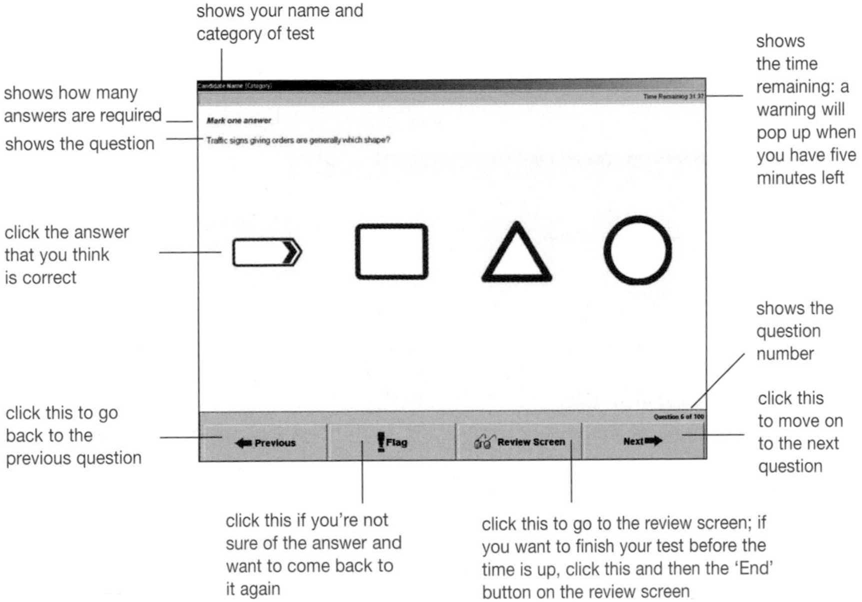
shows your name and category of test
shows how many answers are required
shows the question
click the answer that you think is correct
click this to go back to the previous question
shows the time remaining: a warning will pop up when you have five minutes left
shows the question number
click this to move on to the next question
click this if you're not sure of the answer and want to come back to it again
click this to go to the review screen; if you want to finish your test before the time is up, click this and then the 'End' button on the review screen
Mark one answer
Traffic signs giving orders are generally which shape?
Previous
Flag
Review Screen
Next

To answer, you need to click the box beside the answer you think is correct. If you change your mind and don't want that answer to be selected, click it again. You can then choose another answer.

Take your time and read the questions carefully. You're given 115 minutes for this part of the test, so relax and don't rush. Some questions will take longer to answer than others, but there are no trick questions. The time remaining is displayed on screen.

You may be allowed extra time to complete the test if you have special needs and you let us know when you book your test.

You'll be able to move backwards and forwards through the questions and you can also 'flag' questions that you'd like to look at again. It's easy to change your answer if you want to.

Try to answer all the questions. If you're well prepared, you shouldn't find them difficult.

Before you finish this part of the test, if you have time, you can use the 'review' feature to check your answers. If you want to finish your test before the full time, click the 'review' button and then the 'end' button on the review screen. When you click the review button, you'll see the following screen.

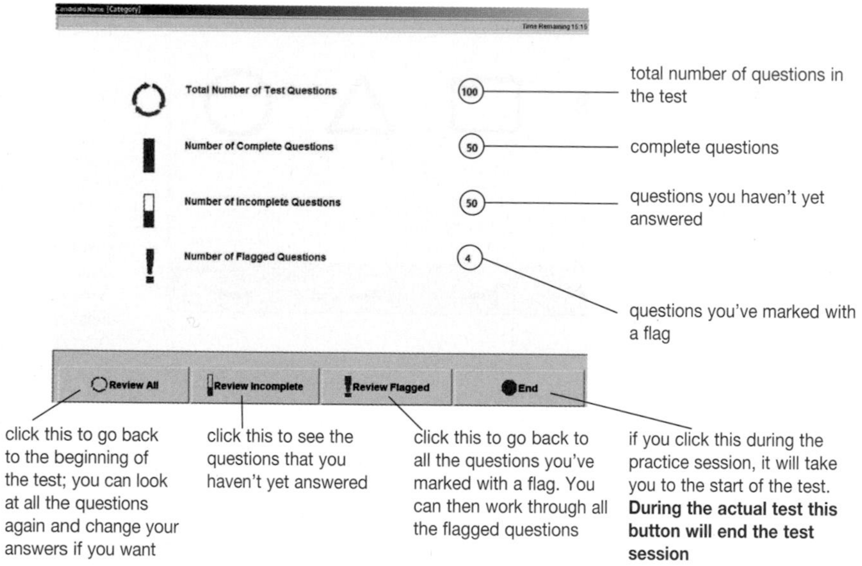

Total Number of Test Questions
100
total number of questions in the test
Number of Complete Questions
50
complete questions
Number of Incomplete Questions
50
questions you haven't yet answered
Number of Flagged Questions
4
questions you've marked with a flag
Review All
Review Incomplete
Review Flagged
End
click this to go back to the beginning of the test; you can look at all the questions again and change your answers if you want
click this to see the questions that you haven't yet answered
click this to go back to all the questions you've marked with a flag. You can then work through all the flagged questions
if you click this during the practice session, it will take you to the start of the test. **During the actual test this button will end the test session**

Hazard perception

The hazard perception part of the test consists of a series of computer-generated image (CGI) video clips, shown from a driver's point of view. You'll be using a mouse for this part of the theory test.

Before you start this part of the test you'll be shown a short CGI video that explains how the test works and gives you a chance to see a sample clip. This will help you to understand what you need to do. You can play this video clip again if you wish.

During the test, you'll be shown 19 CGI video clips. Each clip contains one or more developing hazards. You should press the mouse button **as soon as you see** a hazard developing that may need you, the driver, to take some action, such as changing speed or direction. The earlier you notice a developing hazard and make a response, the higher your score. There are 20 hazards for which you can score points.

Your response won't change what happens in the scene in any way. However, a red flag will appear on the bottom of the screen to show that your response has been noted.

Before each clip starts, there'll be a 10-second pause to allow you to see the new road situation.

The hazard perception part of the test lasts about 25 minutes. For this part of the test no extra time is available, and you can't repeat any of the clips – you don't get a second chance to see a hazard when you're driving on the road.

Trial questions

We're constantly checking the questions and clips to help us decide whether to use them in future tests. After the hazard perception part of the test, you may be asked to try a few trial questions and clips. You don't have to do these if you don't want to, and if you answer them they won't count towards your final score.

Customer satisfaction survey

We want to ensure our customers are completely satisfied with the service they receive. At the end of your test you'll be shown some questions designed to give us information about you and how happy you are with the service you received from us.

Your answers will be treated in the strictest confidence. They aren't part of the test and they won't affect your final score or be used for marketing purposes. You'll be asked if you want to complete the survey, but you don't have to.

The result

You should receive your result at the test centre within 10 minutes of completing the test. This result will also include the score you've obtained for this module.

The theory and hazard perception tests can be taken in any order, and at different sittings, but you must pass both parts to pass the theory test. Should you be required to take the Driver CPC case studies test, you may sit this test either before or after completing both parts of the theory test. You must pass both the theory test and, if applicable, the case studies test before you can take the practical test.

Why do I have to pass both parts of the theory test before I can take my practical test?

As the theory test is one test made up of two parts, you're required to pass both parts to pass the test. These parts are made up of multiple choice questions and hazard perception clips, and can be taken at different times.

What's the pass mark?

To pass the multiple choice part of the theory test, you must answer at least 85 out of 100 questions correctly.

To pass the hazard perception part of the test, you must reach the required pass mark of 67 out of a possible 100. There are 19 clips with 20 scorable hazards.

If I don't pass, when can I take the test again?

If you fail your test, you've shown that you're not fully prepared. You'll have to wait at least three clear working days before you take the theory test again.

Good preparation will save you time and money.

After the theory test

Once you've passed both parts, a pass letter will be sent to you in the post. Keep this safe, as it contains the pass certificate number that you'll need to quote when you book your practical test.

This pass certificate number has a life of two years from the date that you passed the first part of the theory test. This means that you have to take and pass the practical test for the category of vehicle you wish to drive within this two-year period. If you don't, you'll have to take and pass the theory test again before you can book your practical test.

If you're intending to drive professionally, you'll also need to pass the Driver CPC case study test. The case study test is taken on screen at a test centre.

Each case study is based on a real-life scenario that you may encounter in your working life. The case study method tests your knowledge and basic understanding by examining how you put your skills into practice.

Questions are based around these scenarios, and you'll be asked to answer them in different ways – by

- selecting from multiple choice answers
- clicking an area of a photograph or image.

There will be between six and eight case studies, each with between five and ten questions. The test, including the introductory screens, will last for one hour and 30 minutes.

More information specifically relating to gaining the Driver CPC qualification can be found in *Driver CPC – the official DVSA guide for professional goods vehicle drivers* and *Driver CPC – the official DVSA guide for professional bus and coach drivers*. Each book also contains an example case study to help you understand how the case study test will appear on screen.

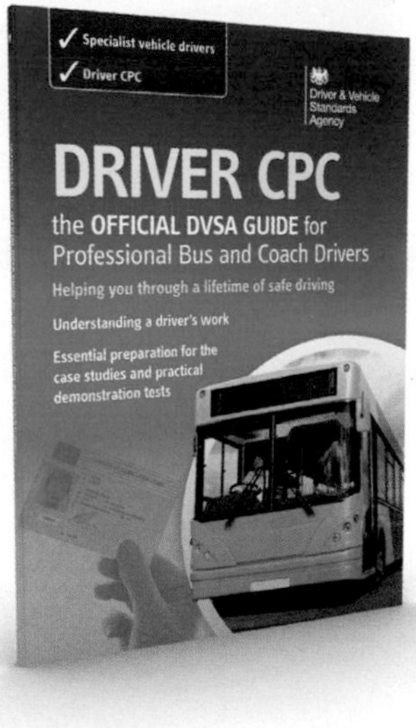

Your practical driving test

Your next step is to prepare for and take your practical driving test. If you intend to drive a large vehicle as a profession, you must take two practical tests to gain the Driver CPC qualification. They are

- the Licence Acquisition Practical Driving Test
- the Driver CPC Practical Demonstration Test.

To help you, and in addition to the two Driver CPC publications mentioned earlier, DVSA has produced *The Official DVSA Guide to Driving Goods Vehicles* and *The Official DVSA Guide to Driving Buses and Coaches,* which contain the official syllabuses for the LGV and PCV practical tests. The information provided in these publications will help you understand the various test modules, develop your skills and prepare you to become a safe and confident driver of large vehicles.

These books include valuable advice about taking the practical tests, and outline the skills you need to show and the faults you should avoid. They also cover legal requirements, different types of vehicle and general driving techniques.

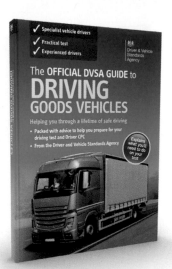

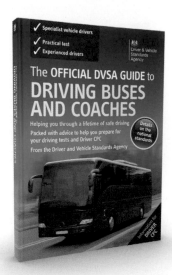

Using the questions and answers sections

Sections 1–15 contain revision questions for the multiple choice part of the theory test. These are very similar to the questions you'll actually be asked in the test and cover the same topics.

For easy reference, the questions are divided into topics. Although this isn't how you'll find them in your test, it's helpful if you want to look at particular subjects.

At the start of each topic, before the questions, there are a few pages of useful information to help you learn more about each topic.

The questions are in the left-hand column with a choice of answers below.

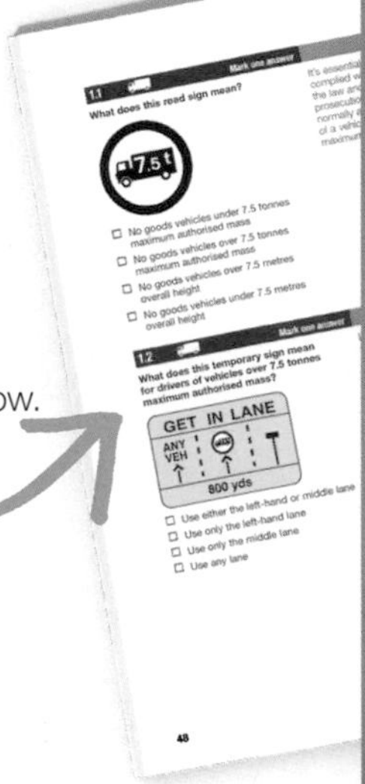

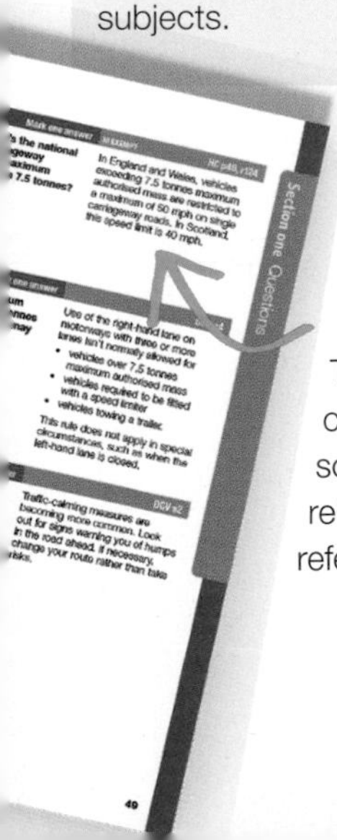

On the right-hand side of the page there's a brief explanation to help you understand the question. There'll also be some advice on correct driving procedures and some short references to the relevant source materials. These refer to the books on page 12.

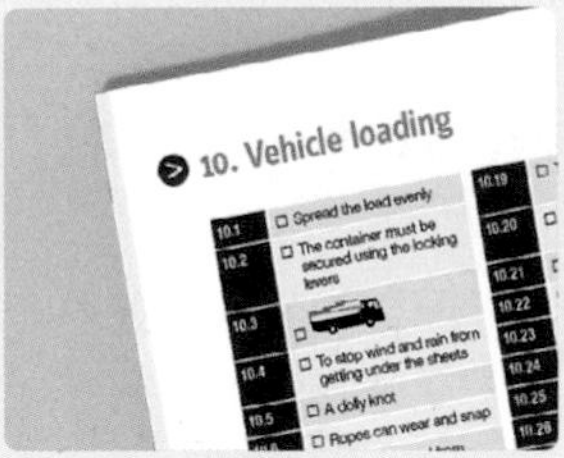

The correct answers are at the back of the book in section 16.

Don't just learn the answers; it's important that you know **why** they're correct. The revision materials listed on pages 12–14 will help you gain the knowledge and understanding that you need to drive a large vehicle safely and competently.

Taking exams or tests is rarely a pleasant experience, but you can make your test less stressful by being confident that you have the knowledge to answer the questions correctly.

Make studying more enjoyable by involving friends and relations. Take part in a question-and-answer game. Test those 'experienced' drivers who've had their licence a while: they might learn something too!

Some of the questions in this book won't be relevant to theory tests taken in Northern Ireland. These questions are marked **NI EXEMPT**.

Questions specific to a single category

Most of the questions refer to drivers of all large vehicles, indicated by two icons – a lorry and a bus – beside the question numbers. However, there are some questions that are only relevant to a specific group of vehicles.

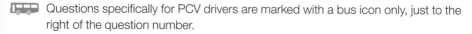

Questions specifically for PCV drivers are marked with a bus icon only, just to the right of the question number.

Questions specifically for LGV drivers are marked with a lorry icon only, just to the right of the question number.

Using this book to learn and revise

We're all different. We like different foods, listen to different music and learn in different ways.

This book is designed to help you learn the important information that you'll need for the large-vehicle theory test in a variety of different formats, so you can find a way of learning that works best for you.

Features

- A summary, at the start of each section, of what you'll learn.

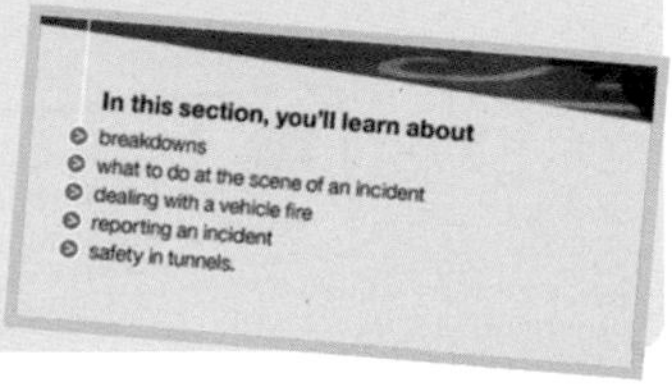
In this section, you'll learn about
- breakdowns
- what to do at the scene of an incident
- dealing with a vehicle fire
- reporting an incident
- safety in tunnels.

- All the key information presented in bite-size chunks with clear headings.

- Images to help you relate the information to the real world.

room as you pass them. They co
ised by your vehicle. Remember t
drain covers in the road.

- Diagrams and tables to help make information clear and summarise key points.

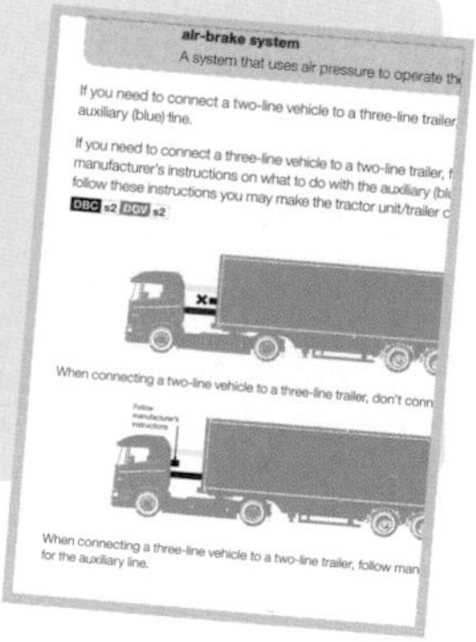
air-brake system
A system that uses air pressure to operate th
If you need to connect a two-line vehicle to a three-line trailer
auxiliary (blue) line.
If you need to connect a three-line vehicle to a two-line trailer, f
manufacturer's instructions on what to do with the auxiliary (bl
follow these instructions you may make the tractor unit/trailer c
DBC s2 DGV s2
When connecting a two-line vehicle to a three-line trailer, don't conn
When connecting a three-line vehicle to a two-line trailer, follow man
for the auxiliary line.

- Links and QR codes to online videos and interactive activities, to further increase your knowledge and skills. Scan the QR code on your smartphone (you'll need a QR code reader app) to access the online content.

- Links to other relevant publications, like *The Official Highway Code, The Official DVSA Guide to Driving Buses and Coaches* and *The Official DVSA Guide to Driving Goods Vehicles.*

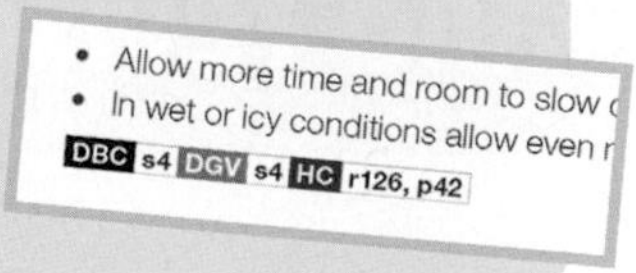

- Tips containing useful extra information about driving safely.

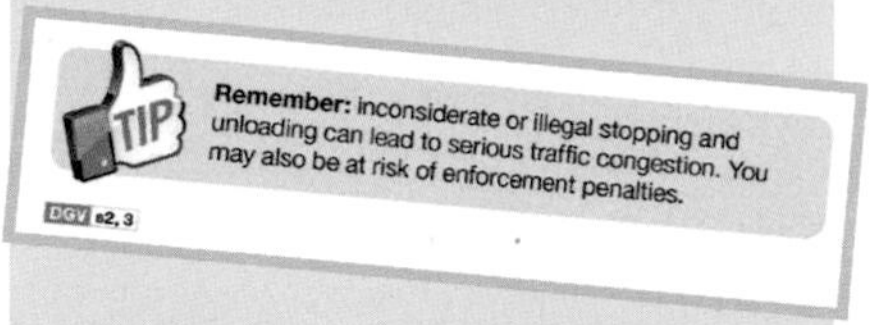

- A summary, at the end of each section, of what you'll need to know and be able to do to meet the National Driving Standards for lorry and bus drivers.

- Pages for your own notes, with suggested things to think about.

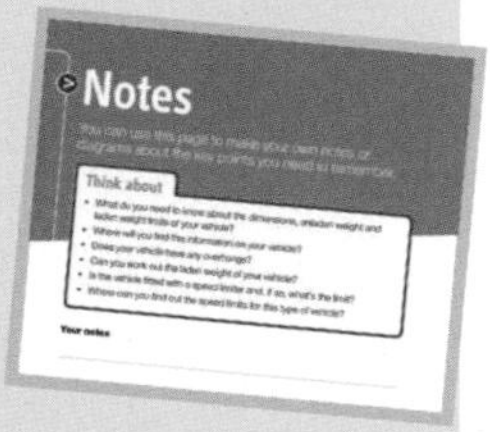

- Ideas to discuss with your driving instructor and practise when driving.

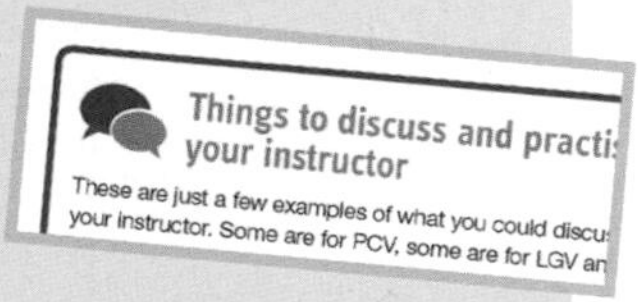

- Self-assessment – revision questions like the ones you'll get in the test.

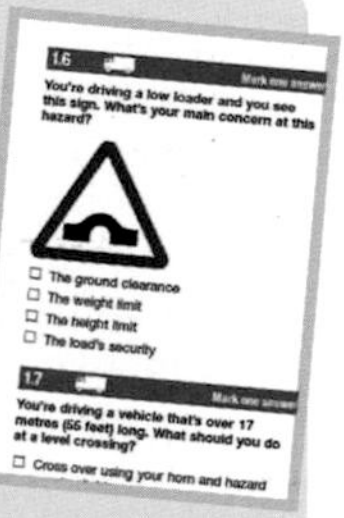

- Icons to show text meant for PCV drivers only (shown by a bus), and LGV drivers only (shown by a lorry).

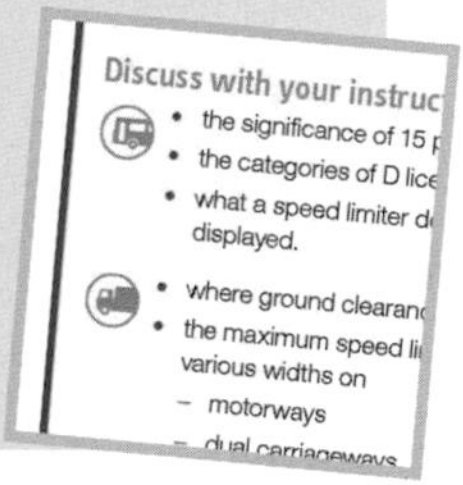

The theory test is just one part of the process of learning to drive a PCV or LGV. You need to learn the facts, but it's important to understand how they relate to real driving, particularly if this is to be your paid profession.

The combination of knowing driver theory and having good practical driving skills won't only help you pass your test; it will also make you a safer driver for life.

What kind of learner are YOU?

Ask yourself these questions

- Why are you doing this? What's motivating you?
- How have you learned best in the past? What helped you to remember what you needed to know?
- What are your strengths and weaknesses as a learner?

Think about the way that you learn best. You could try any combination of the following ideas.

I remember what I see or read

- Create flashcards with important facts or statistics
- Make diagrams and charts
- Use mind maps
- Use colour coding
- Watch the DVSA short films
- Make your own notes
- Cross-reference information using a variety of books, eg *The Official Highway Code*
- Draw your own diagrams to show key information.

I remember best when I physically do something

- Short study sessions
- Do things – create models or diagrams; make lists
- Use props
- Try the interactive activities
- Watch and copy what your driving instructor does
- Mime or act out different driving moves.

I remember what I hear

- Repeat rules out loud
- Use a voice recorder to make recordings of key information
- Work with others and discuss things
- Watch and listen to the DVSA video content.

Top tips

Remember your motivation

Think about the reason you're learning to drive a PCV or LGV. Is it for work? For a hobby? To fulfil a lifelong dream? Remind yourself, from time to time, of your motivation for learning. Don't give up!

Relate to your personal experience

Information is more memorable when it's linked to what you already know. Try to picture yourself in the position of the driver. The 'Think about' sections throughout the book can help you think about how the ideas would work in real life.

Use mnemonics

Mnemonics are little sayings, stories or techniques that help you remember something. A classic example is '**R**ichard **O**f **Y**ork **G**ave **B**attle **I**n **V**ain', which you can use to remember the colours of the rainbow (**r**ed, **o**range, **y**ellow, **g**reen, **b**lue, **i**ndigo, **v**iolet). You can use similar techniques to memorise statistics, facts or information for your driving career.

Question format

However you choose to learn the content, make certain you're familiar with the format of the test and how the questions will be presented. Go through the self-assessment questions in each chapter and see if you can answer them. Mark any you struggle with and try them again at a later date.

Plan your study

Set yourself timelines and targets. Try to set aside dedicated time for study, when you're feeling awake and are unlikely to be interrupted. The environment in which you study is important – try to find an area where you can concentrate.

Getting help

Think about the people you can speak with to ask questions, get advice or share experiences about driving – such as your driving instructor, parents, friends or colleagues at work.

Taking your test

Don't rush into the theory test before you're ready. You need to be confident with the information, and have enough practical experience to give you a deep understanding of the information too.

Section one

Vehicle weights and dimensions

In this section, you'll learn about

- vehicle size
- loading your vehicle
- vehicle markings
- speed limiters.

Vehicle weights and dimensions

To drive safely, you must know your vehicle's dimensions, weight and speed limits, as well as any weight, height, width or length restrictions on your planned route.

Vehicle size

Know your vehicle's dimensions and use them to plan your journey

You **MUST** know your vehicle's dimensions so that you can avoid restrictions and obstructions.

- There should be a plate in the cab of your vehicle that shows its height.
- The vehicle plate will show other information, such as the **axle weight limits**.

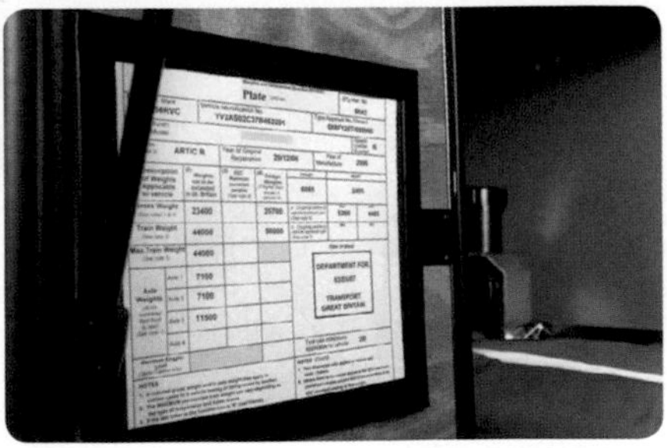

Definition

axle weight limit

Limits laid down for maximum permitted weights carried by each axle.

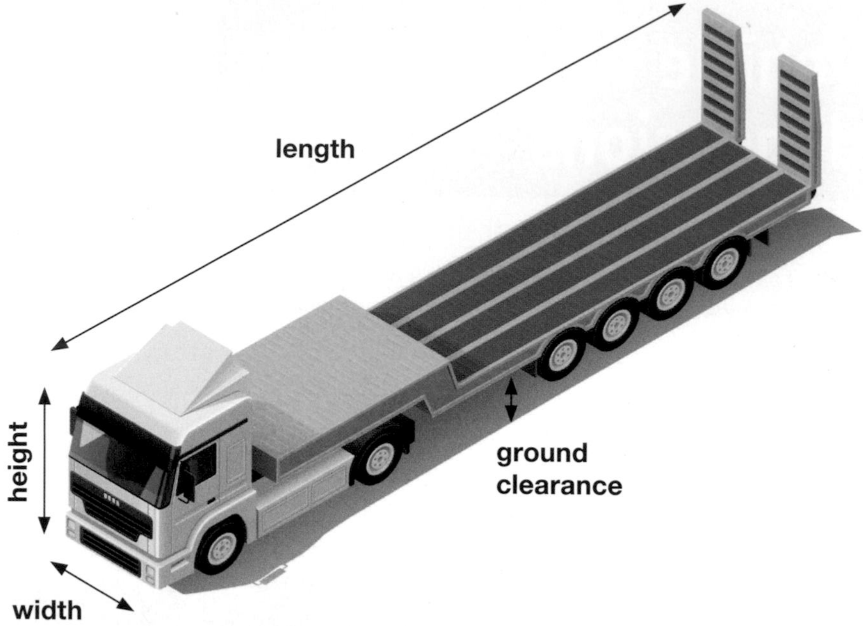

Plan your route carefully to avoid obstructions. Weight, height, width, length, and ground clearance are all important dimensions that you should know. Knowing your vehicle's dimensions will help you to safely negotiate

- overhead cables
- level crossings
- bridges that are low, humped, arched or narrow.

Look for any dimension warning signs during the journey. You **MUST** obey any signed height or weight restriction.

- An unmarked bridge will be at least 5 metres (16 feet 6 inches) in height.
- Lower bridges will be marked and should have advance warning signs.

Some weight restrictions apply to large goods vehicles only and not to passenger-carrying vehicles. A plate beneath a restriction sign will indicate any exemptions.

If your vehicle strikes a railway bridge, you **MUST** tell the police and the railway authority immediately, giving all the details they need.

- Use the telephone number on the bridge's identification plate.
- If there's no plate, use 999.

When driving a passenger-carrying vehicle, you're responsible for knowing the maximum authorised mass (MAM) of the vehicle and staying within this. The weight difference between an **unladen** and **laden** coach may be as much as 7 tonnes.

Definition

unladen
Without a load.

laden
With a load.

The vehicle's unladen weight is displayed on the nearside bodywork as well as on the vehicle's plate. Use this information to help you calculate the vehicle's laden weight.

DBC s3

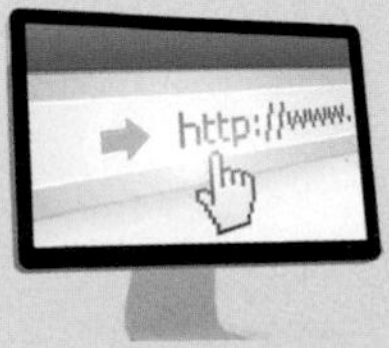

Watch the 'Driver CPC for buses' video on DVSA's YouTube channel.

youtube.com/dvsagovuk

Overhead cables

There may be warning bells strung across the road below overhead cables. If your vehicle disturbs these bells you **MUST** stop and seek advice.

If the height of your load is more than 5.25 metres (17 feet 6 inches), you must tell the telephone companies where you're planning to drive before you start your journey.

Turning

Large vehicles often need to straddle lanes when turning at roundabouts or junctions.

- Be aware of the turning circle that your vehicle needs.
- Try to make sure the rear or trailer wheels don't run over kerbs when you're manoeuvring.

DBC s3 DGV s2

When turning, check your mirrors for cyclists or motorcyclists who may be trying to filter past your vehicle. They may be unaware of the danger they are in from your rear or trailer wheels as you turn.

Beware of overhangs on your vehicle, especially when turning. **Overhangs** increase your risk of colliding with

- pedestrians
- **street furniture**
- traffic signals and signs
- walls or buildings
- other vehicles.

Definition

overhang

The parts of a vehicle that extend beyond the wheelbase, normally described as front overhang and rear overhang.

street furniture

Objects and equipment installed on streets and roads for various purposes. Examples include bus shelters, bins, bollards and traffic lights.

Loading your vehicle

Stopping distances

Your vehicle needs more room and time to stop when laden with cargo or when carrying passengers and their luggage.

Allow for the effect that this weight will have on your vehicle's stopping distance.

- Leave more space between your vehicle and the one in front.
- Allow more time and room to slow down or stop.
- In wet or icy conditions allow even more time and stopping distance.

DBC s4 DGV s4 HC r126, p42

Staying in control

Always drive so that your vehicle and load are stable and you are in control.

Large vehicles are most stable when being driven in a straight line. Any of these things can make your vehicle less stable

- sudden acceleration
- sudden braking
- harsh steering
- bends or corners.

If you lose control of the vehicle, it may cause

- damage to the load
- injury to passengers
- skidding
- a collision.

DBC s2, 3 DGV s2

Restraining loads

Use the right **restraints** for your load.

There are many types of restraints to use on your vehicle.

Straps or ropes

Battens and chocks

Chains

Nets

Twist locks

> Definition
>
> **restraints**
> Anything used to hold a load in place, such as straps or chains.

When loading your vehicle

- use the correct restraints for the load that you're carrying
- check that the restraints are secure and undamaged before you set off.

Unloading legally

Some **restriction notices** and road markings may prevent you from stopping and unloading your vehicle.

- Before you stop or unload your vehicle, check for any notices or road markings.
- Some restrictions are only in force at certain times of day.
- Other restrictions may be in force permanently.

Definition

restriction notice
A plate or sign at the side of the road showing when you may or may not park or wait on the section of road indicated.

Watch the 'Driver CPC for lorries' video on DVSA's YouTube channel.

youtube.com/dvsagovuk

TIP

Remember: inconsiderate or illegal stopping and unloading can lead to serious traffic congestion. You may also be at risk of enforcement penalties.

DGV **s2, 3**

Vehicle markings

Look for the information you need

Your vehicle should have plates that tell you what you need to know about its dimensions and weight limits.

- Large vehicles have a plate in the cab that shows the height.
- The vehicle plate will show other dimensions, such as the axle weight limits and unladen weight.
- Buses and coaches will also show the maximum number of passengers that may be carried.
- Buses and coaches display the unladen weight on the nearside bodywork. The driver can use this information when calculating the laden weight of the vehicle.

Oversized loads

If your vehicle is carrying an **oversized load**, it **MUST** be escorted. If the load overhangs by more than 2 metres (6 feet 6 inches), triangular projection markers **MUST** be displayed.

Definition

oversized load

A load that exceeds the normal or legal size limits for a road, meaning that special arrangements (such as permits or police escorts) need to be made.

Heavy vehicles must display safety markings

Vehicles over 7500 kg maximum **gross weight** or trailers over 3500 kg maximum gross weight **MUST** display red and yellow rear markings.

Definition

gross weight

The total weight of the vehicle and its load.

DGV s2, 4 **HC** r225, p117

Speed limiters

You need to know the speed limits that apply to the vehicle you're driving, and always drive within them.

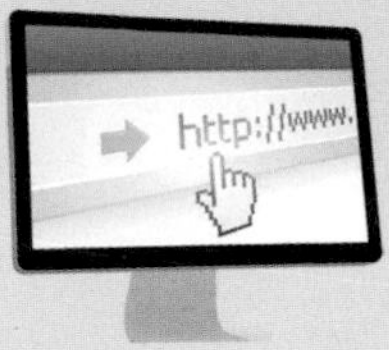

Find speed-limit information for your vehicle at this website.

www.gov.uk

Most large vehicles are fitted with a speed limiter, which will control the vehicle's maximum speed. There should be a notice showing your vehicle's set limit in the cab area.

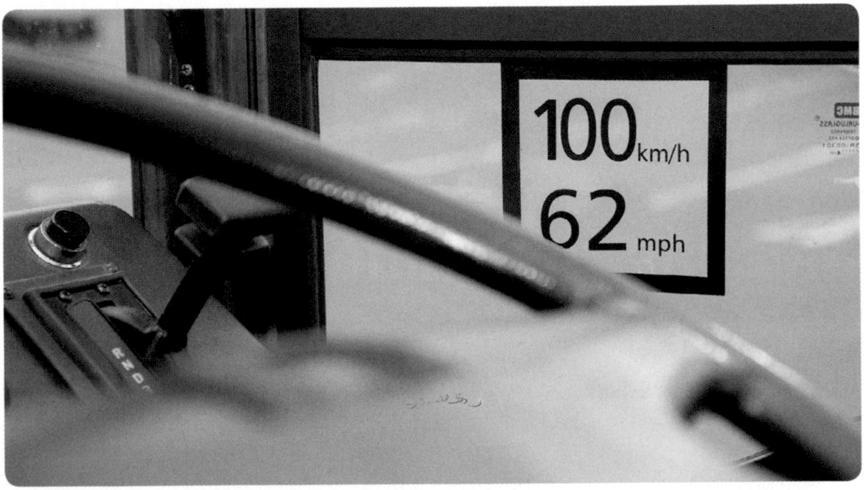

Remember: allow for your set speed limit when overtaking. Don't overtake if your set limit won't let you do this quickly and safely.

If there's a problem with your speed limiter, remember that these can only be repaired at an authorised speed-limiter centre.

DBC s4 DGV s4 HC p40

Obey temporary speed restrictions

Lanes at roadworks often have temporary speed restrictions.

- Temporary speed restriction signs in red circles are **mandatory**.
- These are often enforced by cameras.

Definition

mandatory
Must happen, by law.

Some lanes at roadworks will also have temporary weight limits.

- Know the laden weight of your vehicle.
- Look for and obey the temporary lane restriction signs.

DBC s3 **DGV** s2

Meeting the standards

The National Driving Standards set out the skills, knowledge and understanding that DVSA believes are required to be a safe and responsible driver. If you know, understand and are able to do the things described in the standard, then you'll not only be in a great position to pass your test but will also be well on your way to becoming a safe driver for life.

You can view the national standards for driving buses and coaches, and for driving lorries, at **www.gov.uk**

You must be able to

plan your route, taking into account the location of any

- height
- width
- length
- weight
- access

restrictions that apply to the vehicle you're driving

steer the vehicle safely and responsibly in all road and traffic conditions, paying attention to any restrictions on

- height
- width
- length
- weight
- ground clearance

take into account the effect of the road camber on the position of the vehicle, so that you don't collide with street furniture such as bus stops or street lights.

You must know and understand

the height, width, length and weight of the vehicle you're driving

what the vehicle's payload is, or how to calculate it based on its maximum authorised mass and its unladen weight

how to find your vehicle's maximum permitted gross axle weights

that a vehicle that has compliant maximum permitted gross axle weights at the start of a journey may become illegal as part-loads are offloaded.

Notes

You can use this page to make your own notes or diagrams about the key points you need to remember.

Think about

- What do you need to know about the dimensions, unladen weight and laden weight limits of your vehicle?
- Where will you find this information on your vehicle?
- Does your vehicle have any overhangs?
- Can you work out the laden weight of your vehicle?
- Is the vehicle fitted with a speed limiter and, if so, what's the limit?
- Where can you find out the speed limits for this type of vehicle?

Your notes

Things to discuss and practise with your instructor

These are just a few examples of what you could discuss and practise with your instructor. Some are for PCV, some are for LGV and some are for both. Read more about vehicle weights and dimensions to come up with your own ideas.

Discuss with your instructor

- the significance of 15 passengers
- the categories of D licences and what they mean
- what a speed limiter does and where its set limit should be displayed.

- where ground clearance may be a problem for a low loader
- the maximum speed limits for lorries carrying various loads and of various widths on
  - motorways
  - dual carriageways
- the markings on the back of lorries over 7.5 tonnes. What colour are they and what do they mean?

- signs showing height and weight restrictions for PCV and LGV
- what to do if your vehicle collides with a bridge.

Practise with your instructor

- driving on routes that include mini-roundabouts and turns into narrow roads
- planning your routes to avoid low bridges.

1.1 Mark one answer HC p106

What does this road sign mean?

- ☐ No goods vehicles under 7.5 tonnes maximum authorised mass
- ☐ No goods vehicles over 7.5 tonnes maximum authorised mass
- ☐ No goods vehicles over 7.5 metres overall height
- ☐ No goods vehicles under 7.5 metres overall height

It's essential that all limits are complied with to avoid breaking the law and facing possible prosecution. Weight restrictions normally apply to the plated weight of a vehicle, often referred to as the maximum authorised mass (MAM).

1.2 Mark one answer HC p113, KYTS p131

What does this temporary sign mean for drivers of vehicles over 7.5 tonnes maximum authorised mass?

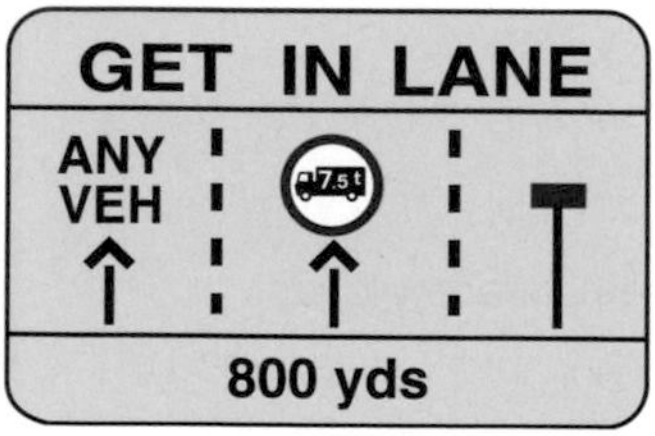

- ☐ Use either the left-hand or middle lane
- ☐ Use only the left-hand lane
- ☐ Use only the middle lane
- ☐ Use any lane

Lanes at roadworks often have weight restrictions to keep larger vehicles in the left-hand lane. The lanes through the roadworks are often narrow and therefore not wide enough for large vehicles to pass each other safely.

Always look for restriction signs at roadworks. They're usually placed well in advance to give you time to move safely into the correct lane in good time.

1.3 Mark one answer | NI EXEMPT | HC p40, r124

In England and Wales, what's the national speed limit on a single carriageway road for a rigid lorry with a maximum authorised mass of more than 7.5 tonnes?

- ☐ 30 mph
- ☐ 40 mph
- ☐ 50 mph
- ☐ 60 mph

In England and Wales, vehicles exceeding 7.5 tonnes maximum authorised mass are restricted to a maximum of 50 mph on single carriageway roads. In Scotland, this speed limit is 40 mph.

1.4 Mark one answer | DGV s4

You're driving a lorry with a maximum authorised mass of more than 7.5 tonnes along a three-lane motorway. When may you use the right-hand lane?

- ☐ To overtake slower lorries
- ☐ When the left-hand lane is closed
- ☐ If you drive at more than 60 mph
- ☐ If you're travelling unladen

Use of the right-hand lane on motorways with three or more lanes isn't normally allowed for

- vehicles over 7.5 tonnes maximum authorised mass
- vehicles required to be fitted with a speed limiter
- vehicles towing a trailer.

This rule does not apply in special circumstances, such as when the left-hand lane is closed.

1.5 Mark one answer | DGV s2

Where are you most likely to ground a low loader?

- ☐ On rural roads
- ☐ On traffic-calming humps
- ☐ On yellow rumble strips
- ☐ On hatched road markings

Traffic-calming measures are becoming more common. Look out for signs warning you of humps in the road ahead. If necessary, change your route rather than take risks.

1.6 Mark one answer — DGV s2

You're driving a low loader and you see this sign. What's your main concern at this hazard?

- ☐ The ground clearance
- ☐ The weight limit
- ☐ The height limit
- ☐ The load's security

You have a responsibility to be aware of the dimensions of your vehicle, including ground clearance. When planning your route, you'll have to consider whether the roads you intend to use have any limits that would apply to your vehicle, such as

- weight
- length
- height
- ground clearance.

1.7 Mark one answer — DGV s2

You're driving a vehicle that's over 17 metres (55 feet) long. What should you do at a level crossing?

- ☐ Cross over using your horn and hazard warning lights
- ☐ Stop before the crossing and phone the signal operator
- ☐ Increase your speed to clear the crossing quickly
- ☐ Stop before the crossing and look both ways before going on

If your vehicle is over 17 metres (55 feet) long and you wish to cross a level crossing, you must stop before the crossing and telephone the signal operator.

1.8 Mark one answer — DGV s2

What type of fastening or restraint should you use when carrying a load of steel?

- ☐ Chains
- ☐ Straps
- ☐ Ropes
- ☐ Sheeting

You're responsible for the safety of the load you're carrying. At no time should the load endanger other road users. It's vital that you make sure your load is secure and safely distributed on your vehicle. How you do this will depend on the

- bulk
- weight
- type of vehicle you're driving (flat bed, curtain side, etc)
- nature of the load.

1.9 Mark one answer DGV s2

You're driving a lorry carrying a load that's 4.5 metres (14 feet 9 inches) wide. What's the maximum speed you're allowed to drive on a motorway?

- ☐ 10 mph
- ☐ 20 mph
- ☐ 30 mph
- ☐ 40 mph

The speed limits imposed on vehicles carrying abnormal loads can be frustrating for other drivers, but you mustn't be tempted to exceed them. Remember, the more weight you're carrying, the longer it will take you to stop safely.

1.10 Mark one answer DGV s2

What's the speed limit on a dual carriageway when you're carrying a load that's 4.5 metres (14 feet 9 inches) wide?

- ☐ 15 mph
- ☐ 25 mph
- ☐ 35 mph
- ☐ 45 mph

Vehicles carrying loads between 4.3 metres (14 feet 1 inch) and 5 metres (16 feet 6 inches) wide are subject to lower speed limits. The limit depends on the type of road. On a motorway it's 40 mph (64 km/h), on a dual carriageway it's 35 mph (56 km/h), and on all other roads it's 30 mph (48 km/h).

1.11 Mark one answer HC p117

What colour are the markings on the rear of lorries over 7.5 tonnes maximum authorised mass?

- ☐ Red/white
- ☐ Red/yellow
- ☐ Black/yellow
- ☐ Black/white

All vehicles over 7.5 tonnes maximum authorised mass must have markings on the rear of the vehicle. These markings are rectangular and are coloured red and yellow. They're there to inform other road users of the characteristics of your vehicle. The markings should be kept clean so that they can be seen clearly at all times, especially at night and in poor visibility.

1.12 Mark one answer DGV s2

When this vehicle transporter turns, how will the overhanging top deck move?

- ☐ Through a greater arc than the cab
- ☐ Through a lower arc than the cab
- ☐ Through a smaller arc than the cab
- ☐ Through the same arc as the cab

As the top deck is longer than the cab and trailer, it will take up more room as the vehicle turns; you must make sure that you've allowed for the wider swing of the deck. Even if your cab is well clear, the overhanging deck could hit obstructions such as telegraph poles or traffic signs as you go round a corner.

1.13 Mark one answer DGV s2

You're the driver of an articulated car transporter. How will the overhanging top deck move as you turn corners?

- ☐ Through a smaller arc than the cab
- ☐ Through a shorter arc than the cab
- ☐ Through a lower arc than the cab
- ☐ Through a greater arc than the cab

The longer top deck needs a wider turning circle than the cab below it. To avoid hitting lampposts, telegraph poles, etc, you must allow for this overhang when turning.

1.14 Mark one answer DGV s2

When would the driver of a car transporter need to be most aware of the front overhang of the trailer?

- ☐ When overtaking
- ☐ When turning
- ☐ When loading
- ☐ When braking

The long overhang at the front of a car transporter can cause problems where street furniture, such as lampposts and traffic signs, is sited close to junctions. Particular problems may be encountered when you're turning right and there are 'Keep left' bollards in the middle of the road you're turning into. Plan your route carefully to avoid such hazards.

1.15 Mark one answer — DGV s3, HC p115

You're scheduled to make a delivery. You arrive at your destination during the morning rush hour. The road is edged with double red lines. What do they mean?

☐ Unload only within a 'white box' area
☐ Unload only within a 'red box' area
☐ Delay your delivery until after the rush hour
☐ Limit your stop to a maximum of 30 minutes

White boxes allow you to unload at any time, but during the day the length of stay is restricted. You should check nearby signs for the specific times. At other times, when it's not so busy, there may not be any restrictions on parking.

1.16 Mark one answer — NI EXEMPT — DGV s2, HC p40, r124

You're driving a lorry with a maximum authorised mass of more than 7.5 tonnes. What's the national speed limit for your vehicle on a dual carriageway in England and Wales?

☐ 40 mph
☐ 50 mph
☐ 60 mph
☐ 70 mph

The speed limit for lorries is lower in Scotland than it is in England and Wales, so be aware – especially if you drive in the border areas.

1.17 Mark one answer — DGV s2

Which axle configuration is most effective at preventing petrol tankers from rolling over?

☐ Tandem axles with double wheels
☐ Tandem axles with air suspension
☐ Tri-axles with single wheels
☐ Tri-axles with double wheels

The type of suspension fitted to a vehicle will influence its resistance to 'roll-over'. Modern tri-axle semi-trailers fitted with single wheels on each side extend the tracking width available, making this the most stable configuration.

1.18 Mark one answer — DGV s2

What warning does the driver of a petrol tanker have that 'rear wheel lift' is about to become 'roll-over'?

☐ The steering becomes heavy
☐ The driver has very little warning
☐ The air pressure drops
☐ The diff-lock engages

The transition from 'rear wheel lift' to 'roll-over' is more rapid on vehicles equipped with air suspension systems. Make sure you take advantage of any extra training that may be available to drivers of this type of vehicle.

1.19 Mark one answer — DGV s2

What effect could a load of hanging meat carcasses cause when you drive around corners or bends?

- ☐ The wave effect
- ☐ The camber effect
- ☐ The gravity effect
- ☐ The pendulum effect

As you turn a corner, the hanging meat carcasses will all swing to one side of your vehicle, adding extra pulling force towards the outside of the curve. The faster you turn a corner, the greater this force will be. After the corner, the carcasses will continue to swing back and forth, like pendulums, making the vehicle unstable.

1.20 Mark one answer — DGV s3

Why do many drivers park at overnight stops with their rear doors close to another lorry?

- ☐ To keep the load safe
- ☐ To ensure a clear path
- ☐ To keep 'same company' lorries together
- ☐ To stop the theft of their fuel

Load security is extremely important. Make sure you park legally and, preferably, in a well-lit area. Some lorry parks are patrolled regularly by the police or security services.

1.21 Mark one answer — HC p117

What should be fitted to a lorry with a maximum authorised mass of more than 7500 kg?

- ☐ 
- ☐ 
- ☐ 
- ☐ 

Motor vehicles over 7500 kg maximum authorised mass (MAM) and trailers over 3500 kg MAM should have these markings fitted to the rear of the vehicle/trailer.

1.22 Mark one answer — DGV s2

How far can a load overhang at the rear before you must use projection marker boards?

- ☐ 1 metre (3 feet 3 inches)
- ☐ 1.5 metres (5 feet 0 inches)
- ☐ 2 metres (6 feet 6 inches)
- ☐ 2.9 metres (9 feet 5 inches)

Projection marker boards must be fitted if the load on your vehicle overhangs by more than 2 metres (6 feet 6 inches). This will help other road users who may not notice the projecting load.

1.23 Mark one answer DGV s2

When must you notify telephone companies that you're moving a high load?

- ☐ When the load's height exceeds 4.00 metres (13 feet)
- ☐ When the load's height exceeds 4.30 metres (14 feet 2 inches)
- ☐ When the load's height exceeds 5.00 metres (16 feet 6 inches)
- ☐ When the load's height exceeds 5.25 metres (17 feet 6 inches)

You should tell telephone companies about your intended route when planning the movement of loads over 5.25 metres (17 feet 6 inches) high. You should tell them in plenty of time before making the journey.

1.24 Mark one answer DGV s2

How wide can a load be before you must use side markers?

- ☐ 2.0 metres (6 feet 6 inches)
- ☐ 2.9 metres (9 feet 5 inches)
- ☐ 3.5 metres (11 feet 5 inches)
- ☐ 4.3 metres (14 feet 2 inches)

Side markers must be displayed if your load is over 2.9 metres (9 feet 5 inches) wide. Make sure that they're clearly visible, at both the front and rear, and that they indicate the actual width of the projection.

1.25 Mark one answer DGV s2

Triangular projection markers are required when your load is wider than 2.9 metres (9 feet 5 inches). What colour are these markers?

- ☐ Black/yellow
- ☐ Red/yellow
- ☐ Black/white
- ☐ Red/white

The marker boards should be red and white. They must be kept clean and independently lit at night and in poor visibility, so that other road users can see them.

1.26 Mark one answer DGV s2

How should you secure an ISO steel cargo container onto your vehicle or trailer?

- ☐ Using battens and chocks
- ☐ Using straps
- ☐ Using twist locks
- ☐ Using ropes

If you're carrying a steel ISO (International Standards Organization) cargo container, ropes or straps won't be strong enough to take the strain. This type of load requires a special type of restraint using twist locks.

1.27 Mark one answer DGV s2

You're driving an articulated lorry on a narrow road. There's a left-hand bend ahead. Why may you need to move out before driving around the bend?

- ☐ To leave more room for braking
- ☐ To prevent anyone from overtaking
- ☐ To make room for the trailer cutting in
- ☐ To make sure oncoming drivers see you

You should always be aware of the amount of room your trailer needs when it's going around bends and corners. If you need to go onto the other side of the road, make sure there's no oncoming traffic before you move out.

1.28 Mark one answer DBC s3, DGV s2

What does this sign mean?

- ☐ Two-axled trailers are prohibited
- ☐ The bridge is too narrow for vehicles over 7.5 tonnes
- ☐ The bridge has an axle weight limit of 7.5 tonnes
- ☐ Vehicles over the weight shown are prohibited

Always look out for road signs, but be especially aware of those that refer to large or heavy vehicles. Get into the habit of checking for signs at junctions. There might be an indication on the junction layout sign. Before you turn, make sure that the road you're using doesn't have any restrictions for the vehicle you're driving.

1.29 Mark one answer DBC s3, DGV s2, HC p109, KYTS p26

What does this sign mean?

- ☐ Slippery road
- ☐ Double bend
- ☐ Overhead electrified cable
- ☐ Cable laying ahead

It's essential to know the height of your vehicle before setting off. This should be clearly marked, usually in the cab, and visible from the driving position. Look out for restrictions that you may not have seen on a map – they may be temporary.

1.30	Mark one answer	KYTS p23

What does this sign mean?

- ☐ No vehicles over 14 feet 6 inches (4.4 metres) wide
- ☐ No vehicles over 14 feet 6 inches (4.4 metres) high
- ☐ Road humps 14 feet 6 inches (4.4 metres) apart
- ☐ No vehicles over 14 feet 6 inches (4.4 metres) long

Always be aware of the height of the vehicle you're driving, particularly if you drive different vehicles. If you use a sat-nav, it may show you the height and weight limits on UK roads.

1.31	Mark one answer	HC p108, KYTS p12

What does this sign warn of?

- ☐ Low bridge ahead
- ☐ Incident ahead
- ☐ Tunnel ahead
- ☐ Accident blackspot ahead

If you're driving a high vehicle, make sure that there's enough headroom available for you to go through the tunnel. There may be additional signs showing height and width restrictions.

1.32 Mark one answer

HC p108, KYTS p12

What does this sign mean?

- ☐ The length of a tunnel
- ☐ The length of a low bridge
- ☐ The distance to a tunnel
- ☐ The distance to a low bridge

As well as a height restriction, the tunnel may have a restriction on the available width of headroom for high vehicles. The advance warning gives you the opportunity to find another route if your vehicle is too high or wide to drive through the tunnel.

1.33 Mark one answer

DBC s3, DGV s2, KYTS p24–26

When must you take extra care if you're driving a vehicle more than 3.0 metres (10 feet) high?

- ☐ When driving over narrow bridges
- ☐ When driving up steep hills
- ☐ When driving near airports
- ☐ When driving under overhead cables

You must take care when approaching any hazard where height is limited. Know the height of your vehicle and what you can drive under safely. If you aren't sure that it's safe, take another route.

1.34 Mark one answer

DBC s3, DGV s2

What's the minimum height of an unmarked bridge?

- ☐ 4.5 metres (15 feet)
- ☐ 4.7 metres (15 feet 6 inches)
- ☐ 4.8 metres (16 feet)
- ☐ 5.0 metres (16 feet 6 inches)

If a height isn't shown on the bridge, the headroom (in the UK) will be at least 5 metres (16 feet 6 inches).

1.35 Mark one answer

NI EXEMPT

DBC s3, DGV s2

Who must you tell when your vehicle collides with a bridge?

- ☐ The police
- ☐ The local authority
- ☐ Your local garage
- ☐ The fire service

Hitting a bridge with your vehicle can have serious consequences for road and rail traffic. You must tell the police about the incident immediately. If the bridge is a railway bridge, call the railway authority as well as the police to report the incident.

1.36 Mark one answer — DBC s3, DGV s2

Your vehicle has collided with a railway bridge. What information must you give the railway authority when you telephone them?

- ☐ The make and model of your vehicle
- ☐ The type of bridge
- ☐ The vehicle's height
- ☐ The bridge number

The railway authority needs to know immediately if one of their bridges has been hit. You'll need to tell them the number of the bridge so they can identify it. The railway authority will then take action to prevent railway passengers from being put at risk.

1.37 Mark one answer — DBC s3, DGV s2, KYTS p26

What do bells hanging across the road warn drivers of?

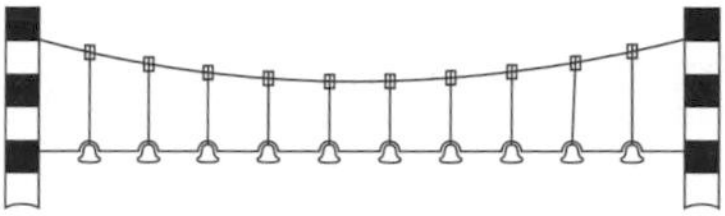

- ☐ Weight restriction ahead
- ☐ Trams crossing ahead
- ☐ Overhead electric cables ahead
- ☐ Railway level crossing ahead

If your vehicle touches bells suspended over the road, they'll obviously ring. This will warn you that your vehicle exceeds the safe travelling height beneath electrified overhead cables ahead. You must stop immediately and either find another route or take advice.

1.38 Mark one answer — HC p106, KYTS p18

What does this sign mean?

- ☐ The width of the road is 6 feet 6 inches (2 metres)
- ☐ No vehicles over 6 feet 6 inches (2 metres) wide
- ☐ No vehicles over 6 feet 6 inches (2 metres) high
- ☐ Trailer length must not exceed 6 feet 6 inches (2 metres)

You must always be aware of the size of your vehicle. Look out for road signs that show a width restriction. There should be an indication of this at the entrance to the road.

Don't get into a situation where you have to reverse out of a narrow road because you haven't seen a sign.

1.39 Mark one answer — HC p102, KYTS p90–91

What does this motorway sign mean?

- ☐ 11 tonnes weight limit
- ☐ Stop: all lanes ahead closed
- ☐ Leave the motorway at the next exit
- ☐ Lane ahead closed

Warning lights show when there's danger ahead. This includes

- lane closures
- incidents
- fog
- icy roads.

1.40 Mark one answer — HC p102, KYTS p90–91

You're driving on a motorway and you see this sign. What does it mean?

- ☐ End of restriction
- ☐ End of crawler lane
- ☐ End of weight limit
- ☐ End of hard shoulder

Look out for variable-message warning signs telling you about

- lane closures
- speed limits
- hazards.

When you're clear of the restriction, you'll see a sign telling you that it has ended.

1.41 Mark one answer — DBC s3, DGV s2

In which of these places might there be a restriction on vehicles over a certain length?

- ☐ On motorways
- ☐ At freight terminals
- ☐ In road tunnels
- ☐ On dual carriageways

Look out for restrictions on long vehicles. Length restrictions are found where there isn't much room to turn (for example, in a tunnel) or where there's a risk of grounding.

1.42 Mark one answer — DBC s3, DGV s2, KYTS p25

At this roundabout, you want to take the fourth exit. What limit is in force on this road?

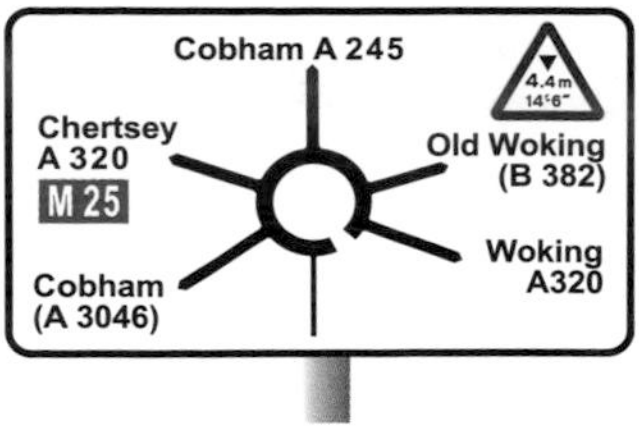

- ☐ Height
- ☐ Weight
- ☐ Width
- ☐ Length

Look out for signs giving you advance warning of height restrictions. As a professional driver, you should always know the height of your vehicle or load. If you aren't sure of the safe height, stop and check.

1.43 Mark one answer — DBC s3, DGV s2

You're driving a large vehicle on a narrow road with passing places. What will you need to be most aware of?

- ☐ Its length
- ☐ Its roof height
- ☐ Its ground clearance
- ☐ Its weight

If your vehicle is too long to fit into a passing place, you may need to wait opposite one. This will allow a following or approaching driver to pass. When you use this type of road, you need to plan and look well ahead to avoid meeting another road user at an inappropriate place.

1.44 Mark one answer — DBC s4, DGV s4, HC r126

What type of road surface will increase your stopping distance by up to ten times?

- ☐ Bumpy
- ☐ Icy
- ☐ Worn
- ☐ Wet

To prevent your wheels from locking, you'll need to brake very gently on ice. Give yourself as much room and time as possible to stop. If your front wheels lock, you can't steer, and that will mean you can't avoid any obstacle in your path.

1.45 Mark one answer DBC s4, DGV s4, HC r126

By how much can your stopping distance increase in icy weather?

- ☐ 2 times
- ☐ 4 times
- ☐ 10 times
- ☐ 20 times

It's vital that you give yourself enough separation distance when driving on icy roads – ten times as much as in normal conditions. You should brake gently to reduce the risk of losing control.

1.46 Mark one answer DBC s2, DGV s3, HC p40

What's the maximum speed allowed for a vehicle towing a trailer on a motorway?

- ☐ 40 mph
- ☐ 50 mph
- ☐ 60 mph
- ☐ 70 mph

Vehicles towing caravans or trailers on motorways are limited in two ways

- the maximum speed is 60 mph
- on motorways with more than two lanes, towing vehicles aren't allowed in the outside lane, unless other lanes are closed.

1.47 Mark one answer DBC s4, DGV s4

What should you check when you've just overtaken a motorcyclist on a motorway?

- ☐ Your speedometer
- ☐ Your left-hand mirror
- ☐ Your right-hand mirror
- ☐ The road ahead

When overtaking with a long vehicle, it's very important to be aware of the position of the road user you're passing. Check the nearside mirror to make sure you've passed safely before returning to the left. Be especially careful when overtaking motorcyclists, as they may be affected by the draught from your vehicle.

1.48 Mark one answer DBC s2, DGV s2

You're driving a large vehicle in a straight line. When is it most stable?

- ☐ During harsh acceleration
- ☐ During gentle braking
- ☐ During gentle acceleration
- ☐ During harsh braking

A vehicle is most stable when travelling in a straight line under gentle acceleration. Sudden acceleration, steering or braking can cause severe loss of control. For maximum control, keep your driving as smooth as possible.

1.49 Mark one answer KYTS p17

What does this sign mean?

- ☐ Warning of lorry crossing a one-way road
- ☐ No entry for vehicles over 32 feet 6 inches (9.9 metres) long
- ☐ No entry for vehicles over 32.6 tonnes
- ☐ Warning of lorry straight ahead

You need to know the length of your vehicle as well as its weight, height and width. Places where the length of your vehicle may be relevant are

- road tunnels
- level crossings
- ferries
- bridges.

1.50 Mark one answer DBC s6, DGV s6

You're driving a long vehicle. What should you do when you want to turn right at a painted mini-roundabout?

- ☐ Make sure none of your wheels touch the roundabout
- ☐ Avoid going over the roundabout if possible
- ☐ Go ahead, and turn right at the next junction
- ☐ Carefully mount the left-hand kerb to make more room

When turning at a mini-roundabout, it's unlikely that you'll be able to avoid going over the marked area. You should, however, avoid going over the roundabout where possible.

1.51 Mark one answer DBC s3, DGV s2

What should you consider when planning a route for your journey?

- ☐ Weight restrictions
- ☐ Speed cameras
- ☐ Speed-limit changes
- ☐ One-way streets

You should plan your route to avoid vehicle width, height and weight restrictions that would affect your vehicle and load. Planning to avoid busy times can also reduce journey times and fuel costs.

1.52 — Mark one answer — DBC s6, DGV s6

Your vehicle is more than 3 metres (9 feet 10 inches) high. Where is this information displayed?

- ☐ On the windscreen
- ☐ On the weight plate
- ☐ In the driver's cab
- ☐ In the engine bay

It's a legal requirement that information about a vehicle's height can be seen by the driver from their seat.

It's important to know the height of your vehicle so that you can avoid any height restrictions on your route.

1.53 — Mark one answer — DBC s3, DGV s2

The height of your vehicle is 4.2 metres (14 feet). What should you do as you approach this bridge?

- ☐ Keep to the centre of the arch and give way to oncoming traffic
- ☐ Drive through slowly, keeping to the left of the marked limits
- ☐ Keep to the centre of the arch and take priority over oncoming traffic
- ☐ Drive through quickly, keeping to the left of the marked limits

The headroom under bridges in the UK is at least 5 metres (16 feet 6 inches) unless marked otherwise. However, this might refer only to the highest point of an arch.

If your vehicle hits a bridge, you must report the incident to the police. If a railway bridge is involved, you must also report it to the railway authority.

1.54 — Mark one answer — DBC s3, DGV s2

Unless otherwise shown, what's the minimum headroom under bridges in the UK?

- ☐ 4.0 metres (13 feet)
- ☐ 4.1 metres (13 feet 4 inches)
- ☐ 5.0 metres (16 feet 6 inches)
- ☐ 5.5 metres (18 feet)

Every year, there are hundreds of incidents in the UK where vehicles hit railway or motorway bridges. Most involve buses, coaches and lorries. Make sure your route doesn't take you under a bridge that's too low for your vehicle. If you hit a bridge, you'll be putting yourself and any passengers at risk, and you could cause a major disruption to road and rail transport.

1.55 — Mark one answer — DBC s4, DGV s4

Where can you park large vehicles at night without needing to use parking lights?

- ☐ In an off-road parking area
- ☐ On a road with a 20 mph speed limit
- ☐ At least 10 metres (32 feet) away from any junction
- ☐ In most lay-bys

Off-road lorry and coach parks are often well lit and patrolled by police or security firms. Never leave your vehicle unlit on a public road or in a lay-by after dark.

1.56 — Mark one answer — DBC s4, DGV s3

When should you allow for the effects of a speed limiter?

- ☐ When you're cornering
- ☐ When you're braking
- ☐ When you're overtaking
- ☐ When you're changing gear

Forward planning is important whenever you consider overtaking another road user. You should carefully assess the speed of the vehicle you intend to overtake. Remember that your vehicle's speed is limited and this could considerably increase the distance and time needed to complete the manoeuvre safely.

1.57 — Mark one answer — DBC s4, DGV s3

When should you consider the effect a speed limiter will have on your vehicle?

- ☐ When braking
- ☐ When changing gear
- ☐ When overtaking
- ☐ When reversing

Plan well ahead before overtaking. Be aware that a speed limiter may cause you difficulties when overtaking another vehicle, particularly when climbing a hill.

1.58 Mark one answer DBC s3, DGV s2, HC p109

What does this sign mean?

- ☐ Hump bridge
- ☐ Risk of grounding
- ☐ Uneven road
- ☐ Road liable to subsidence

If you see this sign, be aware that there's a danger of grounding. This can happen where there's a pronounced bump in the road, such as at a level crossing or a hump bridge.

1.59 Mark one answer DBC s3, DGV s2, HC p109

What should be your main concern as you approach this sign?

- ☐ Your vehicle's height
- ☐ Your vehicle's width
- ☐ Your vehicle's weight
- ☐ Your vehicle's length

At hump bridges, there's a risk of long vehicles grounding. If you aren't sure whether the floor of your vehicle will clear the bridge, find another route.

1.60 Mark one answer DBC s2, DGV s2

You're driving a long vehicle on a two-lane road. What should you do if you want to turn left into a narrow side road?

- ☐ Keep well to the left on the approach
- ☐ Move out to the right immediately before turning
- ☐ Keep to the left and turn later than normal
- ☐ Straddle the lanes to make more room for the turn

When you're driving a long vehicle, you'll often need to straddle lanes to turn into narrow side roads. When you do this, watch out for smaller vehicles – especially motorcycles and cycles – trying to squeeze past on your inside.

1.61 Mark one answer — DBC s1, DGV s2

Where must the repair of a speed limiter be carried out?

- ☐ At an authorised speed-limiter centre
- ☐ At any DVSA site
- ☐ At any dealer centre
- ☐ At the depot by a mechanic

Speed limiters may only be repaired by authorised speed-limiter centres. They'll ensure that all the connections are sealed and that the system is tamperproof.

1.62 Mark one answer — DBC s2, DGV s3

Where can the axle weight limits be found?

- ☐ On the vehicle plate
- ☐ On the operator's licence
- ☐ On the wheel rims
- ☐ On the weighbridge printout

Individual axle weights are shown on the vehicle plate. This can be found in a prominent place on both the vehicle and trailer.

1.63 Mark one answer — DBC s3

By approximately how much would 15 passengers increase the weight of your vehicle?

- ☐ 0.5 tonnes
- ☐ 1 tonne
- ☐ 2.5 tonnes
- ☐ 3 tonnes

You must be aware of the maximum authorised mass (MAM) of your vehicle. This refers to its total weight, including both passengers and luggage. Fifteen passengers would add approximately 1 tonne to the weight of your vehicle. You should also allow for any luggage that they may be carrying.

1.64 Mark one answer — DBC s3

As a guide, how many passengers equal 1 tonne?

- ☐ 15
- ☐ 20
- ☐ 25
- ☐ 30

Your bus will move away more slowly and handle differently when fully laden with passengers and their luggage. You'll have to make allowances for this, especially when moving off uphill.

1.65 Mark one answer DBC s3

How would you know that a weight-limit sign doesn't apply to the bus you're driving?

- ☐ By a plate fitted beneath the weight-limit sign
- ☐ By the colour of the weight-limit sign
- ☐ By a plate attached to the vehicle
- ☐ By a certificate carried by the driver

Some weight restrictions apply to large goods vehicles only and not to passenger-carrying vehicles. Look for a plate beneath a restriction sign that indicates this.

Road signs show weight restrictions in various ways. You should make yourself familiar with all of them, so that you're in no doubt about their meaning and relevance.

1.66 Mark one answer DBC s3

You're driving a bus on a local service. When can you use this lane?

- ☐ Between 4.00 pm and 6.30 pm only
- ☐ Before 4.00 pm and after 6.30 pm only
- ☐ At any time of the day
- ☐ Any time except Saturdays

As a local-service bus driver, you may use the bus lane at any time. The restrictions apply to other road users, to keep the lane clear for you at peak times. Don't be tempted to speed when driving up the inside of slow-moving or stationary traffic. Be alert for pedestrians who may be trying to cross the road.

1.67 Mark one answer HC p40

What's the national speed limit for buses and coaches on a dual carriageway?

- ☐ 55 mph
- ☐ 60 mph
- ☐ 65 mph
- ☐ 70 mph

Don't be tempted to drive on the limiter when using dual carriageway roads. The speed limit for passenger-carrying vehicles is 60 mph. Be considerate to faster-moving traffic by not using the middle or outside lanes unnecessarily.

1.68 Mark one answer DBC s7

What does a D1 category licence allow you to drive?

- ☐ Buses with a maximum of 16 passenger seats
- ☐ Buses with a maximum of 24 passenger seats
- ☐ Buses with a maximum of 32 passenger seats
- ☐ Buses with a maximum of 48 passenger seats

The D1 category allows you to drive any bus with 9 to 16 passenger seats for hire or reward.

1.69 Mark one answer DBC s3

Where can a driver find their vehicle's unladen weight recorded?

- ☐ On the dashboard of the vehicle
- ☐ On the driver's duty roster
- ☐ On the side of the vehicle
- ☐ On the depot noticeboard

The driver of a passenger-carrying vehicle needs to know the vehicle's limits. These include its weight, height, width and length, and its seating and standing capacity. The unladen weight should be displayed on the nearside of the vehicle.

1.70 Mark one answer DBC s1, 7

What category of licence is required to drive an articulated bus?

- ☐ D
- ☐ D+E
- ☐ D1
- ☐ D1+E

A full category D licence entitles you to drive all buses and coaches, including the articulated type.

A full category D licence on its own doesn't cover you to drive buses or coaches drawing detachable trailers of more than 750 kg; for this you'll need a D+E licence.

1.71 Mark one answer DBC s3

The front of your bus overhangs well past the front wheels. Why should you allow for this when cornering?

- ☐ The steering will be much heavier
- ☐ Your speed will be more difficult to control
- ☐ You might hit something close to the road
- ☐ You'll need to brake earlier than normal

The front of a bus or coach often extends well beyond the front wheels. Make allowances for this when turning. Try to avoid overhanging the pavement where possible, but if you have to do this, look out for fences, posts, traffic signs, etc. Be especially careful where pedestrians are present; for example, in bus stations.

1.72 Mark one answer DBC s3

Your bus has a speed limiter fitted. What other related item must it have?

- ☐ An audible warning device for the driver
- ☐ A plate in the cab showing the limited speed
- ☐ A warning sign on the back of the bus
- ☐ A manual override switch for emergencies

A bus equipped with a speed limiter must also be fitted with a plate showing the speed setting. This must be fitted in an obvious position, where it can be clearly seen.

1.73 Mark one answer DBC s3

A speed limiter is fitted to a bus. Where must the speed-limiter setting be displayed?

- ☐ In the driver's cab
- ☐ On the nearside of the vehicle
- ☐ On the rear of the vehicle
- ☐ On the driver's side at the front of the vehicle

If a speed limiter is fitted to a vehicle, there must be a notice showing the speed at which it's set, clearly displayed in the driver's cab.

1.74 Mark one answer DBC s3

What does the term 'maximum authorised mass' (MAM) mean?

- ☐ The weight of your vehicle, with passengers but no luggage
- ☐ The weight of your vehicle, without passengers or luggage
- ☐ The weight of your vehicle, with luggage but no passengers
- ☐ The weight of your vehicle, with both luggage and passengers

Maximum authorised mass (MAM) means the total weight of a vehicle or trailer, including the maximum load that can be carried safely when it's being used on the road. It's also known as gross vehicle weight (GVW) or permissible maximum weight. You're responsible for making sure the MAM isn't exceeded when the vehicle is fully loaded.

1.75 Mark one answer DBC s1

How do speed limiters normally work?

- ☐ By applying the service brake
- ☐ By applying the secondary brake
- ☐ By reducing the fuel supplied to the engine
- ☐ By changing to a lower or higher gear

Most speed limiters work by taking a speed signal from the tachograph and transmitting it to the engine management system. This reduces the fuel supply to the engine, and allows the speed of the vehicle to be controlled to specific limits. The speed at which a limiter is set must be shown on a plate clearly displayed in the cab. You should be aware of the effect a speed limiter will have on your vehicle, especially when overtaking.

1.76 Mark one answer DBC s3

What's a vehicle's turning circle?

- ☐ The number of turns of the steering wheel between locks
- ☐ The amount of space needed for the vehicle to turn
- ☐ The amount by which the vehicle overhangs kerbs
- ☐ The distance by which a vehicle cuts corners

You should be familiar with the room your vehicle needs to make turns or carry out a manoeuvre. You'll need to consider this as you approach junctions and road layouts. If you're driving a new or temporary vehicle, familiarise yourself with its characteristics before you drive on public roads.

Section two

Drivers' hours and rest periods

In this section, you'll learn about

- obeying driving limits
- following tachograph rules
- keeping the correct records
- tiredness
- vehicle security.

Drivers' hours and rest periods

Obeying driving limits

When you drive any large vehicle professionally, you **MUST** follow the EU drivers' hours rules and regulations. These tell you

- how long you can drive before you must have a break
- the minimum time you must allow for each rest break.

There are also **domestic** rules about drivers' hours. These apply when you drive a vehicle on UK journeys that are specifically **exempted** from the EU rules.

Definition

domestic
Within a country.

exempted
Not required to comply with a rule or regulation, or not covered by a rule or regulation.

If you don't follow the EU or domestic rules, you can expect heavy fines and you may lose your driving licence (see **www.gov.uk**).

DBC s3 DGV s3

Following tachograph rules

Types of tachograph

Tachographs can be digital or analogue.

- Digital tachographs record your activity in the memory chip on a **driver smart card**.

- Analogue tachographs record your activity on a round paper chart.

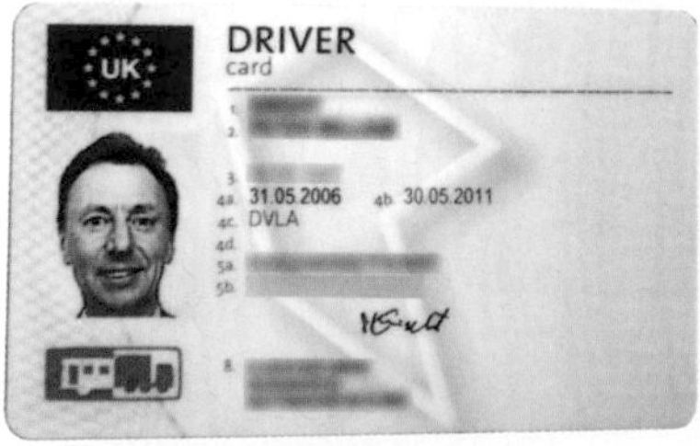

tachograph

A device fitted to a vehicle that records its speed, the distance driven and the driver's activity.

driver smart card

The card used in a digital tachograph. Each driver has their own card, which can be used in any vehicle fitted with a digital tachograph.

You're not allowed to have more than one valid driver smart card. Like a bank card, you may only use the card that has been issued to you.

A driver smart card is valid for up to five years. It's your responsibility to renew it before it runs out.

TIP

A driver smart card can record information for 28 days. If you use it for longer than this, some information will be written over and lost. To avoid that happening, regularly **download** and store your driver records.

download

To transfer data to a local system from a remote system (e.g. a driver smart-card chip).

Your driver smart card may provide a complete record of your activity over the previous 28 days. However, you may need to use what's on your driver smart card plus some of your stored records.

The paper charts that you put in an analogue tachograph cover 24 hours. You **MUST** complete a new chart every day. You **MUST** carry with you the chart for the current day and those for the previous 28 days.

If an **enforcement officer** asks you to produce your charts, driver records or driver smart card, you **MUST** do this.

Definition

enforcement officer
Anyone who's responsible for enforcing the law, such as the police, traffic officers or Driver and Vehicle Standards Agency (DVSA) officers.

It's a serious offence to tamper with the seal of a tachograph or falsify any record of a driver's activity with **intent to deceive**. If found guilty, you can be fined an unlimited amount or imprisoned for up to two years, or both.

Definition

intent to deceive
The will or purpose with which one acts to defraud or cheat.

Using your tachograph

A tachograph has four modes, each with its own symbol.

	driving
	doing other work
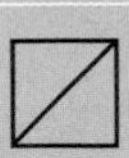	**periods of availability**
	taking a break or rest

periods of availability

Under drivers' hours regulations, waiting time that's known about in advance. Full details of the rules about periods of availability are available in the DVSA guides 'Rules on drivers' hours and tachographs' (GV262 and PSV375).

Under 'other work', you should record any time you spend on

- vehicle walk-round checks
- loading or unloading
- travelling to join or leave the vehicle when asked to by your employer
- training courses.

You should have your own driver smart card or tachograph charts. This is your own personal work record.

TIP If you need to leave the vehicle for a while and you think someone else might use it, take your chart with you and record this time on the back of the chart, or remove your driver smart card.

If you're travelling with a co-driver, always use your own chart or driver smart card to record when you drive, rest or do other work.

If you lose your driver smart card

It's up to you to tell DVLA (DVA in Northern Ireland) if you lose your driver smart card. DVLA will supply you with a replacement card but there will be a fee to pay. In Northern Ireland, you'll need to collect and sign for your replacement card at a local vehicle licensing office or DVA test centre.

If your card is lost, stolen or becomes faulty, you must

- take a tachograph printout at the start and end of each day, showing when you start and finish
- write down the times you spend in each of the four modes (see page 75).

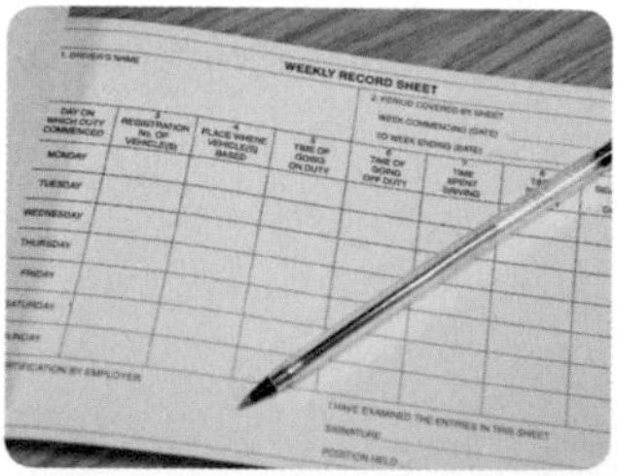

You can do this for up to 15 days. If you don't have a new card by then, you **MUST** stop driving vehicles that use digital tachographs until you have one.

If your tachograph becomes faulty during a journey, you **MUST** keep a **manual record** of everything you do until the end of the journey. If you won't be able to return to base within a week of when your tachograph stops working properly, you **MUST** have it repaired while you're away.

Definition

manual record
A record that a driver keeps of their activities when the vehicle's tachograph isn't working or a record kept by a driver working under domestic drivers' hours.

DBC s3 DGV s3

Keeping the correct records

Your tachograph creates a record of your hours of work and the time you spend driving, resting or doing other work.

- You **MUST** carry with you records for the previous 28 days.
- You **MUST** make sure other people can read your tachograph records.
- You **MUST** allow enforcement officers to inspect your tachograph records at the roadside.

Employers **MUST** keep tachograph records and printouts for at least one year from the date of their use. These **MUST** be

- in chronological order
- in a legible form.

DBC s3 DGV s3

Tiredness

The rules on taking breaks or rest between periods of driving are clear and you **MUST** follow them. If you don't, you might face a heavy fine or even lose your licence.

Don't wait for your next official break if you feel tired or can't concentrate.

- Open a window or turn down the heating so that you don't get drowsy.
- Stop as soon as it's safe to do so.

DBC s3 **DGV** s3

See the Think! road safety website for more information about driving when tired.

http://think.direct.gov.uk/fatigue.html

Vehicle security

Always keep your vehicle secure, especially when you carry a valuable load such as alcohol or tobacco. If you regularly carry valuable goods like these, follow this advice:

- Try to vary the route you take. This makes it harder for criminals to guess where you'll be and when.
- Ask your local crime prevention officer for advice on how to avoid being targeted by criminals.
- If you leave your vehicle for any reason during the journey, be aware of what's going on around you, especially if you use the same route each time.

Always lock your vehicle when leaving it, even if just for a short time. Fit a kingpin or drawbar lock to your trailer if you're going to leave it unattended.

At night, try to park in a secure area and, if possible, block your rear doors by parking near a wall or close to another parked vehicle.

Think carefully before you give a stranger a lift and don't talk about what you're carrying with anyone you don't know well.

You **MUST** stop if a police or enforcement officer asks you to do so. If you're asked to leave your vehicle by an official who isn't in uniform, always request to see their **warrant card** first.

Definition

warrant card

An official document that identifies the holder as having legal authority to perform certain functions.

DGV s3

Meeting the standards

You must be able to

assess whether drivers' hours regulations allow you to drive
follow your organisation's procedures when you're legally unable to drive
operate the tachograph correctly – when it's legally required for the service being performed
continuously monitor whether you're tired while driving and take suitable action.

You must know and understand

the principles of the drivers' hours legislation and how they apply to you
how to operate tachograph equipment
that employers who cause their drivers to exceed drivers' hours are guilty of an offence, but so are the drivers themselves.

Notes

You can use this page to make your own notes or diagrams about the key points you need to remember.

Think about

- Do you know the EU and domestic regulations on drivers' hours?
- What sort of tachograph will you use?
- Do you need a driver smart card?
- How will you store your driver records?
- Do you know how to make manual records if your tachograph becomes faulty?
- How can you keep your vehicle and its load safe and secure?

Your notes

Things to discuss and practise with your instructor

These are just a few examples of what you could discuss and practise with your instructor. Read more about drivers' hours and rest periods to come up with your own ideas.

Discuss with your instructor

- the differences between EU and domestic rules on drivers' hours
- using a tachograph. For example,
  - what it records and the symbols used
  - the number of charts you should keep and their condition
  - the distance of journey over which you must use a tachograph
  - what to do when an enforcement officer keeps your chart
  - the consequences of breaking tachograph regulations
- what to do in the event of an emergency that means you'll have to go over your normal drivers' hours.

Practise with your instructor

- what you'll do if you become tired when driving
- using a tachograph, either analogue or digital
- parking your vehicle in a secure place, as if for an overnight stop.

2.1 Mark one answer DGV s3

Why are goods vehicle drivers' hours of work controlled?

- ☐ To encourage vehicle sympathy
- ☐ To help road safety
- ☐ To ensure fair road use
- ☐ To improve vehicle security

Drivers need to be alert at all times. Fatigue is a recognised danger for professional drivers, so working conditions are governed by EU rules. These set a maximum driving time and minimum requirements for rest and break periods. Setting limits on the length of time a driver can work combats fatigue and so helps road safety.

2.2 Mark one answer DGV s3

What does controlling goods vehicle drivers' hours help to improve?

- ☐ Fuel economy
- ☐ Road safety
- ☐ Traffic calming
- ☐ Journey planning

EU drivers' hours regulations are a series of rules that specify maximum driving time and minimum rest and break periods. These rules are in place to help road safety. Drivers who break them are subject to heavy fines and could lose their licence to drive lorries.

2.3 Mark one answer DGV s3

What must you do when driving under the rules for domestic drivers' hours?

- ☐ Keep a written record of hours worked
- ☐ Only record any driving off public roads
- ☐ Keep a written record of driving time only
- ☐ Always use a vehicle fitted with a tachograph

Domestic rules apply to certain journeys within Great Britain that aren't subject to EU rules. Under domestic rules you must keep a written record of your hours of duty when driving goods vehicles.

2.4 Mark one answer DGV s3

You must have enough tachograph charts with you for your journey. When do you need to start a new chart?

- ☐ Every 10 hours
- ☐ Every 24 hours
- ☐ Every 36 hours
- ☐ Every 48 hours

If you're using a vehicle equipped with an analogue tachograph, you'll need at least one chart for every 24 hours. Your employer should supply you with enough tachograph charts for your entire journey.

2.5 Mark one answer DGV s3

What period of time does one tachograph chart cover?

- ☐ 24 hours
- ☐ 48 hours
- ☐ 5 days
- ☐ 7 days

Your tachograph chart is a legal document; it's a record of your work covering a rolling 24-hour period. Drivers who break the rules are subject to heavy fines and could lose their vocational licence. Altering your tachograph chart with intent to deceive is against the law and could lead to a prison sentence. Similar penalties exist for those who permit such offences.

2.6 Mark one answer DGV s3

What should you do if your tachograph chart becomes dirty or damaged?

- ☐ Continue with the same chart and enter the details in writing
- ☐ Use a spare chart and destroy the damaged one
- ☐ Use a spare chart and attach it to the damaged one
- ☐ Continue to use the chart

If your current tachograph chart becomes damaged, you should start another and then attach it to the damaged one. Your records must be clear and up to date at all times. It's sensible to carry more tachograph charts than you think you'll need for your journey. Then you'll be able to use a spare if one becomes dirty or damaged.

2.7 Mark one answer DGV s3

During your break, your vehicle will be moved by another person. What should you do with the tachograph chart?

- ☐ Leave the chart in the vehicle and record the changes on the back
- ☐ Put in a new chart on your return to the vehicle
- ☐ Switch to rest mode to record the break
- ☐ Remove the chart and make a manual record of the break period

If your vehicle is likely to be used by another person while you're away from it, you should take your tachograph chart with you. Your break from driving should be entered on the reverse of the chart.

2.8 Mark one answer DGV s3

You've been driving a lorry without a break for four-and-a-half hours. Under EU rules, you must now take a break. How long must this break be?

- ☐ 30 minutes
- ☐ 35 minutes
- ☐ 40 minutes
- ☐ 45 minutes

If you're driving under EU rules, you mustn't drive continuously for more than four-and-a-half hours without taking a break. If you've driven continuously for four-and-a-half hours, you must take a break of at least 45 minutes. Include your stops in the timetable when planning your journey.

2.9 Mark one answer DGV s3

You're driving a lorry on a motorway and you start to feel drowsy. What should you do when there are no service areas or exits for some distance?

- ☐ Stop on the hard shoulder and rest
- ☐ Open the window and turn down the heating
- ☐ Slow down and use the hazard warning lights
- ☐ Increase your speed to get to the next service area sooner

During very cold weather, it's tempting to have the heating in the cab turned on full. Be aware that this may help to make you feel drowsy, especially after you've eaten a meal. Open the window slightly to let in some fresh air until it's safe for you to stop and rest.

2.10 Mark one answer DGV s3

You're driving a lorry. During the journey you begin to feel ill and are unable to concentrate. What should you do?

- ☐ Stop in a safe place and seek help
- ☐ Continue your journey and keep your windows open
- ☐ Increase your speed to finish your work earlier
- ☐ Keep stopping at regular intervals for rest

If you become unwell, it will affect your ability to concentrate. You must be fully alert and ready for any hazards that might occur while you're driving. Stop in a safe place and call for help. You may have to call out a relief driver to complete the journey for you.

2.11 Mark one answer DGV s3

What should you do to prevent your load from being stolen?

- ☐ Give a lift to a stranger
- ☐ Make sure all doors and windows are locked
- ☐ Discuss your load with members of the public
- ☐ Have wind deflectors fitted

Be careful of giving lifts to strangers; some employers actively discourage it. Allowing strangers in your cab or telling people about your load could put the security of your vehicle and load at risk, and may put you in danger.

2.12 Mark one answer DGV s3

How can you reduce the likelihood of theft of your load when leaving your vehicle overnight?

- ☐ Park with the rear doors close to another vehicle
- ☐ Park with the rear doors well away from another vehicle
- ☐ Park with the front doors well away from another vehicle
- ☐ Park with the front doors close to another vehicle

Theft of loads from commercial vehicles is a serious problem. You should take steps to safeguard your load when leaving the vehicle overnight. Simply parking with the rear doors close to another vehicle or a wall can act as an effective deterrent to thieves.

2.13 Mark one answer DGV s3

What security precautions should you take if you're carrying a high-value load and have to park and sleep overnight in the cab?

- ☐ Lock the doors but leave a window open for ventilation
- ☐ Make sure the doors and windows are secure
- ☐ Stay at the same location regularly
- ☐ Park in a quiet, unlit, non-residential area

Both your load and your vehicle can be a target for thieves. Taking simple precautions, such as securing the windows and doors, can deter thieves and help ensure your own safety.

2.14 Mark one answer DGV s3

You're often involved in the carrying of high-value goods. What security measures can you adopt?

- ☐ Vary your routes and rest stops
- ☐ Always discuss details of your load
- ☐ Give lifts to anyone for added security
- ☐ Keep your journeys to a strict routine

When carrying high-value goods, you can become a target for thieves. Avoid developing a set routine or pattern. Vary your routes whenever possible to make it difficult for thieves to predict when and where your cargo can be intercepted.

2.15 Mark one answer DGV s3

Where should you park your trailer if you have to leave it unattended?

- ☐ In a public car park
- ☐ On the public highway
- ☐ On secure premises
- ☐ In a quiet residential area

Theft of vehicles is common. You're responsible for the safety and security of your vehicle and trailer. Try to avoid leaving any trailer unattended unless it's on approved secure premises.

2.16 Mark one answer DGV s3

When should your trailer be fitted with a kingpin or drawbar lock?

- ☐ When it's being driven on a motorway
- ☐ When it's being driven abroad
- ☐ When it's being used partially loaded
- ☐ When it's left unattended

A kingpin or drawbar lock is a visible and effective deterrent to thieves wishing to steal an unattended trailer.

2.17 Mark one answer DGV s3

Before starting driving, which of the following should you complete on the centre field of your tachograph chart?

- ☐ The starting point of your day's journey
- ☐ Details of the goods carried
- ☐ The name and address of your employer
- ☐ The amount of daily rest taken prior to starting the shift

Before starting your journey, you must record a number of items on your tachograph chart. One of these is where the journey begins.

2.18 Mark one answer DGV s3

During your working day, you change to another vehicle with the same type of tachograph. How should you keep your tachograph record up-to-date?

- ☐ Use the chart that's already in the other vehicle
- ☐ Take the chart with you and use it in the other vehicle
- ☐ Record your driving hours in a record book
- ☐ Install a new chart in the other vehicle

If you change vehicles during the working day, you should take your chart with you and use it in the next vehicle. This isn't always possible, however, as charts produced by different manufacturers may not be interchangeable. In this case, you should use another chart, making sure that all the information for the day is recorded.

2.19 Mark one answer DGV s3

Which of the following is good practice in keeping your lorry and its load safe?

- ☐ Parking in a quiet area, out of sight
- ☐ Making sure the back doors are accessible
- ☐ Parking in a well-lit lorry park
- ☐ Showing your papers to anyone who stops you

Load security is one of the many responsibilities of the driver. When choosing a site to park your vehicle overnight, you should always look for a location that's legal and well lit. Many allocated lorry parks are patrolled by the police or security firms.

2.20 Mark one answer DGV s3

You're driving a lorry at night. What can you do to help yourself stay alert?

- ☐ Eat a heavy meal before setting off
- ☐ Keep plenty of cool, fresh air moving through the cab
- ☐ Keep the cab warm and comfortable
- ☐ Drive faster to get to your destination sooner

Make sure you have enough rest before you start a night shift. Driving at night can be tiring and you must be able to stay alert for the whole of your shift.

Making sure that there's enough fresh air in the cab can help you to stay alert. Stale, warm air can dull your senses and cause drowsiness.

2.21 Mark one answer DGV s3

How can you reduce the risk of your lorry or trailer being stolen?

- ☐ Fit an alarm and immobiliser
- ☐ Park alongside another vehicle or a wall
- ☐ Use the same route and stop in the same places for your rest periods
- ☐ Park in quiet areas, away from other vehicles

You're responsible for your vehicle, so you should reduce the risk of it being stolen. Plan any lengthy stops for places where your vehicle will be safest, such as a well-lit rest stop at night. Also have an alarm and immobiliser fitted to the vehicle by a security specialist.

2.22 Mark one answer DGV s3

You're planning to carry high-value goods on a regular basis. Whose advice should you seek?

- ☐ Other drivers in your area
- ☐ Your local crime-prevention officer
- ☐ Other operators in your area
- ☐ Your local road-safety officer

High-value cargoes are often targeted by thieves. Operators are recommended to seek advice from their local crime-prevention officer to help prevent their vehicles and loads from being stolen.

2.23 Mark one answer DGV s3

What period of time makes up a driver's week?

- ☐ 00.00 hours Monday to 24.00 hours the following Sunday
- ☐ 00.00 hours Sunday to 24.00 hours the following Saturday
- ☐ Any 7 consecutive days
- ☐ Any 56 hours driven

A driver's week is defined as a period from 00.00 hours on Monday to 24.00 hours the following Sunday.

2.24 Mark one answer DGV s3

While you're driving, you notice your tachograph isn't working. What should you do?

- ☐ Stop immediately and don't drive until it's been repaired
- ☐ Report it to the nearest police station
- ☐ Phone the vehicle testing authority and report the fault
- ☐ Continue your journey but make a manual record

If you can't return to base within a week of the tachograph becoming defective, it must be repaired during the journey. While it's broken, you must keep a manual record.

2.25 — Mark one answer — DBC s3, DGV s3

What can happen to drivers who break EU tachograph regulations?

- ☐ They can be given three warnings
- ☐ They can have their passport withdrawn
- ☐ They can be heavily fined
- ☐ They can be let off if they're new to using tachographs

The driver must take responsibility and follow the drivers' hours and tachograph rules. Failure to do so can result in legal action and penalties – such as a fine.

2.26 — Mark one answer — DBC s3, DGV s3

What must you do when using a vehicle fitted with an analogue tachograph?

- ☐ Carry enough approved charts
- ☐ Use damaged charts if they're clean
- ☐ Use dirty charts if they're undamaged
- ☐ Reuse charts twice

Make sure you carry enough approved tachograph charts for your journey. Store your spare charts in a plastic wallet to keep them clean and undamaged.

2.27 — Mark one answer — DBC s3, DGV s3

The analogue tachograph on your vehicle becomes faulty. How long do you have before you must have it repaired?

- ☐ One day
- ☐ Three days
- ☐ One week
- ☐ Two weeks

If the tachograph on your vehicle becomes faulty, you should take it to an approved tachograph repairer as soon as possible. If you can't return to your base within a week of finding the fault, it must be repaired while you're away. While the tachograph is faulty or broken, you must keep a manual record of your activities.

2.28 Mark one answer DBC s3, DGV s3

An enforcement officer keeps one of your tachograph charts. Who should sign the back of the replacement chart?

- ☐ You, the driver
- ☐ Your transport manager
- ☐ The vehicle owner
- ☐ The officer

When an enforcement officer keeps a record chart, the driver should ask the officer to sign the back of the replacement chart. They'll need to give their name, telephone number and the number of charts they've kept. The replacement chart must be used to continue the journey. You should always carry more blank charts than you think you'll need.

2.29 Mark one answer DBC s3, DGV s3

An enforcement officer keeps your analogue tachograph records. How should they endorse the replacement charts?

- ☐ With their name
- ☐ With their service number
- ☐ With their home address
- ☐ With their date of birth

If your records are kept by an enforcement officer, you should ask the officer to endorse the replacement charts with their name and telephone number. The enforcement officer should also record the number of charts they keep. Alternatively, they can give you a receipt for the charts.

2.30 Mark one answer DBC s3, DGV s3

Why should you carry spare tachograph charts?

- ☐ As a defence against a speeding prosecution
- ☐ To record when you've been in a traffic delay
- ☐ For recording extra loading duties and overtime
- ☐ To replace the original chart if it gets dirty

Your employer should supply enough approved charts for your journey. They should include spares in case any get damaged or are taken by an authorised inspecting officer.

2.31 Mark one answer DBC s3, DGV s3

Under EU rules, a driver must take regular breaks from driving. What maximum period of continuous driving is allowed?

- ☐ 3 hours
- ☐ 4 hours
- ☐ 4.5 hours
- ☐ 5.5 hours

It's essential that you don't become drowsy through driving for excessively long periods. EU rules are in place to prevent this. You must take a break after a maximum of 4.5 hours of continuous driving. Your work details must be available for inspection by enforcement staff. Make sure that you know the rules for the journey and the type of vehicle you're driving.

2.32 Mark one answer — DBC s3, DGV s3

You've been driving non-stop since 5.00 am. The time is now 9.30 am. Under EU rules, what length of break must you take?

- ☐ At least 15 minutes
- ☐ At least 30 minutes
- ☐ At least 45 minutes
- ☐ At least 60 minutes

You must take an uninterrupted break of 45 minutes after four-and-a-half hours of driving. This break may be replaced by two shorter breaks of at least 15 and 30 minutes, taken in that order, during the four-and-a-half hours. During any break, you mustn't drive or do any other work.

2.33 Mark one answer — DBC s3, DGV s3

Under EU rules, what's the maximum daily driving time allowed?

- ☐ 9 hours, extended to 11 hours on three days of the week
- ☐ 10 hours, extended to 11 hours on two days of the week
- ☐ 9 hours, extended to 10 hours on two days of the week
- ☐ 10 hours, extended to 11 hours on three days of the week

You're allowed to extend your daily driving time twice a week, to 10 hours a day. A 'day' is generally any 24-hour period that starts when you begin driving (or other work) again, after the last daily or weekly rest period.

2.34 Mark one answer — DBC s3, DGV s3

Under EU rules, your minimum daily rest is 11 hours. On three days of the week this may be reduced to what length of time?

- ☐ 7 hours
- ☐ 8 hours
- ☐ 9 hours
- ☐ 10 hours

Under EU rules, you must have a minimum daily rest of 11 consecutive hours. A reduced daily rest period is any period of rest of at least 9 hours, but less than 11 hours.

2.35 Mark one answer — DBC s3, DGV s3

Under EU rules, your daily rest can be reduced to 9 hours. For how many days per week is this allowed?

- ☐ One day
- ☐ Two days
- ☐ Three days
- ☐ Four days

Under EU rules, you must have a minimum daily rest of 11 consecutive hours. However you may reduce this to 9 hours for up to three days in any one week.

2.36 Mark one answer — DBC s3, DGV s3

Under EU rules, how long should your normal daily rest period be?

- ☐ 8 hours
- ☐ 11 hours
- ☐ 13 hours
- ☐ 14 hours

Drivers' hours and rest periods are controlled in the interests of road safety. A driver's daily rest period should normally be 11 hours, but this may be reduced to 9 hours for three days in one week. It's a serious offence to break these rules and anyone doing so is liable to a heavy fine and even imprisonment.

2.37 Mark one answer — DBC s3, DGV s3

Under EU rules, what's the normal weekly rest period that must be taken?

- ☐ 40 hours
- ☐ 41 hours
- ☐ 42 hours
- ☐ 45 hours

The working week is defined as from 00.00 hours on Monday to 24.00 hours on the following Sunday. When taking the weekly rest period, a daily rest period must normally be extended to at least 45 consecutive hours.

2.38 Mark one answer — DBC s3, DGV s3

A vehicle fitted with an analogue tachograph has two drivers. How should they use the tachograph?

- ☐ Share the same tachograph chart
- ☐ Use a separate tachograph chart for every driving period
- ☐ Use their own tachograph chart
- ☐ Not use the tachograph for such duties

Your tachograph chart is your personal work record and should only reflect the hours that you drive or do other work. The law says that these charts should be held on file by your employer for at least one year. Enforcement officers can ask for charts to be handed over for inspection.

2.39 Mark one answer — DBC s3, DGV s3

You're making a journey with a co-driver. When the other person is driving, how should you show this time?

- ☐ As a daily rest period
- ☐ As a weekly rest period
- ☐ As a break in daily driving
- ☐ As driving time

You may only record a break when you're not doing any other type of work. You're allowed to take a break on a double-manned vehicle while the other crew member is driving, but any break must be a minimum of 15 minutes.

2.40 Mark one answer DBC s3, DGV s3

What does a tachograph record?

- ☐ Load weight
- ☐ Driving time
- ☐ Fuel consumption
- ☐ Engine temperature

The tachograph is a tool designed to help you be a safe and responsible driver. The tachograph records details of your journey, including time driven, speed and rest periods.

2.41 Mark one answer DBC s3, DGV s3

Which symbol on your tachograph shows your break/rest period?

☐ 

☐ 

☐ 

☐ 

A tachograph allows you to select the mode or task that you're undertaking. Time spent on that task is then recorded automatically. Each task has a different symbol. You need to know the meaning of each, so that your records are correct. The modes are: driving, doing other work, on duty and available for work, and taking a break or rest.

2.42 Mark one answer DBC s3, DGV s3

What does this tachograph chart symbol mean?

- ☐ Driver at rest
- ☐ Chart not required
- ☐ Other work
- ☐ Driving

Each activity has a different symbol. You should know what they mean, so that you can select the correct one. You're responsible for recording all your activities correctly.

Some tachographs don't have a 'driving' mode switch. These tachographs will automatically record driving time whenever the vehicle is moved, whatever mode the switch is set to.

2.43 Mark one answer DBC s3, DGV s3

What is the 'mode' switch on a tachograph used to record?

- ☐ Who's driving
- ☐ Illness
- ☐ Weekly rest periods
- ☐ Other work

As you change activities during the day, you should change the mode switch on the tachograph to record each activity. Failure to operate the mode switch could get you into trouble with your employer or the authorities.

2.44 Mark one answer DBC s3, DGV s3

At the end of your working week, you've driven a total of 56 hours. Under EU rules, what's the maximum number of hours you can drive in the following week?

- ☐ 34
- ☐ 36
- ☐ 38
- ☐ 40

If you've driven a total of 56 hours in any one week, you can only drive for 34 hours in the following week. Keep your own record to make sure that you don't exceed these hours.

2.45 Mark one answer DBC s3, DGV s3

Your vehicle is fitted with an analogue tachograph. What should you do if you need to exceed the normal drivers' hours due to an emergency?

- ☐ Continue using the same tachograph chart and write an explanation on the back
- ☐ Remove the tachograph chart and make a manual record of the rest of the journey
- ☐ Continue using the same tachograph chart, without any explanation or record
- ☐ Remove the tachograph chart and tell your employer the reason

In an emergency, you may exceed the drivers' hours rules. As long as road safety isn't put at risk, you can drive for longer to deliver your passengers, your vehicle or its load to safety. In these circumstances, you should note all the reasons on the back of your tachograph chart.

2.46 Mark one answer DBC s3, DGV s4

What's most likely to cause tiredness?

- ☐ Making frequent and regular stops
- ☐ Taking driving breaks on board the vehicle
- ☐ Having insufficient breaks from driving
- ☐ Using vehicles with automatic gearboxes

Tiredness will affect your concentration. Don't allow yourself to become tired through not taking proper breaks or rest periods. As a professional driver, you have a responsibility either for goods or for passengers, as well as for overall road safety. Make sure your vehicle is well ventilated, so you don't become drowsy between rest periods.

2.47 Mark one answer DBC s3, DGV s4

What should you do if you start to feel tired or unable to concentrate while you're driving?

- ☐ Stop as soon as it's safe to do so
- ☐ Wind down a window and carry on
- ☐ Switch on the radio and complete your journey
- ☐ Speed up to get to your destination sooner

If you start to feel tired, you should stop as soon as it's safe to do so, even if you aren't due a break.

Make sure that you get enough sleep before you're due to work, especially if you're on an early shift.

2.48 Mark one answer DBC s3, DGV s4

What should you do if you feel tired after driving for two-and-a-half hours?

- ☐ Slow down to a safer speed
- ☐ Reduce your planned driving time to three hours
- ☐ Stop as soon as it's safe to do so
- ☐ Take a less busy route

Most incidents happen as a result of a lapse in concentration. Don't let this happen to you. If you start to feel tired, you won't perform as well as you should. Your reactions will slow down, and your anticipation and judgement of hazards will become flawed. If you feel tired, stop and rest as soon as it's safe to do so.

2.49 Mark one answer — DBC s3, DGV s4

What should you do if you suddenly become tired while you're driving on a motorway?

- ☐ Stop on the hard shoulder and rest
- ☐ Leave by the next exit and find a place to stop
- ☐ Stop on the next slip road and rest
- ☐ Stop on the verge of the motorway and rest

If you're driving for long distances on a motorway, have plenty of rest stops. Many incidents have been caused by drivers falling asleep at the wheel. If you feel yourself becoming tired, you need to leave the motorway and find a safe place to stop; usually this would be the next service area.

2.50 Mark one answer — DBC s3, DGV s4

Where can you stop if you feel tired while you're driving on a motorway?

- ☐ On the hard shoulder
- ☐ At a service station
- ☐ On a slip road
- ☐ In a deceleration lane

Travelling long distances on a motorway can be boring. Looking at the same thing for a long period of time can make you feel tired and affect your concentration. If you start to feel tired, leave the motorway by the next exit and stop in a safe place to rest. Ideally you should use a service area, where you can rest and take some refreshment before you restart your journey.

2.51 Mark one answer — DBC s3, DGV s3

What should you do if you have to leave your vehicle unattended for a very short time?

- ☐ Leave a note in the window explaining you'll be back soon
- ☐ Leave the keys available in case of obstruction
- ☐ Keep the engine running but lock the doors
- ☐ Secure the vehicle and lock the doors

Take all the precautions you can to remove opportunities for theft. Lock your vehicle, especially when making deliveries and leaving the vehicle unattended.

2.52 Mark one answer DBC s3, DGV s3

The time is 10.00 am. You've been driving non-stop since 6.00 am. Under EU rules, what's the longest you may now drive without a break?

- ☐ 15 minutes
- ☐ 30 minutes
- ☐ 40 minutes
- ☐ 45 minutes

The maximum driving period under EU rules is 4 hours 30 minutes. After this, you must take a break of at least 45 minutes.

Planning your route will allow you to take your statutory rest periods in a safe place, such as a service area, where you can get food, drink and a rest.

2.53 Mark one answer DBC s3, DGV s3

Under EU rules, you may drive for up to nine hours a day. What maximum may this be increased to on two days of the week?

- ☐ 9.5 hours
- ☐ 10 hours
- ☐ 11 hours
- ☐ 11.5 hours

You're permitted to extend the daily limit of nine hours to 10 hours twice a week. Don't drive for more than the maximum hours allowed, as you risk a heavy fine and may lose your licence.

2.54 Mark one answer DBC s3, DGV s3

Under EU rules, you can drive for a maximum of nine hours a day. On how many days each week can this be extended to 10 hours?

- ☐ One day
- ☐ Two days
- ☐ Three days
- ☐ Four days

Under EU rules, your normal daily driving time mustn't exceed nine hours. This nine-hour period is defined as the time between

- any two daily rest periods, or
- a daily rest period and a weekly rest period.

You're allowed to extend these hours to 10 hours twice a week.

2.55 Mark one answer DBC s3, DGV s3

How many days does a driver's smart card normally cover?

- ☐ 7
- ☐ 14
- ☐ 21
- ☐ 28

The smart card will record information covering a period of about 28 days. If you use it beyond this period, some of the recorded information will be overwritten.

2.56 Mark one answer DBC s3, DGV s3

Under EU drivers' hours regulations, a 45-minute break must be taken after 4.5 hours of driving. What shorter breaks can this be split into?

- ☐ One of 10 minutes, plus one of 35
- ☐ One of 15 minutes, plus one of 30
- ☐ One of 20 minutes, plus one of 25
- ☐ One of 40 minutes, plus one of 5

Under EU regulations, you must take a break of at least 45 minutes after four-and-a-half hours of driving. You may choose to take this break during the driving period. In this case, it can be split into two breaks of at least 15 and 30 minutes, taken in that order.

2.57 Mark one answer DBC s3, DGV s3

You're driving under EU tachograph regulations and lose your smart card. When must you inform the relevant authority?

- ☐ Within 5 days
- ☐ Within 7 days
- ☐ Within 14 days
- ☐ Within 28 days

As a professional driver, you have a responsibility to report any loss or theft of your digital smart card. You must inform the nearest relevant authority within seven days.

2.58 Mark one answer DBC s3, DGV s3

How long is a digital tachograph driver smart card valid?

- ☐ One year
- ☐ Three years
- ☐ Five years
- ☐ Ten years

Digital tachograph driver smart cards are valid for a maximum of five years. You should receive a reminder about three months before the expiry date. However, it's your responsibility to make sure that you apply for a new card at least 15 days before the old one expires.

2.59 Mark one answer NI EXEMPT DBC s3, DGV s3

Which authority must you contact if your tachograph card is lost or stolen?

- ☐ The police
- ☐ The Driver and Vehicle Licensing Agency
- ☐ A tachograph centre
- ☐ The Driver and Vehicle Standards Agency

Driver cards for digital tachographs are issued by the Driver and Vehicle Licensing Agency in Swansea. In Northern Ireland, the cards are issued by the Driver and Vehicle Agency. Tell them if your card is lost or stolen.

2.60 Mark one answer NI EXEMPT DBC s3, DGV s3

Where can you get a replacement driver tachograph card?

- ☐ The Driver and Vehicle Licensing Agency
- ☐ The Driver and Vehicle Standards Agency
- ☐ Any MOT test centre
- ☐ A tachograph centre

Driver cards for digital tachographs are issued by the Driver and Vehicle Licensing Agency. In Northern Ireland, the cards are issued by the Driver and Vehicle Agency. They can send you a replacement card, for a fee.

2.61 Mark one answer DBC s3, DGV s3, HC p126

A driver is convicted of obstructing an enforcement officer in the course of their duties. Under EU drivers' hours regulations, what's the maximum fine they can receive?

- ☐ Unlimited
- ☐ £3000
- ☐ £4000
- ☐ £5000

EU drivers' hours regulations say that any driver who fails to comply with, or obstructs, an enforcement officer in the course of their duties can face an unlimited fine.

2.62 Mark one answer DBC s3, DGV s3

When are you allowed to alter your tachograph record?

- ☐ If there are two or more drivers
- ☐ If your journey is over 50 miles
- ☐ At no time
- ☐ When you have no spare charts

Altering drivers' hours records or tampering with a tachograph with intent to deceive is a criminal offence. You, and anyone else involved, could go to prison. You could also lose your licence.

2.63 Mark one answer DBC s3, DGV s3

You're driving under EU drivers' hours. How is a week defined under these rules?

- ☐ Between 00.00 hours on Monday and 24.00 hours the following Sunday
- ☐ Any seven-day period
- ☐ Between 00.00 hours and 24.00 hours six days later
- ☐ A working period of 56 hours

The rules define a week as a period between 00.00 hours on Monday and 24.00 hours the following Sunday. You must make sure that you don't exceed any weekly driving limit.

2.64 Mark one answer — DBC s3, DGV s3

Which type of digital tachograph card is used by an approved calibration centre when they're recalibrating the tachograph?

- ☐ Control card
- ☐ Company card
- ☐ Driver card
- ☐ Workshop card

The workshop card is available only to approved calibration centres. There are three other types of digital tachograph card:

- driver card, used by the driver
- company card, used by an operator
- control card, used by enforcement authorities.

2.65 Mark one answer — DBC s3, DGV s3

Which digital tachograph card is only available to enforcement authorities?

- ☐ Control card
- ☐ CPC card
- ☐ Company card
- ☐ Workshop card

Various types of smart card are used in digital tachograph systems:

- driver card, used by drivers
- company card, used by operators
- workshop card, available only to approved calibration centres
- control card, available only to Driver and Vehicle Standards Agency (DVSA) officials and the police, for carrying out enforcement.

2.66 Mark one answer — NI EXEMPT — DBC s3, DGV s3

Drivers must have a driver smart card for use in digital tachographs. Where can you obtain a replacement smart card in Great Britain?

- ☐ The Driver and Vehicle Licensing Agency
- ☐ Highways England
- ☐ A tachograph calibration centre
- ☐ The Post Office

In Great Britain, a replacement smart card can only be obtained from the Driver and Vehicle Licensing Agency (DVLA). In Northern Ireland, they're available from the Driver and Vehicle Agency (DVA).

2.67 Mark one answer — DBC s3, DGV s3

Who's responsible for the issue of tachograph charts to a bus or lorry driver?

- ☐ The driver's employer
- ☐ The Driver and Vehicle Standards Agency
- ☐ The authorised calibration centre
- ☐ The local MOT testing centre

The driver's employer is responsible for the issue of tachograph charts. The driver must ensure that the correct information is recorded on the chart.

2.68 Mark one answer DBC s3, DGV s3

How can you find out when an analogue tachograph was last recalibrated?

- ☐ From a date on the tachograph chart
- ☐ By contacting the vehicle's manufacturer
- ☐ By checking the vehicle's service record
- ☐ From a plaque on or near the tachograph

An analogue tachograph must be checked every two years and recalibrated every six years. A plaque on or near the tachograph will show when it was last checked.

2.69 Mark one answer DBC s3, DGV s3

When must an analogue tachograph be recalibrated?

- ☐ Every two years
- ☐ Every four years
- ☐ Every six years
- ☐ Every eight years

When an analogue tachograph is installed and calibrated, an installation plaque is fixed near the tachograph. This shows the date of the most recent tachograph calibration. Under EU rules, this must take place every six years.

2.70 Mark one answer DBC s3, DGV s3

Your vehicle breaks down during a journey. You continue by driving in another vehicle with the same type of tachograph. What must you do with your tachograph chart?

- ☐ Leave it in the broken-down vehicle
- ☐ Take it with you for security, but use a new chart in the new vehicle
- ☐ Telephone the testing authority for permission to drive without a chart
- ☐ Take it with you, and use it in the new vehicle

When changing vehicles, you should also record certain pieces of information, including the closing odometer reading, the registration number of the new vehicle, the odometer start reading and the time of the vehicle change.

2.71 Mark one answer DBC s3, DGV s3

Under EU rules, what's the maximum driving time allowed in any two consecutive weeks?

- ☐ 85 hours
- ☐ 90 hours
- ☐ 100 hours
- ☐ 105 hours

Under EU rules, the maximum number of hours that you can drive in any two consecutive weeks is 90 hours. These don't have to be split evenly, but the total amount of driving time in any one week mustn't exceed 56 hours. Don't exceed your driving hours; heavy fines can be given to drivers who break the law.

2.72 Mark one answer DBC s3, DGV s3

How often must an analogue tachograph be checked at an approved calibration centre?

- ☐ Every year
- ☐ Every two years
- ☐ Every five years
- ☐ Every six years

An analogue tachograph must be checked every two years and recalibrated and sealed every six years. This must be done at an approved calibration centre. Digital tachographs, unlike analogue ones, must be recalibrated every two years.

2.73 Mark one answer DBC s3

You're driving a bus with 18 seats on an excursion from London to Scotland. Which drivers' hours rules should you use?

- ☐ AETR only
- ☐ Domestic
- ☐ EU only
- ☐ EU and AETR

If you're driving a bus on an excursion (that is, not a regular journey) and your vehicle has 18 or more seats, you must use EU rules and record the details.

2.74 Mark one answer DBC s3

After driving continuously for the maximum period under EU rules, a bus driver must take a break. What's the minimum duration of this break?

- ☐ 15 minutes
- ☐ 30 minutes
- ☐ 45 minutes
- ☐ 60 minutes

After a driving for a period of no more than 4.5 hours, you must take a break of at least 45 minutes, which mustn't be interrupted.
If you're carrying passengers, they'll probably also be grateful for a break. Taking breaks at the correct time will keep you and your passengers safe and comfortable.

2.75 Mark one answer DBC s3

When may a driver take their rest period in a parked vehicle?

- ☐ If it's fitted with a bunk
- ☐ If a smoke alarm is fitted
- ☐ If the vehicle's in an authorised coach park
- ☐ If there are no passengers on board

Some vehicles are fitted with sleeping accommodation for the driver. If your vehicle has this facility, you're allowed to take your daily rest period there, provided your vehicle is stationary.

2.76 Mark one answer DBC s3

What should you do if you're asked to leave your bus by an official who isn't in uniform?

- ☐ Comply with the request
- ☐ Ask to see a warrant card
- ☐ Refuse to leave the vehicle
- ☐ Invite the official aboard

If you're asked to leave your vehicle by an official who isn't in uniform, ask to see their warrant card. The official is likely to be an enforcement or police officer, but don't presume this.

2.77 Mark one answer DBC s2

What must the driver do when they leave their bus unattended?

- ☐ Check that the tachograph chart is removed
- ☐ Check that the gear lever is in reverse
- ☐ Check that the gear lever is in first
- ☐ Check that the parking brake is applied

Make sure that your vehicle is safe whenever you leave it unattended. Always stop the engine and apply the parking brake.

2.78 Mark one answer DBC s3

When are bus operators required to use tachographs for regular journeys?

- ☐ When the journey is more than 10 km
- ☐ When the journey is more than 20 km
- ☐ When the journey is more than 50 km
- ☐ When the journey is more than 80 km

Operators of buses used for regular journeys of more than 50 km must use tachographs.

Section three
Braking systems

In this section, you'll learn about

- types of braking system
- using your brakes properly
- connecting the air lines
- maintenance and inspection
- draining air tanks in freezing conditions
- driving with anti-lock brakes.

Braking systems

Types of braking system

There are three types of braking system on a large vehicle: the service brake, the secondary brake and the parking brake.

The **service brake** is your vehicle's main braking system. Your foot control works the service brake to slow your vehicle and stop it safely. Your service brake may have an **anti-lock braking system (ABS).**

Definition

anti-lock braking system (ABS)
A braking system, fitted to most modern vehicles, that prevents the wheels locking up under braking and helps to avoid uncontrolled skidding.

The **secondary brake** is there in case your service brake doesn't work. It may be combined with the service brake's foot control or with the parking brake.

The secondary brake normally works on fewer wheels than the service brake. This means it will have a reduced level of performance compared with the service brake.

You must always set the **parking brake** (handbrake) when you leave your vehicle unattended. A parking brake must be mechanical and is usually a hand control. Your parking brake may also be your secondary brake but unless your service brake isn't working you should only use this when your vehicle is stationary.

DBC s2 DGV s2

Using your brakes properly

To stay in control when you brake, make sure that you

- start braking in good time
- brake when travelling in a straight line, wherever possible
- brake in a controlled manner.

If you use your brakes for a long time on long or steep downhill slopes, you may notice that they don't work so well. This is known as 'brake fade' and it happens when the brakes overheat.

Large vehicles often come with an **endurance braking system**, or **retarder**. A retarder is helpful on long downhill slopes, where it can restrict the vehicle's speed without you using the service brake. This helps to stop the brakes overheating.

Definition

endurance braking system/retarder
A system that slows the vehicle independently of the brakes.

TIP

If your vehicle doesn't have a retarder, use a low gear to help control your speed on long or steep hills. This will reduce how much you need to use the brakes and make brake fade less likely.

Some steep hills have an escape lane for traffic travelling downhill. The escape lane is for use in the event of brake failure and is designed to safely slow or stop the vehicle. This is done with the use of an uphill gradient and/or a gravel-filled arrester bed either alongside or adjacent to the carriageway.

Jack-knifing and trailer swing

If you brake hard when driving an articulated vehicle or when towing a drawbar trailer, you may cause **jack-knifing** or **trailer swing**. Either one is more likely if you're not travelling in a straight line when you brake.

Jack-knifing is more likely to happen if the vehicle is unladen. When you brake, the trailer's tyres can't grip the road. The trailer will be unable to stop and will swing around the **tractor unit** or push it aside.

Trailer swing is more likely when you brake with a drawbar trailer, but it can also happen with an articulated vehicle. All of these can cause trailer swing

- braking hard when you turn a corner or when on a bend
- using too much steering while travelling at speed
- having incorrectly adjusted brakes on either the tractor unit or the trailer.

Definition

jack-knifing
Jack-knifing is where a tractor/trailer combination pivots so far around the coupling that the combination can't be driven.

trailer swing
When the trailer swings out of line independently of the tractor unit.

tractor unit
A large goods vehicle that pulls a trailer; the drive unit of an articulated vehicle.

DBC s2 DGV s2

Connecting the air lines

When you draw a trailer, you must connect brake lines between the trailer and the tractor unit. Braking systems can have two or three lines.

Three-line system	Two-line system
Service line – yellow	**Service line – yellow**
Emergency line – red	**Emergency line – red**
Auxiliary line – blue	

Air-brake systems usually have two lines.

service line
Transmits the service brake signal from the tractor unit to the trailer.

auxiliary line
Transmits the secondary brake signal from the tractor unit to the trailer (where fitted).

emergency line
Supplies air to the trailer's air reservoirs.

air-brake system
A system that uses air pressure to operate the brakes.

If you need to connect a two-line vehicle to a three-line trailer, don't connect the auxiliary (blue) line.

If you need to connect a three-line vehicle to a two-line trailer, follow the vehicle manufacturer's instructions on what to do with the auxiliary (blue) line. If you don't follow these instructions you may make the tractor unit/trailer combination dangerous.
DBC s2 DGV s2

When connecting a two-line vehicle to a three-line trailer, don't connect the auxiliary line.

When connecting a three-line vehicle to a two-line trailer, follow the manufacturer's instructions for the auxiliary line.

View this DVSA guide and see why understanding your brake connections is vital.

www.gov.uk/government/publications/understanding-your-braking-connections-when-using-a-trailer

Maintenance and inspection

Correct any fault in your braking system before you drive the vehicle. Take note of any pressure warning lights or buzzers, or any changes that you might notice in the 'feel' of the brake pedal.

With air-assisted hydraulic brakes, a brake pedal that feels 'hard' could mean a loss of vacuum or a fault in the **vacuum pump**.

Definition

vacuum pump
An engine-driven vacuum pump that creates the vacuum for air-assisted hydraulic brake systems.

If the brake air-pressure warning light comes on while you're driving, you must stop and get the fault put right straight away.

DBC s2 DGV s2

Draining air tanks in freezing conditions

In cold weather, any moisture in an air-brake system can cause ice to form in valves and pipes. This may lead to low air pressure and/or brake failure.

Some air-brake systems have automatic drain valves to remove moisture from the system, but you may need to do this yourself on other types. In cold weather, always drain a **manually drained brake system** at the end of each day of driving.

Definition

manually drained air tanks
Compressed air tanks that require manual draining of water via a drain tap located on the underside of the tank.

DBC s2 DGV s2

Driving with anti-lock brakes

Vehicles with anti-lock brakes (ABS) should have a warning light on the dashboard. This light should come on with the ignition and go out when the vehicle reaches about 6 mph.

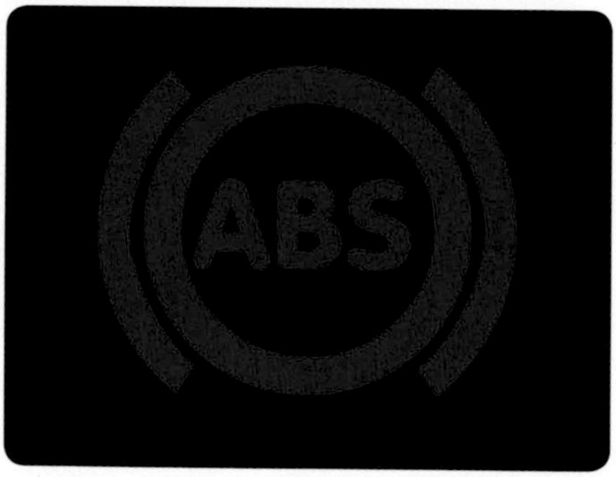

On some vehicles, the ignition switch doesn't operate the ABS warning light. In these cases there should be a 'check' switch on the dashboard for you to use.

It's an offence to drive with an ABS system that doesn't work properly.

DBC s2 DGV s2

Meeting the standards

You must be able to

use braking systems safely and effectively

stop the vehicle safely and under control in an emergency

use the vehicle's endurance braking system (retarder) when needed.

You must know and understand

that larger vehicles may need a greater distance to stop

how to apply a safe, systematic approach when stopping

the principles of the various endurance braking systems (retarders) that may be fitted to large vehicles, such as

- electric
- engine driven
- exhaust brakes.

Notes

You can use this page to make your own notes or diagrams about the key points you need to remember.

Think about

- How do you operate your vehicle's secondary brake?
- Can you remember how to avoid losing control while braking?
- How should you use your endurance braking system or retarder, if you have one?
- Where can you find information about connecting your tractor unit's air lines to two- and three-line trailers?
- What signs can show a fault or problem with your brakes?
- Does your brake system drain automatically or do you need to know how to do this yourself?
- How can you check your ABS is working properly?

Your notes

Things to discuss and practise with your instructor

These are just a few examples of what you could discuss and practise with your instructor. Some are for PCV, some are for LGV and some are for both. Read more about braking systems to come up with your own ideas.

Discuss with your instructor

- the impact of various weather conditions on your braking
- the disadvantages of progressive braking when driving a bus.

- what could cause an articulated lorry to jack-knife
- how to stop a lorry safely with and without anti-lock braking.

Practise with your instructor

- making short stops while driving uphill
- carrying out an emergency stop in a safe place
- driving down steep hills.

3.1 Mark one answer DGV s2

When is trailer swing most likely to occur on a lorry and drawbar combination?

- ☐ While you're braking on a bend
- ☐ While you're using an endurance brake
- ☐ While you're steering at slow speed and fully loaded
- ☐ While you're moving off at an angle

The forces that act on your vehicle if you brake while cornering can lead to instability and may cause trailer swing. You should try to plan ahead so that you brake smoothly and in good time as you approach a bend.

3.2 Mark one answer DGV s2

Your vehicle is fitted with a 'diff-lock'. When would you normally use it?

- ☐ While driving on straight roads
- ☐ While towing an empty trailer
- ☐ While driving on muddy construction sites
- ☐ While uncoupling a trailer

The differential allows the drive wheels to rotate at different speeds; this is necessary for the vehicle to negotiate corners and bends. On a slippery surface, however, it can cause problems: if one wheel has less traction, it can spin, resulting in a complete loss of drive. The 'diff-lock' prevents the wheels from turning independently and so helps to stop this happening.

3.3 Mark one answer DGV s2

What colour is the auxiliary line on a three-line braking system?

- ☐ Red
- ☐ Blue
- ☐ Green
- ☐ Yellow

If you're driving an articulated vehicle or a trailer combination, it's vital that you understand the rules that apply to coupling and uncoupling the brake lines. If you take a practical driving test with a trailer, you'll be expected to demonstrate this during the test. The lines are colour-coded: red is the emergency line, blue is the auxiliary line and yellow is the service line. They must be connected strictly in accordance with the correct procedure.

3.4 Mark one answer — DGV s2

The emergency line is common to both two-line and three-line brake systems. What colour is it?

- ☐ Red
- ☐ Blue
- ☐ Black
- ☐ Yellow

The red emergency line is common to both two-line and three-line braking systems. Always set the parking brake before disconnecting any brake line.

3.5 Mark one answer — DGV s2

Air-brake systems usually have two lines. Which additional line is fitted on a three-line system?

- ☐ Emergency
- ☐ Service
- ☐ Electrical
- ☐ Auxiliary

The blue (auxiliary) line isn't used when connecting to a two-line system. Follow the manufacturer's instructions about what to do with this third line.

3.6 Mark one answer — DGV s2

In frosty weather, what precaution could a lorry driver take to prevent moisture from freezing in the air-brake system?

- ☐ Drain the air tanks daily
- ☐ Cover the air tanks with a blanket
- ☐ Keep the engine at high revs when starting
- ☐ Pump the brakes

You should make sure that you drain the air tanks daily to avoid moisture in the system. Most modern vehicles have an automatic draining system, which should be checked regularly.

3.7 Mark one answer — DGV s2

How can a lorry driver avoid brake fade?

- ☐ Ensure that the air tanks are drained before journeys
- ☐ Check that the air pressure is correct
- ☐ Make sure that the parking brake is applied before stopping
- ☐ Select an appropriate gear before downhill gradients

Using the brakes to control a lorry's speed as it's driven down a hill will cause them to get hot. If they overheat, they can become less effective – a condition known as brake fade. By selecting a low gear, the engine's resistance will slow the lorry. This reduces the demands on the brakes, making them less likely to overheat.

3.8 Mark one answer DGV s2

When are exhaust brakes most efficient?

- ☐ At high engine speed, while in low gears
- ☐ At low engine speed, while in high gears
- ☐ During stop/start town work
- ☐ While travelling at a constant speed on a motorway

An exhaust brake alters the engine's exhaust flow, using it to help control the vehicle's speed. It's most efficient when the engine is running at a high speed and the vehicle is in a low gear, such as when descending a long hill. Using the exhaust brake can relieve the service brakes, preventing them from overheating and fading.

3.9 Mark one answer DGV s2

What's the principal braking system on a lorry called?

- ☐ The endurance brake
- ☐ The service brake
- ☐ The parking brake
- ☐ The jake brake

The service brake is usually operated by the brake pedal. It's used to control the speed of the vehicle and to bring it to a halt safely. It may also incorporate an anti-lock braking system.

3.10 Mark one answer DGV s2

You're driving an articulated lorry. What could happen if you change to a lower gear while you're going too fast?

- ☐ The vehicle could jack-knife
- ☐ The engine could stall
- ☐ The brakes could fail
- ☐ The trailer could uncouple

Severe braking or selecting a gear too low for your road speed can cause the tractor unit to slow more quickly than the trailer. The trailer can then pivot around the coupling (fifth wheel), causing the vehicle to jack-knife. Jack-knifing is more likely to occur with an unladen vehicle, particularly on a bend or curve.

3.11 Mark one answer DGV s2

When is an articulated vehicle most at risk of jack-knifing?

- ☐ When it's fitted with an endurance brake (retarder)
- ☐ While it's manoeuvring slowly
- ☐ While it's under heavy braking
- ☐ When it's fully loaded

Heavy braking can cause an articulated vehicle to become unstable and may even lead to jack-knifing; this is more likely to occur when the trailer is unladen.

3.12 Mark one answer DGV s2

Your lorry is stuck in snow. You use the diff-lock to move off. When should you switch the diff-lock off?

☐ Only after selecting top gear
☐ Once the engine has warmed up
☐ As soon as the vehicle is moving
☐ As soon as the snow has cleared

You must always disengage the diff-lock as soon as the vehicle is moving. The differential allows the rear wheels to revolve at different speeds, and this allows the vehicle to follow a curved path. Attempting to turn with the diff-lock engaged could result in the vehicle continuing straight on.

3.13 Mark one answer DGV s2

Your tractor unit has three air lines. You're connecting to a trailer that has two air-line couplings. What colour is the line you shouldn't connect to the trailer?

☐ Red
☐ Yellow
☐ Black
☐ Blue

When connecting a three-line tractor unit to a two-line trailer, the blue line is the one that shouldn't be connected to the trailer. This is the auxiliary line. Always follow the manufacturer's advice: it may be necessary to reconnect the extra line to the tractor unit.

3.14 Mark one answer DGV s2

You're driving an articulated lorry that has three air lines connected to the trailer. What's the purpose of the red line?

☐ It's the emergency line
☐ It's the service line
☐ It's the auxiliary line
☐ It's the electrical line

The red emergency line is common to both two-line and three-line brake systems. The other colours are

- blue – auxiliary
- yellow – service.

3.15 Mark one answer DGV s2

You're driving a tractor unit fitted with two air lines. You want to couple to a trailer that has three air-line couplings. How should the air lines be connected?

- ☐ The trailer auxiliary line should be left unconnected
- ☐ The trailer service line should be left unconnected
- ☐ Only the service line should be connected
- ☐ Only the auxiliary line should be connected

A two-line system consists of the emergency line (red) and the service line (yellow). You need to understand the rules that apply to safe connection of brake systems and mixing two-line and three-line systems.

3.16 Mark one answer DGV s2

Your lorry is equipped with anti-lock brakes. What's the correct procedure for stopping in an emergency?

- ☐ Apply the footbrake firmly in a pumping action until the vehicle has stopped
- ☐ Apply the footbrake firmly and continuously until the vehicle has stopped
- ☐ Apply the footbrake and handbrake until the vehicle has stopped
- ☐ Apply the handbrake only

When your vehicle has anti-lock brakes and you need to stop in an emergency, you should keep your foot firmly on the brake pedal until the vehicle stops; this will allow the system to work. Although anti-lock brakes contribute to safety, they don't reduce your stopping distance or take away the need to drive with good planning and anticipation.

3.17 Mark one answer DGV s2

What does changing to a lower gear on a long downhill gradient help to avoid?

- ☐ Brake fade
- ☐ Clutch slip
- ☐ Excessive engine revs
- ☐ Tyre wear

Engaging a lower gear and using engine braking will help you to control your speed as you're going downhill. This reduces the demands on the brakes, and helps to prevent them from overheating. This will reduce the likelihood of brake fade.

3.18 Mark one answer DGV s2

You have to drive onto a muddy building site. Why should you switch on your diff-lock?

- ☐ To make your steering lighter
- ☐ To improve your fuel consumption
- ☐ To increase your engine power
- ☐ To make the wheels less likely to spin

Engaging the diff-lock means that the driven wheels are locked together. This reduces the likelihood of wheel spin. Remember to switch off the diff-lock as soon as you're on firm ground again, otherwise you could damage the transmission.

3.19 Mark one answer DBC s2, DGV s2

How will anti-lock brakes improve your vehicle's safety?

- ☐ They allow you to follow other vehicles more closely
- ☐ They help you keep steering control while braking heavily
- ☐ They let you drive faster on wet roads
- ☐ They enable you to brake later than normal

When your steering wheels are locked – as can happen in a skid – you can't steer the vehicle. Anti-lock brakes help you to keep steering control while you're braking heavily, but they aren't a substitute for good forward planning. Anticipation and planning ahead will minimise the risk of skidding more effectively than relying on your braking system.

3.20 Mark one answer DBC s2, DGV s2

What could prevent air pressure from building up in an air-brake system in frosty weather?

- ☐ Moisture in the air may form bubbles in the brake fluid
- ☐ The air will contract, reducing the pressure
- ☐ The dampness may cause valves to rust
- ☐ Moisture drawn in with the air may freeze and cause a blockage

When air is compressed, moisture condenses and collects in the air tanks. This can find its way along the network of pipes connected to the brakes. In frosty weather, the moisture can freeze in the pipes, blocking them completely. On modern vehicles, the air is dried before it's compressed and the air tanks drain automatically to shed any moisture. On older vehicles, the air tanks need draining manually.

3.21 Mark one answer DBC s2, DGV s2

What should you do if the brake air-pressure warning light comes on while you're driving?

- ☐ Stop and get help without delay
- ☐ Report the fault when you return to your depot
- ☐ Build up the pressure by accelerating
- ☐ Drain the air tanks and continue normally

When a warning light or device indicating a loss of brake pressure comes on, you must stop safely and get the fault put right immediately. The safety of you, your load, your passengers and all other road users is at risk.

3.22 Mark one answer DBC s2, DGV s2

You've just started the engine. What must you do if the brake air-pressure warning light is showing?

- ☐ Report it as a fault
- ☐ Keep the parking brake on
- ☐ Switch your engine off
- ☐ Engage the clutch

Even though the warning light is showing, there may be sufficient pressure to release the parking brake. However, you mustn't do this, because there may not be enough air pressure to operate the service brake.

3.23 Mark one answer DBC s2, DGV s2

Your vehicle is fitted with an anti-lock braking system. What's it designed to allow you to do?

- ☐ Drive at faster speeds
- ☐ Brake much later than normal
- ☐ Apply the brakes more quickly
- ☐ Stop safely in an emergency

Anti-lock braking systems sense when the wheels are about to lock. The sensor control releases the brakes and immediately applies them again. It will do this many times per second. This both prevents skidding and enables the driver to steer – something that's impossible when the steering wheels are skidding.

3.24 Mark one answer DBC s2, DGV s2

Your vehicle is fitted with an anti-lock braking system. When should you check that it's working properly?

- ☐ Before each service
- ☐ At the start of each working day
- ☐ At the start of each working week
- ☐ Before every journey

Modern anti-lock braking systems rely on electronics for their operation. A warning light on the dashboard shows that the system is operating. This light will usually go out when the vehicle reaches about 6 mph (10 km/h).

3.25 Mark one answer DBC s2, DGV s2

What will 'pumping' the brake pedal in a vehicle fitted with anti-lock brakes cause?

- ☐ Increased effectiveness
- ☐ Reduced effectiveness
- ☐ Reduced brake wear
- ☐ Increased brake wear

Before driving a vehicle fitted with anti-lock brakes, refer to the handbook. This will give you details of the manufacturer's recommended method of use. Remember, anti-lock brakes will enhance your skills, but they can't replace them.

3.26 Mark one answer DBC s2, DGV s2

You're driving a vehicle fitted with anti-lock brakes. What should you do when braking in an emergency?

- ☐ 'Pump' the brake pedal harshly
- ☐ Apply minimum force to the brake pedal
- ☐ Use the exhaust brake (retarder) before the footbrake
- ☐ Apply firm, continuous pressure to the brake pedal

Applying a firm, continuous brake pressure will enable the anti-lock braking system (ABS) to function properly. However, ABS is becoming increasingly sophisticated, so you should read the vehicle manufacturer's handbook for its advice about using the system.

3.27 Mark one answer DBC s2, DGV s2

What's 'brake fade'?

- ☐ Reduction of air pressure
- ☐ Smooth progressive braking
- ☐ Reduction of braking effectiveness
- ☐ Low hydraulic brake fluid level

Continually using the brakes could cause them to overheat. When that happens, they become less effective. This is called brake fade, and it occurs most often when the driver has to control the vehicle's speed on a long downhill gradient. To help prevent brake fade, use the endurance brake (retarder) and a lower gear to keep the vehicle's speed in check, rather than the wheel-mounted brakes.

3.28 Mark one answer DBC s2, DGV s2

'Brake fade' is a loss of effectiveness of the brakes when they overheat. When would 'brake fade' be most likely to happen?

- ☐ On a long journey
- ☐ On a long downhill gradient
- ☐ On the approach to hazards
- ☐ On a long uphill gradient

Continuous use of the brakes will cause them to overheat, and in extreme cases they'll become ineffective. When you're going downhill, the momentum of your vehicle will cause you to gather speed very quickly. Use the endurance brake (retarder) and stay in a low gear to reduce the need to use the brakes.

3.29 Mark one answer DBC s2, DGV s2

What should you do when driving a large vehicle down a steep hill?

- ☐ Partly apply the parking brake
- ☐ Select the vehicle's highest gear
- ☐ Use the endurance brake
- ☐ Put the gear lever into neutral

When approaching a long hill, you should take note of any early warning signs. Reduce your speed and select the appropriate gear in good time. Using an endurance brake (retarder), if fitted, will help control your speed.

3.30 Mark one answer DBC s2, DGV s2

What should anti-lock brakes let you do when you have to stop in an emergency?

- ☐ Brake more gently
- ☐ Brake much later
- ☐ Maintain steering control
- ☐ Stop over a long distance

Anti-lock brakes are a driver aid, and can help you maintain steering control while braking. However, you mustn't rely on them to get you out of trouble or expect to be able to make sudden direction changes if you're going too fast.

3.31 Mark one answer DBC s2, DGV s2

Your vehicle has anti-lock brakes. How does this affect you when you're braking normally?

- ☐ There's no need to alter the way you brake
- ☐ You'll be able to brake much later
- ☐ You'll need to brake more firmly
- ☐ There's no need to brake so early

Plan well ahead to enable you to brake normally. Don't rely on anti-lock brakes to compensate for deficiencies in your driving.

3.32 Mark one answer — DBC s2, DGV s2

Where would you see an escape lane?

- ☐ Outside a fire station
- ☐ Alongside a bus lane
- ☐ Before a motorway exit
- ☐ Down a steep hill

An escape lane is for use in emergencies, usually when brakes have become hot and faded or failed completely while descending a long, steep hill. A sign will show you the direction of the road, and a chequered area, usually straight ahead, defines the escape route.

3.33 Mark one answer — DBC s2, DGV s2

When would you use an escape lane?

- ☐ When motorways merge
- ☐ When carrying a dangerous cargo
- ☐ When your brakes have failed
- ☐ When emergency repairs are needed

Escape lanes are found on steep downhill sections of road. They're designed to give a 'run-off' area, usually straight ahead, to allow vehicles to stop in an emergency. This is usually when brakes have faded or failed on large vehicles.

3.34 Mark one answer — DBC s2, DGV s2

When can an endurance brake (retarder) be especially useful?

- ☐ While driving down long hills
- ☐ When driving on steep cambers
- ☐ As a way to reduce gear changes
- ☐ To improve fuel consumption

Plan ahead and use your endurance brake (retarder) to help keep your speed in check on long downhill gradients. This can help to prevent your brakes from overheating and fading.

3.35 Mark one answer — DBC s2, DGV s2

You're driving down a long hill. What should you use to prevent the brakes from overheating?

- ☐ The anti-lock braking system
- ☐ The footbrake
- ☐ The secondary brake
- ☐ The endurance brake (retarder)

Endurance brakes or 'retarders' help to control a vehicle's speed. They operate by applying resistance to slow the vehicle without using the wheel brakes, so avoiding brake fade. The methods used include: increased engine braking, exhaust braking and transmission-mounted devices.

3.36 Mark one answer — DBC s2, DGV s2

What system is used to control a vehicle's speed without using the footbrake?

- ☐ A secondary brake
- ☐ An endurance brake (retarder)
- ☐ A differential lock
- ☐ An emergency air system

Large vehicles can be fitted with various types of endurance brake (retarder). Make sure you're familiar with how the system operates on the vehicle you're using. Don't risk a crash or collision through ignorance.

3.37 Mark one answer — DBC s2, DGV s2

When should an endurance brake (retarder) be used?

- ☐ On motorways only
- ☐ When braking quickly
- ☐ When you stop or park
- ☐ On long downhill slopes

An endurance brake can help to control your speed on long downhill slopes without using the service brake. Reducing the use of the service brake helps to prevent the brakes from overheating and reduces the risk of brake fade.

3.38 Mark one answer — DBC s2, DGV s2

Your vehicle is fully loaded. Where should you be braking when dealing with bends?

- ☐ As close to the bend as possible
- ☐ As you start to turn the wheel
- ☐ When halfway around the bend
- ☐ When driving in a straight line

When a vehicle changes direction, forces are applied to the vehicle and its load. Braking at the same time can result in additional forces being introduced, and this can lead to a loss of control. For this reason, braking should be carried out while driving in a straight line, so preventing these forces from acting together.

3.39 Mark one answer DBC s4, DGV s4

A lorry is overtaking you on a two-lane motorway. What should you do if it doesn't have the speed to get past?

- ☐ Continue at the same speed
- ☐ Be prepared to reduce your speed
- ☐ Increase your speed and force the lorry to drop back
- ☐ Brake hard to allow the other driver to cut in

Always be prepared to give way to overtaking lorries or buses. Maintaining your speed will only block the motorway to other traffic unnecessarily. You may find yourself in a similar situation if the limiter on your vehicle doesn't allow you enough speed to successfully overtake another vehicle.

3.40 Mark one answer HC r121

What should you do after driving through a flood?

- ☐ Carry out an emergency stop
- ☐ Drive in a low gear with the footbrake lightly applied
- ☐ Avoid braking until the brakes have dried out
- ☐ Pump the footbrake when approaching hazards

If you have to drive through a flood, do so with caution. Once out of the flood, you need to test your brakes to make sure they're working properly. To do this, check what's behind you, then drive in a low gear with the brakes gently applied.

3.41 Mark one answer DBC s2, DGV s2

What can coasting downhill seriously affect?

- ☐ The air brakes
- ☐ The cooling system
- ☐ The tachograph
- ☐ The electrical systems

Air-brake systems rely on an engine-driven compressor to keep the reservoir tanks at their operating pressure. Coasting downhill and relying on the brakes to control your speed could result in the loss of sufficient air pressure to operate the brakes effectively.

3.42 Mark one answer DBC s1, 2, DGV s2

When does 'brake fade' occur?

- ☐ When the brakes get too hot
- ☐ When the brakes get too cold
- ☐ When the brakes get too dry
- ☐ When the brakes get too wet

Continuous use of the brakes can result in them overheating and losing their effectiveness. This is known as 'brake fade' and it's most likely on long downhill gradients.

3.43 Mark one answer DBC s1, 2, DGV s2

How can you help to prevent 'brake fade'?

- ☐ Use the endurance brake
- ☐ Apply the parking brake
- ☐ Select neutral for a short distance
- ☐ Repeatedly pump the brake pedal

Brake fade occurs when the brakes overheat. Good forward planning and correct use of the gears when descending hills, together with proper use of the endurance brake (retarder), can help to prevent brake fade.

3.44 Mark one answer DBC s1, 2, DGV s2

What causes 'brake fade'?

- ☐ Continuous use of the brakes
- ☐ Repeated pumping of the brakes
- ☐ Loss of air pressure in the system
- ☐ Badly worn brake pads

Continuous use of the brakes on a long downhill gradient can cause them to overheat and become ineffective. This is called brake fade. You should use a low gear to enable the engine's braking effect to help control your speed. Also use the vehicle's retarder, if fitted. As well as reducing the likelihood of brake fade, this helps to keep air pressure in the tanks.

3.45 Mark one answer DBC s2, DGV s2

Where must extra care be taken when using an endurance brake (retarder)?

- ☐ On uneven roads
- ☐ On slippery roads
- ☐ On downhill gradients
- ☐ On uphill gradients

The endurance brake (retarder) slows the vehicle without the use of the wheel-mounted brakes. However, care should be taken in slippery conditions, when the retarder's effect can be powerful enough to cause a skid.

3.46 Mark one answer DBC s2, DGV s2

You're stationary. The brake air-pressure warning light comes on. Why should you keep the parking brake on?

- ☐ Because the vehicle will suddenly roll backwards
- ☐ Because it will cause the air pressure to rise
- ☐ Because the service brake may not stop you
- ☐ Because it will make the warning light go out

Even though a warning device indicates low air-brake pressure, you may still be able to release the parking brake. If you do this and start moving, the service brake may be ineffective and you may not be able to stop.

3.47 Mark one answer DBC s2, DGV s2

You're about to drive a vehicle fitted with air-assisted hydraulic brakes. Why might the brake pedal feel hard when you press it?

- ☐ The vacuum pump isn't working
- ☐ The pedal movement needs adjusting
- ☐ The brakes are locked on fully
- ☐ The brake-fluid reservoir is empty

If the brake pedal is hard to press, this could mean a loss of vacuum or a fault in the vacuum pump. You shouldn't drive the vehicle until the fault has been repaired.

3.48 Mark one answer DBC s2, DGV s2

Your vehicle is fitted with an air-assisted hydraulic braking system. What would warn you that the vacuum pump isn't working?

- ☐ The brake pedal feels spongy when pressed
- ☐ The brake pedal has little resistance
- ☐ The brake pedal feels hard when pressed
- ☐ The brake pedal travels a long way

The pump creates a vacuum in the servo unit. This reduces the amount of pressure you need to apply to the brake pedal. Without this vacuum, it's extremely difficult to press the brake pedal down.

3.49 Mark one answer DBC s2, DGV s2

Your vehicle is fitted with hydraulic brakes. Why might the brake pedal go down too far when it's pressed?

- ☐ There's too much fluid in the braking system
- ☐ The pedal travel requires adjustment
- ☐ The vacuum exhauster isn't working
- ☐ There isn't enough fluid in the braking system

A loose brake pedal would suggest a serious loss of fluid from the hydraulic system. Don't drive your vehicle until the fault has been checked and fixed by a qualified person.

3.50 Mark one answer DBC s2, DGV s2

You're about to drive a vehicle fitted with hydraulic brakes. What should you check before driving the vehicle?

- ☐ The hydraulic-brake fluid level
- ☐ The power-steering fluid level
- ☐ The cooling-system fluid level
- ☐ The windscreen-washer fluid level

It's important to check all fluid levels during your daily checks, but it's especially important to remember the hydraulic fluid reservoir. A loss of fluid could lead to brake failure.

3.51 Mark one answer — DBC s2, DGV s2

Your vehicle is fitted with air-assisted hydraulic brakes. What would warn you that there isn't enough air in the system?

- ☐ An increase in pressure on the air gauge
- ☐ A buzzer or light
- ☐ The exhaust brake won't work
- ☐ Brake fade

If a warning buzzer or light alerts you to a loss of air pressure, you should stop on the side of the road immediately. You should have enough air in reserve to allow you to do this safely. Don't start or continue your journey until the fault has been repaired.

3.52 Mark one answer — DBC s2, DGV s2

Which coloured band on the rev counter makes the best use of engine braking?

- ☐ Blue
- ☐ Red
- ☐ Amber
- ☐ Green

There are usually four coloured bands on the rev counter: green, amber, blue and red. The blue band will allow you to make the best use of engine braking. Try to keep in this band when going down a steep hill.

3.53 Mark one answer — DBC s2, DGV s2

Your vehicle is fitted with air brakes. What does it mean if a brake warning light shows when you've just started the engine?

- ☐ Low air pressure
- ☐ Increased air pressure
- ☐ The parking brake isn't working
- ☐ The air reservoirs are fully charged

It's dangerous to drive a vehicle with low air pressure. To alert you to this, all vehicles are fitted with a warning light and/or buzzer. Don't move your vehicle if this warning shows or sounds.

3.54 Mark one answer — DBC s2, DGV s2

Your vehicle is fitted with air-assisted hydraulic brakes. What fault would you suspect if the brake pedal becomes hard to press?

- ☐ The brake system has a loss of vacuum
- ☐ The brake linings are worn
- ☐ The brake shoes need adjusting
- ☐ The brake system requires more fluid

A fault in the vacuum pump could be the cause of a stiff brake pedal. There could also be a leaking connection allowing air into the vacuum. Don't continue your journey until the fault has been fixed.

3.55 Mark one answer DBC s2, DGV s2

Which component will last longer when you use a retarder?

- ☐ The brake linings
- ☐ The catalytic converter
- ☐ The exhaust system
- ☐ The transmission

By using a retarder, the life of the brake linings is extended. This is because a retarder works by slowing the vehicle without using the wheel-mounted brakes.

3.56 Mark one answer DBC s2, DGV s2

What should you do when an air-pressure warning device activates?

- ☐ Continue to drive the vehicle
- ☐ Drain the air tanks
- ☐ Stop and have the fault put right
- ☐ Pump the brake pedal repeatedly

Air-brake systems are fitted with a warning device that operates if the air pressure in the tanks drops below a safe level. There may be a warning buzzer and/or pressure gauges. You must be aware of the function of all gauges on your vehicle, and check them as you drive.

3.57 Mark one answer DBC s2, DGV s2

You're driving down a snow-covered hill. Why should you take care when using an independent endurance brake (retarder)?

- ☐ Your brakes could overheat
- ☐ Your speed could increase
- ☐ Compressed air could escape
- ☐ The drive wheels could lock

Select an appropriate gear in good time and, if your vehicle has a dashboard-mounted lever, apply the endurance brake (retarder) in stages. Braking too much, too soon may result in locking your drive wheels.

3.58 Mark one answer DBC s2, DGV s2

What should you avoid when using an independent endurance brake (retarder) on slippery roads?

- ☐ The front wheels spinning
- ☐ The drive wheels locking
- ☐ Brake-pad wear
- ☐ Anti-skid road surfaces

The endurance brake (retarder) usually operates by applying resistance, via the transmission, to the vehicle's driven wheels. If the road is slippery, the tyres may lose grip and the wheels will lock.

3.59 Mark one answer DBC s2, DGV s2

You're about to drive an unfamiliar vehicle. What should you do if you think there may be moisture in the air-brake reservoir?

- ☐ Assume the system has automatic drain valves
- ☐ Find out whether you need to drain the system manually
- ☐ Nothing; it's the vehicle owner's responsibility
- ☐ Leave the engine running for a while before driving

Moisture in the braking system can cause serious problems, especially in cold weather, when it could freeze and stop the brakes from working. Most modern vehicles have a system that drains automatically, but make sure you know which system is fitted to any vehicle that you drive.

3.60 Mark one answer DBC s2

You're about to start a journey in very frosty weather. What's the likely cause of low air-brake pressure?

- ☐ Engine temperature too low
- ☐ Weak engine anti-freeze mixture
- ☐ Brake pedal needs adjustment
- ☐ Frozen moisture in the storage tanks

Air-brake systems use air from the atmosphere, which contains moisture. The moisture condenses under pressure, resulting in water in the system. Make sure the air tanks are drained daily to remove any water and to help prevent the system from freezing in cold weather.

3.61 Mark one answer DBC s2

When should an anti-lock braking system (ABS) warning light turn off?

- ☐ When the brakes are used for the first time
- ☐ When the ABS comes into operation
- ☐ When road speed is 6 mph (10 km/h) or more
- ☐ When the secondary braking system is used

Every vehicle fitted with anti-lock brakes must have a warning light in the cab of the vehicle. The warnings may differ between manufacturers, but on all types the warning light should come on when the ignition is switched on, and turn off when the vehicle reaches a speed of 6 mph (10 km/h). If the light stays on, there may be a fault; you should have the brakes checked before continuing your journey.

3.62 Mark one answer DBC s2

What's the most powerful brake on a bus?

- ☐ Secondary brake
- ☐ Anti-lock braking system
- ☐ Endurance brake (retarder)
- ☐ Service brake

The most powerful and effective brake on the vehicle is the service brake, and this should be used in normal circumstances. Well-maintained brakes should apply an even pressure to all the wheels, providing an efficient, controlled stop.

3.63 Mark one answer DBC s2

What should you do when making a short stop, facing uphill?

- ☐ Hold the vehicle on the clutch
- ☐ Hold the vehicle on the footbrake
- ☐ Select neutral and apply the parking brake
- ☐ Apply the parking brake after stopping

As with stopping in any other situation, if you have to make a stop on an uphill gradient, wait until the vehicle has come to a halt before applying the parking brake.

3.64 Mark one answer DBC s2

You're about to move off in a vehicle that has automatic transmission. What must you do before you select 'D' (drive)?

- ☐ Put your foot on the footbrake
- ☐ Signal to move off
- ☐ Alter your seat position
- ☐ Adjust your mirrors

It's important to apply the footbrake before you engage 'D' (or drive), so your vehicle doesn't creep forward or roll back out of control when you're about to move away. This can be dangerous; for example, if there's another road user close behind.

3.65 Mark one answer DBC s2

What's an advantage of progressive braking?

- ☐ Improved passenger safety and comfort
- ☐ Increased air-brake pressure
- ☐ Avoidance of 'brake fade'
- ☐ Longer rest periods

The way you drive matters.
The safety and comfort of your passengers is your priority; if you have the correct attitude when you're driving, your passengers will be assured of a comfortable and pleasant journey.

Good forward planning and anticipation will help you avoid having to turn or brake suddenly.

Section four
The driver

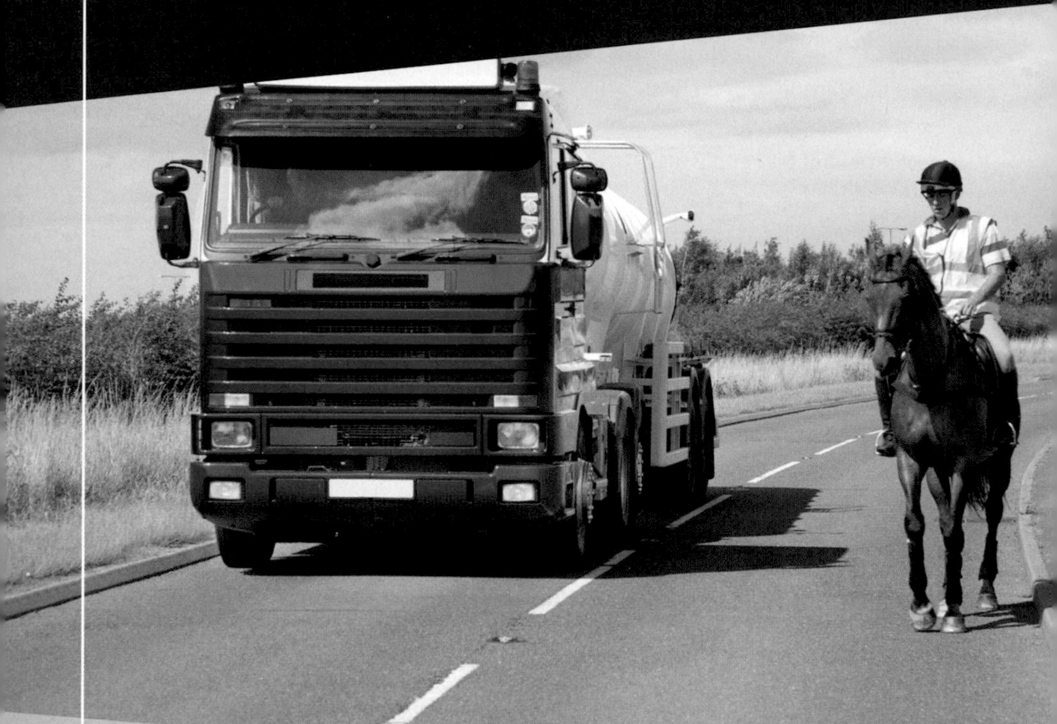

In this section, you'll learn about

- consideration towards other road users
- safety equipment
- mobile phones when driving
- fitness to drive
- medication
- emergency vehicles.

The driver

Consideration towards other road users

Driving to a high standard

When you drive a large vehicle professionally, always drive to a high standard. Be patient and considerate towards other road users.

- Don't block side-road junctions if you have to stop in a line of traffic.
- Don't intimidate others by driving too closely behind their vehicle.
- Don't use your air brakes or horn to frighten other road users.
- Don't park where you'll put other people in danger or get in their way.

Taking care in rural areas

In rural areas, try not to park on the grass verge. Heavy vehicles can damage the verge and you could leave mud and debris on the road as you drive away.

You may need to pass horses and riders on the road, or other animals such as sheep.

- Keep your speed and noise as low as possible.
- Give them as much room as you can.
- Be ready to stop if you need to.

Staying aware of cyclists, motorcyclists and pedestrians

Cyclists and motorcyclists may ride close behind you or between the kerb and your vehicle. Always check your nearside mirrors for riders before you change lanes or make a turn.

Some road users may not understand why you need to straddle lanes on the right before you turn left. Check there's no-one alongside who may be in danger from your rear or trailer wheels as you turn.

Take care at pedestrian crossings. If you're moving slowly in traffic, try to avoid stopping where your vehicle will block a crossing.

DBC s1, 4 **DGV** s1, 4

Safety equipment

Seat belts

If a seat belt is fitted to your vehicle, you **MUST** wear it unless you're

- **exempt for medical reasons**
- travelling less than 50 metres (162 feet); for example, for deliveries or collections.

> Definition
>
> **exempt for medical reasons**
> Have a certificate signed by a medical practitioner stating that some activity is inadvisable, on medical grounds.

You can take off your seat belt when performing a reversing manoeuvre.

Your passengers **MUST** wear any seat belts that are fitted and available for them. The number of passengers you carry **MUST NOT** be more than the number of seats and belts available.

Children **MUST** wear a **child restraint** suitable for their size and age, if one is available, otherwise they **MUST** use the adult seat belt.

Definition

child restraint
Baby seat, child seat, booster seat or booster cushion.

DBC s2 DGV s3 HC r99–102

Legal lettering

Your vehicle **MUST** clearly display all the information classified as 'legal lettering'. This must be where people can see it. This information covers

- seating/standing capacity
- emergency exit location
- fuel cut-off switch (if fitted)
- electrical isolator switch.

Public address systems

If your vehicle has a public address system inside, you should only use this if you're stationary and in a safe place, or if you have a colleague on board who's giving out information during the journey.

DBC s1, 2

Personal protective equipment (PPE)

Your operator should provide you with any protective equipment you need to do your job. It's **your** responsibility to use this properly, for your own safety.

As well as wearing a high-visibility jacket, you may also need

- protective footwear
- heat- or corrosion-resistant gloves
- safety glasses or protective goggles
- protective headgear
- a face mask or breathing apparatus
- hearing protectors or earplugs.

DGV s3

Mobile phones when driving

It's illegal to use a hand-held mobile phone while you drive. It's best to turn your mobile phone off before you set off. Don't pick up or send calls, voicemail messages or texts until you've stopped in a safe, legal place.

If you use hands-free equipment, remember this can also distract you from the road and other road users.

DBC s1 DGV s1

Fitness to drive

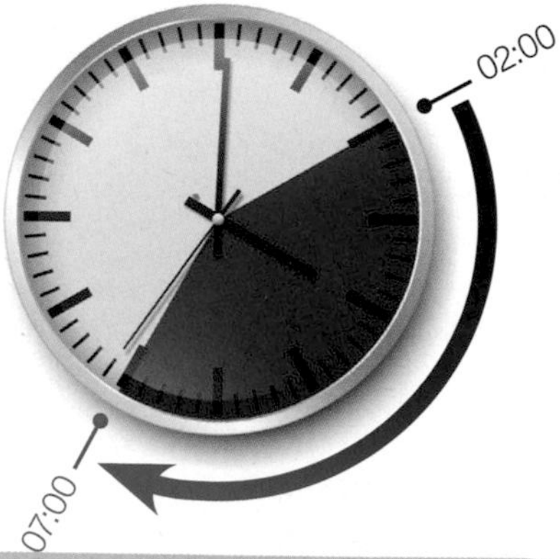

Up to 40% of collisions involving commercial vehicles are thought to involve tiredness. You're most at risk between 2 am and 7 am, as this is when most people's 'body clock' is programmed for sleep.

If you start to feel drowsy or tired while driving, open a window or turn down the heating, then stop as soon as it's safe to do so.

Before you drive, make sure you're fit and well enough to do so.

- Alcohol can remain in your body for 24–48 hours.
- The effects of drugs can last for up to 72 hours.

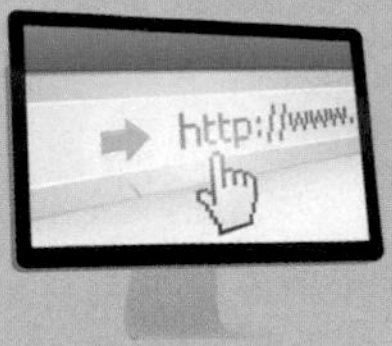

See the Think! road safety information on drink-driving.

http://think.direct.gov.uk/drink-driving.html

See the Think! road safety information on drug driving.

http://think.direct.gov.uk/drug-driving.html

Medication

Many medicines, both prescription and 'off-the-shelf', can cause drowsiness – even some everyday cold or flu remedies.

If your doctor needs to prescribe any medication for you, make sure you tell them that you're a professional driver and ask whether it could affect your ability to drive safely.

Read the label of any medicine carefully. If in doubt, don't drive.

DBC s3 DGV s3

Emergency vehicles

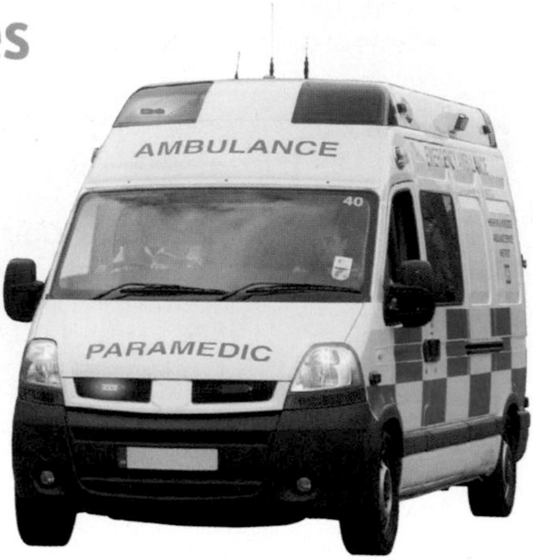

If you see an emergency vehicle showing red, blue or green lights, try to move out of the way to let them pass.

Do this safely and legally, and comply with all traffic lights and signs. Don't mount the pavement or put pedestrians in danger.

If a police officer signals you to stop and pull over, you **MUST** obey them. They'll attract your attention with blue flashing lights and/or a siren.

You **MUST** also obey traffic officers and Driver and Vehicle Standards Agency (DVSA) officers. They'll use hand signals and amber flashing lights on their vehicle.

DBC s4 DGV s4 HC r105–108, 219

HM coastguard

Bomb disposal

Mountain rescue

Blood transfusion

As well as fire, police and ambulance services, other emergency services (including those shown here) use a blue flashing light.

Doctors' vehicles may use green flashing lights when answering an emergency call.

Meeting the standards

You must be able to

assess whether your ability to drive safely and legally is affected or likely to be affected by the use of

- over-the-counter medicines
- prescription medicines
- illegal or controlled substances
- alcohol

assess whether your ability to drive safely and legally is affected by

- your emotional state
- a short- or long-term physical condition
- tiredness

manage your diet and fluid intake, taking shift patterns into account, to make sure that you're fit to drive and to minimise tiredness.

You must know and understand

what the law says about driving while you have illegal or controlled substances or alcohol in your system

that, regardless of any legal limits, the desirable level of alcohol to have in your system is zero

how being tired, before or during your journey, affects your ability to drive safely

how emotional states such as anger, grief, sadness and joy can affect your ability to drive safely.

Notes

You can use this page to make your own notes or diagrams about the key points you need to remember.

Think about

- How can I maintain a high standard of driving, while being patient and considerate?
- How should I drive when I need to pass animals on the road?
- How can I stay aware of cyclists, motorcyclists and pedestrians?
- How many passengers can I legally carry?
- What 'legal lettering' should I show and where should this be on my vehicle?
- What PPE do I need and do I know how to use it properly?
- What safe, legal phone habits should I follow?
- Do I stay aware of my fitness to drive and level of tiredness?
- How will emergency vehicles signal to me and how should I respond?

Your notes

Things to discuss and practise with your instructor

These are just a few examples of what you could discuss and practise with your instructor. Read more about the driver to come up with your own ideas.

Discuss with your instructor

- the rules on wearing a seat belt
- illnesses and disabilities that may prevent you from getting a bus or lorry licence
- emergency vehicles
  - which colour lights the various vehicles use
  - how to behave when you see their flashing lights.

Practise with your instructor

- driving down narrow roads and how to behave towards following traffic
- your ability to ignore your mobile phone. Arrange for someone to call you during your lesson so that you can practise your reaction
- driving in an area with pelican and zebra crossings.

4.1 Mark one answer HC r176

You're approaching a green traffic light and are going straight ahead. What should you do when the traffic ahead, beyond the junction, has stopped and is queuing?

- ☐ Go forward if your vehicle will clear the junction
- ☐ Drive slowly across the junction
- ☐ Maintain your speed and sound your horn
- ☐ Stop across the junction until the traffic clears

Look well ahead and plan your approach to all junctions. Even if the traffic lights are green, don't go forward if queuing traffic will cause you to block the junction. Be aware of the length of your vehicle and how much space it needs to clear the junction.

4.2 Mark one answer DGV s3, HC r99–102

What rules apply to the wearing of seat belts fitted to your lorry?

- ☐ It's not advisable
- ☐ It's advisable
- ☐ It's required by law
- ☐ It's not required by law

If seat belts are fitted to your lorry, you must wear them. Exemptions are allowed for holders of medical exemption certificates and for delivery drivers when travelling less than 50 metres (162 feet).

4.3 Mark one answer HC r180

You're driving this box van and are waiting to turn right. What should you do just before turning?

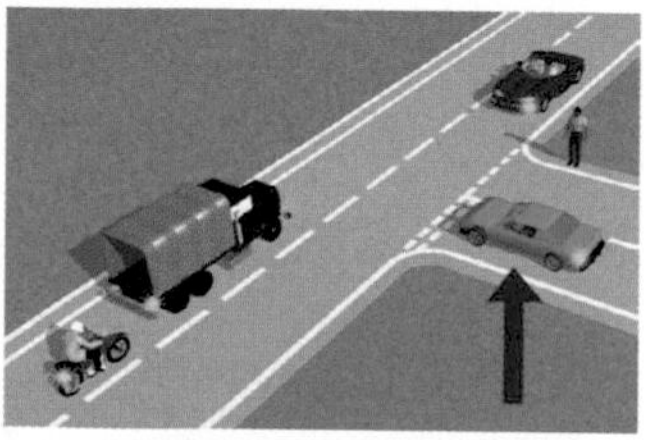

- ☐ Wave the pedestrian across
- ☐ Check your left-hand mirror and blind spot
- ☐ Wave out the green car (arrowed)
- ☐ Check your right-hand mirror and blind spot

Always make a final check in your mirrors and check your blind spots before you turn. Another driver or a motorcyclist may have committed themselves to overtaking traffic behind you before realising you intend to turn.

4.4	Mark one answer	DGV s4

What will you have to deal with when you're driving at night?

☐ Headlight dazzle
☐ Other drivers speeding
☐ Cold weather conditions
☐ Dazzle from shop windows

At night, the headlights of oncoming traffic can dazzle you, reducing how much you can see of the road ahead. Reduce your speed so that you can safely respond to any situation that may arise and make sure you've dipped your headlights.

4.5	Mark one answer	DGV s1

You're driving a lorry in a busy town. A driver pulls out in front of you and you have to brake hard. What should you do?

☐ Overtake as quickly as possible
☐ Stay calm and accept the error
☐ Flash your lights to show your annoyance
☐ Sound your horn and speed up

Some drivers might emerge from a junction when it's not safe to do so. In this situation, understand that other drivers might make mistakes. Don't intimidate them by driving aggressively to teach them a lesson.

4.6	Mark one answer	HC r99

Your lorry is fitted with a driver's seat belt. When may you drive without wearing it?

☐ When you're making deliveries that are less than 50 metres apart
☐ When you're towing at less than 50 mph
☐ When you're working less than 50 hours in a week
☐ When you're less than 50 miles away from your depot

It's compulsory for drivers and passengers to wear seat belts. However, if you're making deliveries that are less than 50 metres apart, you may leave the seat belt undone. It would be impractical to keep fastening and unfastening your seat belt over such short distances.

4.7	Mark one answer	DGV s3

You've been issued with protective clothing. Who's responsible for its use?

☐ You, the driver
☐ The insurance company
☐ The Health and Safety Executive
☐ Your employer

Protective clothing will usually be provided by your company. However, it's your responsibility to make sure that you use it properly.

4.8 Mark one answer HC r116

You're driving on a motorway. What does it mean when a lorry ahead of you switches on its hazard warning lights?

- ☐ There are speed cameras ahead
- ☐ The lorry is about to overtake
- ☐ The lorry is leaving the motorway
- ☐ Traffic further ahead may be stopping

When someone ahead of you switches on their hazard warning lights, it means that they've spotted a hazard that you may not be able to see yet. Slow down and be ready to stop if necessary.

4.9 Mark one answer HC r150

What should you do if you lose your way in heavy traffic?

- ☐ Stop at traffic lights and ask pedestrians
- ☐ Shout to other drivers to ask them the way
- ☐ Drive on until you find a safe place to stop
- ☐ Check a map as you keep moving with the traffic

Driving in heavy traffic needs 100% concentration. If you become lost, find a safe place to stop before checking a map or asking for directions. Don't risk losing concentration by glancing at a map while driving, even if you're in traffic that keeps stopping.

4.10 Mark one answer DBC s1, DGV s1

Where are you most likely to find pedestrians stepping into the road?

- ☐ Motorways
- ☐ One-way streets
- ☐ Wooded areas
- ☐ Shopping areas

In shopping areas, pedestrians frequently want to cross the road. They may be hidden by parked vehicles before stepping out into your path. Keep your speed down and be aware of the risk of pedestrians suddenly walking into the road.

4.11 Mark one answer DBC s4, DGV s4

What's most likely to cause danger to a group of horse riders?

- ☐ Powerful brake lights
- ☐ Leaving plenty of room
- ☐ The noise of your vehicle
- ☐ Reacting too early

The hiss of air-assisted mechanisms, as well as engine noise, can easily startle horses. Keep the noise to a minimum by gentle use of the brakes and, if necessary, stopping and turning your engine off.

4.12 Mark one answer DBC s4, DGV s4

What should you do when passing sheep on a road?

- ☐ Pass quickly and quietly
- ☐ Sound your horn gently
- ☐ Drive very slowly
- ☐ Keep your vehicle moving

Animals can be very unpredictable. You should give them as much room as you can, keep your speed and noise to a minimum to avoid panicking them, and always be ready to stop if necessary.

4.13 Mark one answer DBC s1, DGV s4

When is your nearside mirror most likely to endanger pedestrians?

- ☐ When driving over a speed hump
- ☐ When driving near a hospital
- ☐ When driving close to the kerb
- ☐ When driving past a car park

When you drive too close to the kerb, there's a risk that your nearside mirror could strike the head of a pedestrian. Be aware of this hazard and, if you see pedestrians close to the road, make sure you leave them enough room.

4.14 Mark one answer DBC s1, DGV s1

What can prevent you from holding a bus or lorry driving licence?

- ☐ Heart disorders
- ☐ Dyslexia
- ☐ Skin problems
- ☐ Stomach problems

An illness that could affect you while you're driving, such as a heart condition, may make it unsafe for you to drive. You must tell DVLA if you develop any serious illness or disability that's likely to last more than three months and that could affect your driving. Partial blindness and mental disorders can also prevent you from getting a driving licence for large vehicles.

4.15 Mark one answer DBC s3, DGV s3

You're about to drive a vehicle you've never driven before. What should you consider when adjusting your seat?

- ☐ Your distance from the controls
- ☐ Being able to reach your sat-nav
- ☐ The tension in the seat belt
- ☐ The wear on the seat fabric

Being seated properly is very important when driving long distances. A poor driving position can quickly cause fatigue. You should make sure that you can reach all the controls comfortably.

4.16 Mark one answer DBC s3, DGV s3

Why is it important to adjust your head restraint properly?

- ☐ To make you more comfortable
- ☐ To protect your neck in a collision
- ☐ To help you relax while you're driving
- ☐ To maintain a good driving position

Your head restraint should be adjusted so that, in the event of a collision, it protects your neck from whiplash injury.

4.17 Mark one answer DBC s2, DGV s3

When may you drive without wearing your seat belt?

- ☐ When reversing
- ☐ When carrying out a hill start
- ☐ When making an emergency stop
- ☐ When driving slowly

You can legally remove your seat belt when carrying out a manoeuvre that involves reversing. Don't forget to put it back on again when you've finished.

4.18 Mark one answer DBC s4, DGV s4

You're driving along this road. What should you do if the red van cuts in close in front of you?

- ☐ Accelerate to get closer to the red van
- ☐ Give a long blast on the horn
- ☐ Drop back to leave the correct separation distance
- ☐ Flash your headlights several times

There are times when other road users make incorrect or ill-judged decisions. Try to stay calm and don't retaliate or react aggressively, especially as your vehicle is likely to be larger and more intimidating than the other person's. Always put safety first.

4.19	Mark one answer	DBC s1, DGV s4

You're approaching a large puddle near the left-hand kerb. What should you do if pedestrians are close to the water?

- ☐ Ignore the puddle
- ☐ Brake suddenly and sound your horn
- ☐ Try to avoid splashing the pedestrians
- ☐ Wave at the pedestrians to keep back

The effect of your vehicle driving through a puddle will be to throw water onto the pavement. If there are pedestrians close by, they could be splashed with the water. Be considerate and, if it's safe to do so, avoid driving through the puddle.

4.20	Mark one answer	DBC s4, DGV s4

What should you do if a long, heavily laden lorry is taking a long time to overtake you?

- ☐ Speed up
- ☐ Slow down
- ☐ Hold your speed
- ☐ Change direction

A long lorry with a heavy load will need more time to pass you than a car. It won't be able to accelerate enough to pass you quickly, especially on an uphill stretch of road. Be considerate to the lorry driver – ease off the accelerator and allow the lorry to pass.

4.21	Mark one answer	HC r169

You're driving a slow-moving vehicle on a narrow, winding road. What should you do to let other vehicles overtake you?

- ☐ Wave them past
- ☐ Stop in a safe place
- ☐ Show a left-turn signal
- ☐ Drive more slowly

Don't frustrate other road users by driving for long distances with a queue of traffic behind you. This could cause them to lose concentration and overtake in an unsafe way. Stop where it's safe and let them pass.

4.22	Mark one answer	DBC s1, 4, DGV s1, 4, HC r112

When should you use your vehicle's horn?

- ☐ To warn others of your presence
- ☐ To allow you right of way
- ☐ To greet other road users
- ☐ To signal your annoyance

Only use your horn to warn people of your presence when they may not have seen you. Don't use it to greet others, to show impatience, or to give or claim priority.

Your horn mustn't be used between 11.30 pm and 7.00 am in a built-up area, or when your vehicle is stationary, unless another vehicle poses a danger.

4.23 Mark one answer DBC s4, DGV s4

You're following a car being driven slowly by a learner driver. What should you do if it isn't safe to overtake it?

- ☐ Flash your lights so the driver sees you
- ☐ Be patient and stay well behind
- ☐ Switch your hazard lights on
- ☐ Drive along the centre line of the road

Learner drivers are often nervous. If you stay well back, this will reduce the risk of incidents occurring if they make a mistake, and you won't intimidate them. Remember, you were once in that situation yourself.

4.24 Mark one answer HC r106

What should you do if you're signalled to stop by a police officer in a patrol car?

- ☐ Brake harshly to a stop
- ☐ Drive on until you reach a side road
- ☐ Stop on the left as soon as it's safe
- ☐ Stop immediately wherever you are

If a police officer signals for you to stop, stop as soon as possible in a safe place on the left.

4.25 Mark one answer HC r106

You're being followed by a marked police car. As well as flashing the headlights, what will the police officer do to signal you to stop?

- ☐ Signal with the right indicator
- ☐ Signal with the left indicator
- ☐ Switch on the hazard warning lights
- ☐ Switch on the rear fog lights

When a police officer is following you and wants you to stop, they'll flash the vehicle's headlights and signal with the left indicator. You should stop in a safe place as soon as possible.

4.26 Mark one answer HC r149

Your vehicle is fitted with a hand-held telephone. What should you do before taking or making a call?

- ☐ Find a safe place to stop
- ☐ Reduce your speed to less than 30 mph
- ☐ Steer your vehicle with one hand
- ☐ Be very careful when dealing with junctions

Telephone calls can distract you, meaning that you're not in proper control of your vehicle. If you need to use a telephone when driving, find a safe place to stop first. Better still, use a phone answering service and listen to your messages at the end of your journey.

4.27 Mark one answer HC r149

A mobile telephone is fitted to your vehicle. When should it be used?

- ☐ When you're stopped in a safe place
- ☐ When travelling slowly
- ☐ When on a motorway
- ☐ When in light traffic

It's illegal to use a hand-held phone while driving. When you've stopped in a safe place, you can concentrate on your call or message. Using a phone while driving means you won't have full control of your vehicle. This could result in a collision, with serious or even fatal consequences.

4.28 Mark one answer HC r197

You approach a pelican crossing that goes straight across the road. How must you treat it if it has a central island?

- ☐ As one crossing in daylight only
- ☐ As one continuous crossing
- ☐ As two separate crossings
- ☐ As two crossings during darkness

A pelican crossing that goes straight across the road must be treated as one continuous crossing even if it has a central island. The lights controlling the crossing show to both directions of traffic. You must give way to pedestrians who are still crossing when the amber light is flashing.

4.29 Mark one answer HC r196

What does the flashing amber light mean at a pelican crossing?

- ☐ Stop if you can do so safely
- ☐ Give way to pedestrians already on the crossing
- ☐ Stop and wait for the green light
- ☐ Give way to pedestrians waiting to cross

Pedestrians use push-button controls to activate the signals. Pelican crossings have no red-and-amber stage before green. Instead, they have a flashing amber light. This means you must give way to pedestrians on the crossing, but you may proceed if the crossing is clear.

4.30 Mark one answer HC r195

What should you do at a zebra crossing?

- ☐ Rev your engine to encourage pedestrians to cross quickly
- ☐ Park only on the zigzag lines on the left
- ☐ Always leave the crossing clear in traffic queues
- ☐ Wave pedestrians to cross if you intend to wait for them

When traffic is heavy, plan well ahead so that you don't block crossings. Leave them clear so that pedestrians can cross safely.

4.31 Mark one answer DBC s4, DGV s4

A coach is overtaking you. What should you do when it's safe for the coach to move back to the left?

- ☐ Do nothing and let the driver decide
- ☐ Switch your sidelights on and off
- ☐ Flash your headlights once
- ☐ Flash your headlights twice

When you're being overtaken, the other driver has to decide when it's safe to complete the manoeuvre. Don't give an unofficial signal, however well intended. There may be a hazard you're unable to see from your position.

4.32 Mark one answer HC r219

What should you do when you're being followed by an ambulance showing a flashing blue light?

- ☐ Stop to let it pass as soon as it's safe to do so
- ☐ Accelerate quickly to get away from it
- ☐ Ignore it if possible, unless you're forced to let it pass
- ☐ Brake harshly and immediately to a stop in the road

When an ambulance is displaying a flashing blue light, try to stop in a safe place to allow it to pass. Make sure you don't endanger any other road users or commit an offence; for example, by driving past a red traffic light.

4.33 Mark one answer DBC s4, DGV s4, HC r219

Should you give way to a car showing a flashing green beacon?

- ☐ Yes, because it's a doctor going to an emergency
- ☐ Yes, because it's a fire-crew support vehicle
- ☐ No, because it's a slow-moving vehicle
- ☐ No, because it's a breakdown vehicle

If you see a vehicle with a green flashing light, it will be a doctor on an emergency call, so give way by pulling over and letting the vehicle pass. But don't just stop suddenly – choose a safe place as soon as you can to let the doctor's car pass safely.

4.34 Mark one answer DBC s4, DGV s4, HC r219

What type of emergency vehicle has a green flashing beacon?

- ☐ Fire engine
- ☐ Road gritter
- ☐ Ambulance
- ☐ Doctor's car

A green flashing light on a vehicle means the driver or passenger is a doctor on an emergency call. Give way to them if it's safe to do so. Be aware that the vehicle may be travelling quickly or may stop suddenly.

4.35 Mark one answer HC r195

You stop for pedestrians at a zebra crossing. What should you do if they don't start to cross immediately?

- ☐ Be patient and wait
- ☐ Sound your horn
- ☐ Drive on
- ☐ Wave them to cross

If you stop for pedestrians and they don't start to cross, be patient. Don't wave them across or sound your horn. This could be dangerous if the driver of an approaching vehicle isn't aware of them. The driver may not have seen or heard your signal, making it very dangerous for the pedestrians to start to cross.

4.36 Mark one answer HC r195

Why could it be dangerous to wave people across at pedestrian crossings?

- ☐ There may be another vehicle coming
- ☐ They may be distracted
- ☐ They may not see your signal
- ☐ They may not be ready to cross

You should always try to stop, if it's safe, for people waiting at pedestrian crossings. However, avoid waving them to cross. Another driver may not have seen them, may not have seen your signal or may not be able to stop safely.

4.37 Mark one answer DBC s4, DGV s1

What danger should you be most aware of if you're driving close to the kerb in a busy shopping area?

- ☐ The nearside mirror striking the heads of pedestrians
- ☐ The amount of fuel being used when driving slowly
- ☐ Solid white line markings in the centre of the road
- ☐ Traffic lights that may suddenly change to green

When you need to drive close to the kerb, be aware of the dangers.

- Pedestrians may step off the kerb.
- The nearside mirror may be at a pedestrian's head height.
- Cyclists may be tempted to pass you on your left if you're driving slowly in congested conditions.

4.38 Mark one answer — HC r243

Which of these is an unsafe place to park your vehicle or trailer?

- ☐ At an overnight service area
- ☐ Near the brow of a hill
- ☐ In the yard at a factory
- ☐ In a lay-by in a built-up area

Don't park where you would endanger or inconvenience others. If your choice of parking place obstructs drivers, riders or pedestrians, move to a more suitable area.

4.39 Mark one answer — DBC s3, 4, DGV s3, 4

What must you check before setting out on a journey?

- ☐ Your mirrors are tinted
- ☐ Your mirrors are convex
- ☐ Your mirrors are concave
- ☐ Your mirrors are clean

It's important to know what's happening behind as well as ahead. Your mirrors must always be clean and properly adjusted.

4.40 Mark one answer — DBC s4, DGV s4

A group of schoolchildren are standing close to the left-hand kerb. What should you do as you drive past?

- ☐ Check your offside mirror
- ☐ Check your nearside mirror
- ☐ Switch on your headlights
- ☐ Switch on your hazard warning lights

As you approach the children, consider whether you need to sound the horn as a warning. Always check your nearside mirror as you pass any hazards on the left, so that you know when your vehicle is clear of the hazard.

4.41 Mark one answer — HC r168

You're driving at the legal speed limit. What should you do if a vehicle comes up quickly behind, flashing its headlights?

- ☐ Accelerate to maintain a gap behind you
- ☐ Touch the brake pedal sharply to show your brake lights
- ☐ Maintain your speed and prevent the vehicle from overtaking
- ☐ Allow the vehicle to overtake

Don't try to enforce the speed limit by blocking another vehicle's progress. This is likely to cause the other driver to become more frustrated. Slow down or move over when it's safe and allow the other vehicle to pass.

4.42 Mark one answer | DBC s4, DGV s4

What should you do when a vehicle pulls out in front of you at a junction?

- ☐ Swerve past it and sound your horn
- ☐ Flash your headlights and drive up close behind
- ☐ Slow down and be ready to stop
- ☐ Accelerate past it immediately

Plan ahead and learn to anticipate hazards. Give yourself more time to react to any problems that might occur – such as a vehicle pulling out unexpectedly. Be tolerant of other road users who don't behave correctly. Don't react by flashing your lights, sounding your horn or making gestures, as any show of aggression on the road can quickly lead to a serious incident.

4.43 Mark one answer | DBC s3, DGV s3, HC r95

How is drinking alcohol likely to affect your driving?

- ☐ You'll have faster reactions
- ☐ You'll become colour-blind
- ☐ You'll feel more confident
- ☐ You'll be able to concentrate better

Alcohol can increase confidence to a point where a driver's behaviour might become 'out of character': someone who normally behaves sensibly suddenly takes risks and enjoys it. Never let yourself or your friends get into this situation.

4.44 Mark one answer | DBC s3, DGV s3

What can seriously affect your concentration while driving?

- ☐ Contact lenses
- ☐ Fresh air
- ☐ Tinted windows
- ☐ Tiredness

Tiredness can seriously affect your concentration – even to the point of falling asleep. You always need to be alert so that you can safely deal with the constantly changing environment around your vehicle.

4.45 Mark one answer | DBC s3, DGV s3, HC r95

How does alcohol affect your driving?

- ☐ It speeds up your reactions
- ☐ It increases your awareness
- ☐ It improves your coordination
- ☐ It reduces your concentration

Concentration and good judgement are needed to be a good, safe driver. You're strongly advised not to drink any alcohol before driving. Alcohol can stay in your system for 24–48 hours, so you should also consider your intake the day and night before driving.

4.46 Mark one answer DBC s3, DGV s3

You've been convicted of driving while unfit through drink or drugs. What will increase in cost when you start driving again?

- ☐ Vehicle tax
- ☐ Insurance premiums
- ☐ Vehicle test certificate
- ☐ Driving licence

If you've been convicted of drink- or drug-driving, statistically you're a risk to yourself and others on the road. For this reason, when you're able to drive again, insurance companies may charge you a high premium to insure your vehicle.

4.47 Mark one answer DBC s3, DGV s3

What should you do if you've had a few alcoholic drinks at a party?

- ☐ Have a strong cup of coffee and then drive home
- ☐ Drive home carefully and slowly
- ☐ Go home by public transport
- ☐ Wait a short while and then drive home

Drinking black coffee or waiting a few hours won't make any difference. Alcohol takes time to leave the body, so you should find a way to get home without using your vehicle; for example, by public transport. Depending on how much you've had to drink, you may even be unfit to drive the following morning.

4.48 Mark one answer DBC s3, DGV s3

As a driver, what should you do when you attend a social event?

- ☐ Drink plenty of coffee after drinking alcohol
- ☐ Avoid busy roads after drinking alcohol
- ☐ Avoid drinking alcohol completely
- ☐ Avoid drinking alcohol on an empty stomach

Avoiding alcohol will always be the safest option. There are many factors that influence how alcohol will affect you, such as height, weight and gender. It's better not to risk a single alcoholic drink before you drive.

4.49 Mark one answer DBC s3, DGV s3

It's been eight hours since you last had an alcoholic drink. What effect could the alcohol have on your driving?

- ☐ You may still be over the legal drink-drive limit
- ☐ You'll have no alcohol in your system
- ☐ Your driving won't be impaired
- ☐ You can't be breathalysed

Alcohol can take a long time to leave the body. You may feel all right to drive, but its effect will last for many hours.

4.50 Mark one answer DBC s3, DGV s3

Your doctor has given you a course of medicine. Why should you ask whether it's safe to drive?

- ☐ Drugs make you a better driver by quickening your reactions
- ☐ You'll have to let your insurance company know about the medicine
- ☐ Some types of medicine can cause your reactions to slow down
- ☐ The medicine you take may affect your hearing

Some medicines can have side-effects that affect your driving. Ask your doctor whether you'll be safe to drive while taking the medicine that's been prescribed.

4.51 Mark one answer DBC s3, DGV s3

You've been taking medicine for a few days, and it has made you feel drowsy. Today you feel better, but you still need to take the medicine. When should you drive?

- ☐ Only if your journey is necessary
- ☐ Only at night on quiet roads
- ☐ Only if someone goes with you
- ☐ Only after checking with your doctor

Take care; it's not worth taking risks. Always check with your doctor to be absolutely sure. The medicine may have an effect on you later in the day and for longer than you think.

4.52 Mark one answer DBC s3, DGV s3

What should you do if you aren't sure whether your cough medicine will affect your driving?

- ☐ Ask your doctor for advice
- ☐ Drive a little more slowly
- ☐ Drive if you feel all right
- ☐ Ask a friend for advice

If you're taking medicine or drugs prescribed by your doctor, check that they won't affect your driving. If you forget to ask when you visit the surgery, check with your pharmacist.

4.53 Mark one answer DBC s3, DGV s3

You take some cough medicine given to you by a friend. What must you do before driving?

- ☐ Drink some strong coffee
- ☐ Ask your friend whether taking the medicine affected their driving
- ☐ Check the label to see whether the medicine will affect your driving
- ☐ Make a short journey to see whether the medicine is affecting your driving

Never drive after taking drugs you don't know about. They might affect your judgement and perception – and, therefore, endanger lives.

4.54 Mark one answer DBC s3, DGV s3

What should you do if you're driving along a motorway and become tired?

- ☐ Stop on the hard shoulder at a safe place to rest
- ☐ Leave the motorway at the next exit and take a rest
- ☐ Close your windows and turn up the radio volume
- ☐ Close your windows and set the heating to warm

If you feel the onset of tiredness, you need to take a break. This might be at a motorway service station, but if there isn't one nearby you'll need to leave at the next exit and find a safe place to stop and rest.

4.55 Mark one answer DBC s3, DGV s3

You're about to drive home. What should you do if you feel very tired and have a severe headache?

- ☐ Wait until you're fit and well before driving
- ☐ Drive home, but take painkiller tablets
- ☐ Drive home if you can stay awake for the journey
- ☐ Wait for a short time, then drive home slowly

All your concentration should be on your driving; any pain you feel will distract you. Be safe and wait until you're fit before starting your journey.

4.56 Mark one answer DBC s3, DGV s3

What should you do if you're feeling tired but there's nowhere to stop safely?

- ☐ Increase your speed to find a stopping place more quickly
- ☐ Ensure a supply of fresh air
- ☐ Gently tap the steering wheel
- ☐ Keep changing speed to improve your concentration

If you're travelling on a long journey, plan your route before you leave. This will help you to

- be decisive at intersections and junctions
- plan your rest stops
- know approximately how long the journey will take.

Make sure that the vehicle you're travelling in is well ventilated. A warm, stuffy atmosphere can make you drowsy, which will impair your judgement and perception.

4.57 Mark one answer DBC s3, DGV s3

When will your reactions be slower?

- ☐ When driving while you're tired
- ☐ When driving in fog
- ☐ When driving too quickly
- ☐ When driving in rain

Try to avoid becoming tired by taking plenty of rest stops and allowing fresh air into your vehicle.

4.58 Mark one answer DBC s3, DGV s3

What should you do if you're taking medication that could affect your driving?

- ☐ Seek medical advice
- ☐ Make short journeys only
- ☐ Drive only at night
- ☐ Drink plenty of water

Check all medicines for their side-effects, such as drowsiness. If you're unsure, ask your doctor or pharmacist to explain how you might be affected by the medication.

4.59 Mark one answer DBC s3, DGV s3

What should you do if you feel tired while driving on a motorway?

- ☐ Stop on the hard shoulder for a rest
- ☐ Carry on, but drive slowly
- ☐ Leave the motorway at the next exit
- ☐ Try to complete your journey more quickly

Don't continue to drive if you feel the onset of tiredness. Find a safe place to stop for a rest. Walking around in the fresh air during your break will help but it's no substitute for adequate rest.

4.60 Mark one answer DBC s3, DGV s3

You've driven a long distance and feel tired. Your tachograph shows that you haven't exceeded your driving hours. What should you do in these circumstances?

- ☐ Park in a suitable place and rest
- ☐ Reduce your speed and drive more slowly
- ☐ Carry on driving to use up your hours
- ☐ Increase your speed and reduce your journey time

The smallest lapse in concentration can result in loss of control. Even when your tachograph would allow you to drive for longer, stop somewhere safe and rest if you feel your driving may be affected by your tiredness.

4.61 Mark one answer DBC s3, DGV s3

What might misuse of drugs or alcohol lead to?

- ☐ Better concentration
- ☐ Better eyesight
- ☐ Withdrawal of a driving licence
- ☐ Faster reactions

Misuse of drugs and/or alcohol may lead to the withdrawal of your driving licence. Future insurance premiums will probably increase as well.

4.62 Mark one answer DBC s4, DGV s4

What should you do if a front tyre bursts while you're driving on a motorway?

- ☐ Loosen your grip on the steering wheel
- ☐ Brake firmly to a stop
- ☐ Hold the steering wheel firmly
- ☐ Drive to the next service area

A front tyre bursting will seriously reduce your control of the vehicle. Keep calm and resist the temptation to brake hard or swerve. Hold the steering wheel firmly and try to get the vehicle onto the hard shoulder while allowing it to slow down gradually. Stop as far to the left as possible and switch on your hazard warning lights.

4.63 Mark one answer DBC s4, DGV s4

Your vehicle breaks down on a motorway and you need to call for help. Why may it be better to use an emergency roadside telephone rather than a mobile phone?

- ☐ It connects you to a local garage
- ☐ Using a mobile phone will distract other drivers
- ☐ It allows easy location by the emergency services
- ☐ Mobile phones don't work on motorways

On a motorway, it's best to use a roadside emergency telephone so that the emergency services are able to locate you easily.

4.64 Mark one answer HC r149

When are you most likely to lose concentration while you're driving?

- ☐ When you use a mobile phone
- ☐ When you switch on the windscreen wipers
- ☐ When you switch on the heated rear window
- ☐ When you look at the door mirrors

Using a hand-held mobile phone while driving is illegal. It will distract you from your driving, because your attention is split between the phone call and your driving.

4.65 Mark one answer HC r149

What should you do if you want to make a business call on your mobile phone while you're driving?

- ☐ Make the call but take extra care
- ☐ Use the cruise control while you're making the call
- ☐ Stop in a safe place before making the call
- ☐ Reduce speed while you make the call

It's illegal to use a hand-held mobile phone while you're driving, other than to call 999 or 112 in a genuine emergency. Switch it off before you start driving and use the messaging facility. You can pick up messages when you stop for a rest break. If you must make a call, stop in a safe place before using the phone.

4.66 Mark one answer — HC r149

You're driving on a single carriageway. When may you use your mobile phone?

- ☐ Only when you're receiving a call
- ☐ Only when you're suitably parked
- ☐ Only when you're driving at less than 30 mph
- ☐ Only when you're driving an automatic vehicle

It's illegal to use a hand-held mobile phone while you're driving. Park in a suitable place before receiving or making a call or text.

4.67 Mark one answer — HC r149

What effect will using a mobile phone have on your driving?

- ☐ Your vehicle will be harder to steer
- ☐ Your field of vision will be reduced
- ☐ Your attention will be diverted from the road
- ☐ Your vehicle's electronic systems will be disrupted

Driving today requires all of your attention, all of the time. Any distraction, however brief, is dangerous. This is why it's illegal to use a hand-held mobile phone while you're driving.

4.68 Mark one answer — DBC s4, DGV s4

Your vehicle breaks down on the hard shoulder of a motorway. What should you do if you decide to use your mobile phone to call for help?

- ☐ Stand at the rear of the vehicle while making the call
- ☐ Use the speed-dial function to save time
- ☐ Set the tachograph to 'Other work' before making the call
- ☐ Check your location from the marker posts on the left

In an emergency, time can be of the essence. The emergency services need to know your exact location. Before you phone, get the post number from the nearest marker post on the edge of the hard shoulder. Give this to the emergency services, as it will help them to locate you.

4.69 Mark one answer — HC r149

How should you answer a mobile phone call while you're driving?

- ☐ Reduce your speed wherever you are
- ☐ Stop safely before answering
- ☐ Keep one hand on the steering wheel
- ☐ Slow down and allow others to overtake

No phone call is important enough to endanger someone's life. If you must be contactable when driving, plan your route to include breaks where you can catch up on telephone messages in safety. Always choose a safe place to take a break.

4.70 Mark one answer DBC s4, DGV s4

You're overtaking a lorry. What should you do if you see the driver flash their headlights?

- ☐ Move back to the left when you judge it's safe to do so
- ☐ Take their signal to mean it's safe to move back to the left
- ☐ Stop; there must be something wrong with your vehicle
- ☐ Flash your hazard warning lights and move back to the left

Never presume what someone means when they flash their headlights, as they might be signalling to someone else. Use your own judgement and wait until it's safe to complete your manoeuvre.

4.71 Mark one answer HC r223

A bus has stopped at a bus stop ahead of you. What should you do if its right-hand indicator is flashing?

- ☐ Flash your headlights and slow down
- ☐ Slow down and give way if it's safe to do so
- ☐ Sound your horn and keep going
- ☐ Slow down and then sound your horn

Give way to buses whenever you can do so safely, especially when they signal to pull away from bus stops. Look out for people who have left the bus and wish to cross the road. Also look for people rushing to catch a bus: they may be more concerned about catching the bus than watching for traffic.

4.72 Mark one answer HC r207

You've stopped for an elderly pedestrian who's slowly crossing the road. What should you do if traffic behind you is being held up?

- ☐ Edge forward slowly and make them hurry
- ☐ Stay where you are and allow them to cross in their own time
- ☐ Steer slowly around them to ease the build-up of traffic
- ☐ Get out of your vehicle and wave them across

Elderly pedestrians can be hesitant and may move slowly when crossing the road. Also, their awareness of traffic may be limited because of hearing or visual impairment. Be patient and show courtesy and understanding.

4.73 Mark one answer — HC r169

How should you let other vehicles overtake when you're driving a slow-moving vehicle along a narrow road?

- ☐ Maintain a steady speed
- ☐ Wave them past
- ☐ Give a left-turn signal
- ☐ Pull in when you can

Drivers queuing behind you may make hasty or ill-judged decisions in an effort to overtake. If you see a queue of traffic building up behind, give way as soon as you can do so safely by pulling in to the left.

4.74 Mark one answer — DBC s1, DGV s1, HC r149

How is using a hands-free phone likely to affect your driving?

- ☐ It will improve your safety
- ☐ It will increase your concentration
- ☐ It will reduce your view
- ☐ It will divert your attention

While you're using a phone, your attention will be divided between the call and the road. This means you're unable to take in all that's happening around your vehicle. You need to be concentrating on your driving all of the time, and especially when dealing with a hazard.

4.75 Mark one answer — DBC s1, DGV s1, HC r149

You're driving on the motorway and want to use your mobile phone. What should you do before using it?

- ☐ Reduce your speed to 40 mph
- ☐ Stop on the hard shoulder
- ☐ Move into the left-hand lane
- ☐ Stop in a safe place when you can

Plan your journey and take breaks to keep in touch if necessary. When driving on motorways, you can't just stop to use your mobile phone. You'll need to turn off the motorway or wait until you reach the next service area. Avoid being distracted by switching your phone off while you're driving; use the message facility to listen to any calls when you've parked safely.

4.76 Mark one answer — DBC s3, DGV s3

What effect will drinking alcohol have on your driving?

- ☐ Better forward planning
- ☐ A false sense of confidence
- ☐ Faster reactions
- ☐ Greater awareness of danger

Drinking alcohol will reduce your ability to drive safely. One drink is too many if you're going to drive.

4.77 Mark one answer HC r282

You're driving on a motorway. There's been a collision on the opposite carriageway. What should you do if drivers ahead are slowing to look?

- ☐ Concentrate on the road ahead
- ☐ Slow down to take a look
- ☐ Stop on the hard shoulder
- ☐ Overtake using the hard shoulder

Drivers passing incidents often end up having collisions themselves when they're distracted by the scene. You need to keep your concentration in a situation like this and ignore what's happening on the other carriageway.

4.78 Mark one answer DBC s1, DGV s6

What do you need to be aware of when driving a vehicle fitted with power-assisted steering?

- ☐ It causes less tyre wear
- ☐ It prevents you from oversteering
- ☐ It makes it easier for you to steer
- ☐ It only works at high speeds

Power-assisted steering operates when the engine is running. If a fault develops, much greater effort is required to turn the steering wheel. Don't attempt to drive a vehicle if you're aware of a fault in the power-steering system.

4.79 Mark one answer DBC s1, DGV s1

What should you do if a car driver makes a mistake that causes you to brake?

- ☐ Keep calm and don't retaliate
- ☐ Overtake and sound your horn
- ☐ Drive close behind and sound your horn
- ☐ Flag the driver down and explain the mistake

Sometimes other road users disobey the rules or make errors of judgement. Try to accept this calmly and learn from their mistakes.

4.80 Mark one answer DBC s1, DGV s1

What should you do if you're upset by another driver's behaviour?

- ☐ Stop and take a break
- ☐ Shout abusive language
- ☐ Gesture to them with your hand
- ☐ Follow them, flashing your headlights

Tiredness may make you more irritable than you would be normally. You might react differently to situations because of it. If you feel yourself becoming tense, take a break.

4.81 Mark one answer | DBC s1, DGV s1

What should you do if another driver does something that upsets you?

- ☐ Stay calm and don't retaliate
- ☐ Let them know how you feel
- ☐ Flash your headlights several times
- ☐ Sound your horn

Sometimes other drivers or riders make a mistake. If this happens, try not to let it worry you. Don't react by sounding the horn, flashing your headlights or shouting at the other driver; it won't help the situation. A professional driver will remain calm and stay focused on their driving.

4.82 Mark one answer | DBC s4, DGV s4

You're driving in fast-moving traffic along a motorway. What should you do if there's a stationary queue of traffic ahead?

- ☐ Move to the hard shoulder
- ☐ Change lanes
- ☐ Switch on your rear fog lights
- ☐ Switch on your hazard warning lights

Traffic queues on the motorway are fairly common, often due to the volume of traffic at peak times or to incidents. If you see a queue of stationary traffic ahead, switch on your hazard warning lights for a short while to warn those behind you of a hazard ahead.

4.83 Mark one answer | DBC s6, DGV s6, HC r173

You're turning right onto a dual carriageway from a side road. What should you do if your vehicle is too long for the gap in the central reservation?

- ☐ Move forward and wait in the middle
- ☐ Wait until it's clear in both directions
- ☐ Move out, blocking traffic from the right
- ☐ Edge out slowly so other traffic will see you

When turning right onto a dual carriageway, don't stop in the gap in the central reservation unless your vehicle will fit in the gap without any overhang. When it's busy, consider turning left and using a roundabout further up the road so you don't have to cross the central reservation.

4.84 Mark one answer | DBC s6, DGV s6

You want to turn left at a junction. What affects the position you take up as you approach the junction?

- ☐ The length of your vehicle
- ☐ The axle weight of your vehicle
- ☐ The camber of the road
- ☐ The type of road surface

Plan well ahead and take care if you need to move out or straddle lanes due to the length of your vehicle. Other road users may not understand your reasons for doing this and may try to pass on your left. Check your mirrors before and as you turn. It's better to use extra space on the road you're leaving than to find there's not enough room when you turn in.

4.85 Mark one answer DBC s4, DGV s4

What should you do when you have to travel a long distance?

- ☐ Allow plenty of time for your journey
- ☐ Plan to go at busy times
- ☐ Avoid all national-speed-limit roads
- ☐ Prevent other drivers from overtaking

Always allow plenty of time for your journey, in case of unforeseen problems such as punctures, breakdowns, road closures or diversions. You'll feel less stressed and less inclined to take risks if you're not under time pressure.

4.86 Mark one answer DBC s1, 3, DGV s1, 3

What can you do to help maintain concentration while you're driving?

- ☐ Eat sugary snacks when on duty
- ☐ Have regular meals and rest breaks
- ☐ Don't eat at all when on duty
- ☐ Avoid meals containing bread

Being as healthy and alert as possible will make you safer on the road. Take regular meal and rest breaks, and keep to foods that contain complex carbohydrates, such as bread and vegetables. These release energy slowly and will keep you satisfied for longer than foods with a high sugar content.

4.87 Mark one answer DBC s1, DGV s1

What's the best type of fluid to drink during work time?

- ☐ Bottled water
- ☐ Sugary canned drinks
- ☐ High-caffeine drinks
- ☐ Hot chocolate

Water is the ideal drink, because it quenches your thirst for longer than tea or coffee. It's a good idea to carry water when you're driving, especially in very hot weather. The medical profession recommends that you drink 1.5 to 2 litres of water per day.

4.88 Mark one answer DBC s1, DGV s1

What should you do to help your concentration when you have to drive through the night?

- ☐ Eat lots of snacks while you're driving
- ☐ Eat a big meal in the middle of your shift
- ☐ Eat a meal at the beginning of your shift
- ☐ Eat nothing during your period of work

Concentration will be improved by eating regular, light meals timed to fit in with your scheduled rest periods. If you don't eat at all, your blood sugar levels will drop and you'll feel tired. If you just eat snack foods, most of which are sugary or high in fat, your energy levels will peak and then drop – and eating while you're driving is a dangerous distraction.

4.89 Mark one answer DBC s3, DGV s3

When is there a particular risk of sleep-related incidents?

- ☐ Between 2 am and 7 am
- ☐ Between 11 pm and 2 am
- ☐ Between 6 pm and 11 pm
- ☐ Between 7 am and 2 pm

There's a particular risk when driving between 2 am and 7 am. This is when the 'body clock' is programmed for sleep in most people. There's another, smaller, dip between 2 pm and 4 pm. If you feel that tiredness is affecting your ability to drive, stop in a safe place and rest.

4.90 Mark one answer DBC s3, DGV s3

Approximately what percentage of sleep-related vehicle incidents (SRVIs) involve commercial vehicles?

- ☐ 40%
- ☐ 55%
- ☐ 70%
- ☐ 95%

Research shows that approximately 40% of SRVIs involve commercial vehicles. The consequences are usually very serious because of the size of the vehicle and the fact that there is no braking before the impact.

4.91 Mark one answer DBC s1, DGV s1

Poor eating habits can increase your risk of long-term health problems. What may result from a poor diet?

- ☐ Deafness
- ☐ Epilepsy
- ☐ Lung disease
- ☐ Heart disease

A poor diet can increase your risk of ill health. It may cause obesity, diabetes or heart disease, which can increase your risk of sudden incapacity at the wheel. There's a greater risk of developing a serious illness – and ultimately an early death. Smoking and a lack of exercise will increase the risks.

4.92 Mark one answer DBC s3, DGV s3

What should you do when you've been driving for a long time and are fighting sleep?

- ☐ Stop and rest
- ☐ Open a window
- ☐ Play loud music
- ☐ Stretch your arms

The only effective solution to driver sleepiness is to stop in a safe place and take a break.

4.93 Mark one answer DBC s3, DGV s3

Where do most sleep-related vehicle incidents occur?

- ☐ On motorways
- ☐ In one-way streets
- ☐ On rural roads
- ☐ On tourist routes

Driving at a constant speed for long periods, such as on a motorway, can cause drowsiness. Falling asleep while driving accounts for a significant proportion of vehicle incidents. You should plan and take regular rest stops, but if you feel the onset of tiredness between these breaks, stop and rest in a safe place.

4.94 Mark one answer DBC s3, DGV s3

At what age are men most at risk from sleep-related vehicle incidents (SRVIs)?

- ☐ 30 years and under
- ☐ 31–45 years
- ☐ 46–59 years
- ☐ 60 years and over

SRVIs are more evident in young male drivers, who tend to ignore or deny how they're feeling. The greatest risk is between the hours of 2.00 am and 7.00 am, when the body clock is in a trough. Make sure you take sufficient rest breaks and don't ignore feeling sleepy.

4.95 Mark one answer DBC s4, DGV s4

What should you try to do when your vehicle has a front-tyre blow-out?

- ☐ Hold the steering wheel firmly and slow down gradually
- ☐ Steer to the right-hand side and brake firmly
- ☐ Don't use your brakes under any circumstances
- ☐ Brake hard and steer towards the affected side

With a front-tyre blow-out, you won't be able to steer properly. Keep a firm hold of the steering wheel and slow gradually to a stop. Don't brake hard or steer sharply.

4.96 Mark one answer — DBC s2, DGV s3, HC r99

When can you drive without wearing a seat belt?

- ☐ When you hold a medical exemption certificate
- ☐ When the seat belt is too large for your use
- ☐ When you're driving on a motorway
- ☐ When the speed limit is 30 mph or less

When seat belts are fitted to your vehicle, they must be worn unless you hold a valid medical exemption certificate. People making deliveries or collections in goods vehicles are also exempt, when they're travelling less than 50 metres (162 feet).

4.97 Mark one answer — DBC s3, DGV s3

How will your body clock affect your driving between 2.00 am and 7.00 am?

- ☐ You're more likely to fall asleep at the wheel
- ☐ You're more likely to feel road rage
- ☐ You're more likely to break the speed limit
- ☐ You're more likely to drive too close to the vehicle in front

A significant number of sleep-related vehicle incidents involve commercial vehicles. There's a particular risk when driving between 2.00 am and 7.00 am, because this is when the body clock craves sleep. It's important to take proper rest before, during and between driving duties.

4.98 Mark one answer — DBC s3, DGV s3, HC r97

What's provided specifically to protect you against neck and whiplash injuries?

- ☐ An air-sprung seat
- ☐ Anti-lock brakes
- ☐ A collapsible steering wheel
- ☐ A head restraint

If you're in a collision, a properly adjusted head restraint can help to reduce the risk of neck injury. However, an incorrectly adjusted head restraint could worsen any damage to your neck.

4.99 Mark one answer — DBC s2

When are you allowed to drive a bus without wearing the seat belt?

- ☐ When the seat belt is uncomfortable
- ☐ When it's a lap-only type of seat belt
- ☐ When you're reversing the vehicle
- ☐ When your passengers are children

If your vehicle is fitted with a seat belt, you must wear it unless you're exempt for medical reasons. You may also remove it before a reverse manoeuvre. If the fitting of the belt is uncomfortable and it prevents you from obtaining a safe driving position, report this to your employer.

4.100 Mark one answer DBC s1

What must be clearly displayed on your bus?

- ☐ The route timetable
- ☐ The location of all bus stops
- ☐ The emergency-exit location
- ☐ Your employee number

Make sure that your vehicle displays all the information required by law (the 'legal lettering'): the seating/standing capacity and the locations of the emergency exit, fuel cut-off switch and electrical isolator.

4.101 Mark one answer DBC s4

You're driving on a dual carriageway. When would you use the nearside mirror?

- ☐ To check that the driver's door is closed properly
- ☐ After overtaking, to see whether it's safe to return to the left
- ☐ To see whether any passengers want the next bus stop
- ☐ To look for following traffic moving out to overtake

Use your nearside mirror when you've overtaken another vehicle. This is particularly important on dual carriageways or motorways, where it can take some distance to overtake safely. Look carefully to make sure you're far enough ahead and can safely return to the left lane.

4.102 Mark one answer DBC s4

What should you do while you're driving a bus in a built-up area?

- ☐ Keep junctions clear
- ☐ Reduce your stopping distance
- ☐ Stay ahead of cyclists and motorcyclists
- ☐ Allow less time for passengers to board

Blocking other vehicles' access to side roads is inconsiderate and can cause congestion. You should look at the flow of traffic and be aware of junctions when you're slowing down or stopping.

Section five

Carrying passengers

In this section, you'll learn about

- passenger safety and comfort
- vehicle stability
- passenger care
- older passengers and those with disabilities.

Carrying passengers

Passenger safety and comfort

When you drive a passenger-carrying vehicle, your job is to get your passengers to their destination safely and on time.

- Treat passengers with patience, courtesy and consideration.
- Try to give them a comfortable journey.
- Always make sure they're safe.

Be careful when braking, accelerating or cornering. If you do any of these sharply or harshly you may throw passengers off-balance. This can be uncomfortable and may even injure them.

Give passengers time to sit down safely before you move off. If your vehicle has seat belts, you **MUST** advise passengers to wear them. You can do this with an announcement or by using signs or pictograms.

TIP

When driving at night, switch your inside lights on to help passengers move about safely on the bus.

Always make sure the doors are shut before you move off and don't open the doors again until the bus has stopped completely. Never allow passengers to travel on an open platform.

Using bell signals

When driving a half-cab bus and carrying passengers, a **designated person** must be in charge of the passenger area. If this is an official conductor you must both know the system of bell signals.

Stop – 1 ring	
Move off when safe – 2 rings	
Bus full – 3 rings	
Emergency on bus – 4 rings	

Definition

designated person
Someone who has been nominated for a task.

DBC **s1, 2, 4**

Vehicle stability

When you accelerate, brake or change direction, forces act on your vehicle and its load. Stronger or more sudden changes cause stronger forces. These may cause you to lose control of the vehicle.

- Plan ahead so you approach bends and corners at a speed that lets you go around them carefully.
- Brake or accelerate gently, if possible, when travelling in a straight line.

When driving a double-deck bus or high-floor coach, stay aware of the **camber** of the road. This will tip the top of the vehicle to one side.

Definition

camber
The curvature of a road surface.

camber

no camber

Look out for anything that may hit the top of the vehicle as it tips. This could include

- lampposts
- overhanging buildings
- shop awnings
- traffic signs and signals
- street furniture.

DBC s1, 2, 3

Passenger care

Whenever you can, stop close to the kerb when you pick up or set down passengers. Some passengers will find it hard to get on or off if there's a big gap to step across.

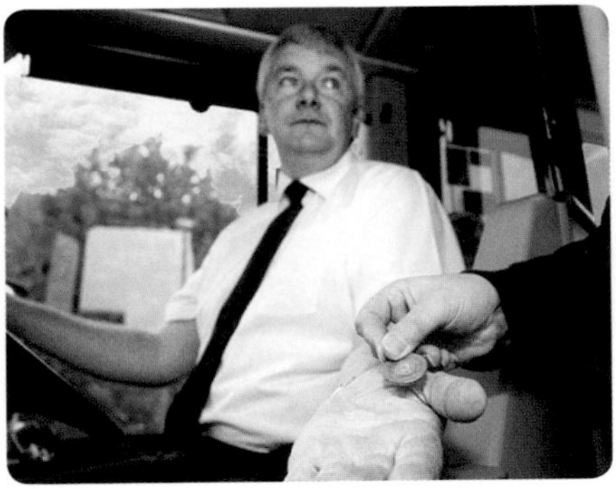

Make eye contact and listen to each passenger. Looking directly at them when you speak is important, especially if they have hearing difficulties and need to lip-read.

- Do give passengers time to sit down before you move off.
- Don't move off while taking fares, giving change or issuing tickets.

DBC s1

Older passengers and those with disabilities

Some older passengers or those with a disability may need extra patience, courtesy and consideration. Be ready to offer assistance if they ask for it.

TIP

Remember that some older passengers may prefer to be independent and help themselves.

Your vehicle may have a **'kneeling' facility**. Use this to make it easier for older passengers and those with a disability to get on or off. Make sure that you've been trained in how to use it. Don't move off until the platform has returned to its travelling position.

Definition

'kneeling' facility

A feature of buses or coaches, where air is let out of the suspension system to allow the exit of the vehicle to be lowered to pavement level.

Vehicles that are wheelchair accessible will have special areas for wheelchairs.

You must not **discriminate** against people with a disability. Make all reasonable attempts to allow them to board the vehicle.

Definition

discriminate

To disadvantage someone because they're different in some way (such as those with a disability).

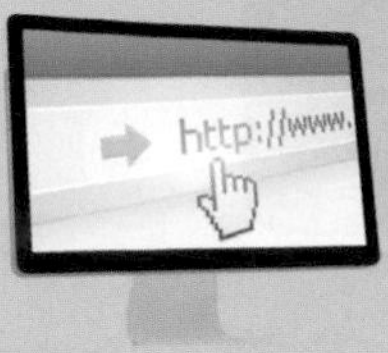

Read about the rights of disabled bus and coach passengers at this website.

citizensadvice.org.uk

You may only refuse to allow someone with a disability to board the vehicle when

- the passenger lifting equipment has broken down
- another wheelchair user already occupies the area suitable for wheelchairs
- the bus is already full to its stated capacity with passengers and their luggage.

Driving a school bus

A school bus will display the yellow and black school bus sign. When driving a school bus, you're permitted to use the vehicle's hazard warning lights when you stop to let children on and off the vehicle.

DBC s1 HC r209, p117

Meeting the standards

You must be able to

enforce the regulations that apply to passengers and their behaviour, and make sure that you don't carry more passengers than

- the vehicle is designed for
- the law allows

manage the effect that passengers may have on your ability to drive safely

make sure of the safety and comfort of passengers at all times.

You must know and understand

current legal requirements for the fitting and use of seat belts

how to deal with the effects of social pressure and distractions caused by passengers

the regulations that apply to passengers and their behaviour.

Notes

You can use this page to make your own notes or diagrams about the key points you need to remember.

Think about

- What can I do to help my passengers be safe and comfortable?
- Do I know the four bell signals?
- How can I be aware of road camber, and of anything that might hit the top of my vehicle?
- Do I stop close to the kerb when I can?
- Do I make eye contact, listen to passengers and speak clearly and politely?
- Do I wait until all passengers are seated before moving off?
- How can I offer help to an older person or someone with a disability, while also letting them be independent if they want to?

Your notes

Things to discuss and practise with your instructor

These are just a few examples of what you could discuss and practise with your PCV instructor. Read more about carrying passengers to come up with your own ideas.

Discuss with your instructor

- the ways that you, as a bus driver, can show good passenger care
- how to provide the best possible service for disabled passengers. For example
  - wheelchair users
  - blind passengers
  - passengers with learning difficulties
- the rules about the use of seat belts on buses.

Practise with your instructor

- accelerating and braking smoothly
- taking bends in the best way for your passengers and to keep the bus stable
- pulling up at stops in the correct position to make boarding as safe as possible for passengers.

5.1 Mark one answer — DBC s1

What's your main responsibility as a bus driver?

- ☐ The safety and comfort of your passengers
- ☐ Keeping to a strict timetable
- ☐ The collecting of fares
- ☐ The issuing of tickets

Your passengers should be your first consideration. It's your responsibility to deliver them to their destination safely and on time. You should also treat passengers courteously at all times.

5.2 Mark one answer — DBC s1

How can a bus driver show care for their passengers?

- ☐ By taking fares while moving, to save time
- ☐ By reaching destinations early
- ☐ By not speaking when taking fares
- ☐ By stopping close to the kerb

At bus stops, stop close to the kerb, where it's safe and convenient for your passengers to get on and off. Accelerating as you move off can easily unsteady a passenger. Try to wait until all passengers are seated or settled before moving off.

5.3 Mark one answer — DBC s1

What's the main reason for using smooth acceleration when you're driving a bus?

- ☐ To reduce wear on the tyres
- ☐ To reduce wear on the engine
- ☐ To improve fuel consumption
- ☐ To improve passenger comfort

Operators often publicise journeys as being comfortable, convenient and fast. You play an important part in delivering this standard of service.

5.4 Mark one answer — DBC s2

How can you avoid harsh braking?

- ☐ Pump the brakes when approaching a bus stop
- ☐ Use only the gears to slow down
- ☐ Use the parking brake just before stopping
- ☐ Plan ahead and take early action on all stops

Your first duty is to your paying customers, who want to reach their destination comfortably and safely. Harsh braking can cause your passengers to be thrown forward on the bus, so avoid this by looking and planning well ahead. Give yourself plenty of time to react to traffic situations.

5.5	Mark one answer	DBC s2

Where are passengers likely to be thrown when a bus driver brakes harshly?

- ☐ To the front of the bus
- ☐ To the rear of the bus
- ☐ To the nearside
- ☐ To the offside

When approaching hazards or bus stops, you should always be aware of how your passengers will be affected by the change in speed. Late, harsh braking as passengers leave their seats can take them by surprise and cause them to fall towards the front of the bus, possibly injuring themselves or others. Wherever possible, avoid harsh braking by planning ahead, giving yourself time to reduce speed smoothly.

5.6	Mark one answer	DBC s2

How can you avoid harsh braking?

- ☐ Gently apply the parking brake
- ☐ Plan ahead and take early action
- ☐ Slow down by using your gears only
- ☐ Pump the brake pedal several times

Always look well ahead. Early planning and anticipation will help you avoid the need to brake harshly. It's important that you give your passengers a smooth journey, especially if any of them are having to stand.

5.7	Mark one answer	DBC s2

When would passengers be most likely to notice weight transfer?

- ☐ When the vehicle is parking
- ☐ When the vehicle is cornering
- ☐ When the vehicle is reversing
- ☐ When the vehicle is overtaking

A smooth ride at all times isn't always easy to achieve. However, scanning ahead for hazards will help you avoid late braking and cornering too fast.

5.8	Mark one answer	DBC s4

Well ahead of you are traffic lights on green. What should you do in case the lights change to red?

- ☐ Accelerate to make sure you can cross before they change
- ☐ Slow down to avoid the need to stop suddenly
- ☐ Accelerate, but warn your passengers you may have to stop
- ☐ Carry on at a constant speed, but be ready to sound your horn

If you're approaching a set of traffic lights and you can see they've been on green for a while, be prepared for them to change. Ease off the accelerator and be ready to come to a gradual stop if you need to. Don't try to beat the red light; it may change as you come closer, causing you to brake late and harshly. Think of your passengers.

5.9 Mark one answer DBC s2

Why should a bus driver avoid stopping harshly?

- ☐ To preserve the tyres
- ☐ To avoid wear on the brakes
- ☐ To keep the passengers safe
- ☐ To stop the suspension bouncing

Passengers are very vulnerable when they're standing in a moving vehicle. Inconsiderate use of the footbrake can easily cause them injury – especially if they're young or infirm. Some passengers may not wait until you've stopped before leaving their seat. If you seem to be in a hurry, they may even leave their seat early because they don't want to feel responsible for holding you up.

5.10 Mark one answer DBC s2

Where may passengers be thrown when a bus takes a bend too fast?

- ☐ Towards the outside of the bend
- ☐ Towards the inside of the bend
- ☐ Towards the front of the bus
- ☐ Towards the rear of the bus

Turning corners or travelling around bends too quickly will push passengers towards the outside of the bend. In addition, the weight of the passengers being transferred to one side of the vehicle will make the bus even more unstable. This will make steering out of the bend more difficult.

5.11 Mark one answer DBC s1

What would contribute to the safety and comfort of your passengers?

- ☐ Giving change on the move
- ☐ Driving with the door open
- ☐ Thinking and planning well ahead
- ☐ Braking hard before reaching a bend

Thinking and planning well ahead will help you to deal competently and safely with hazards and other road users. A safe, comfortable ride, even when you're under pressure, will be appreciated by your passengers.

5.12 Mark one answer DBC s1

What should you be most aware of before moving off from a standstill?

- ☐ Passengers making phone calls
- ☐ Passengers smoking on the bus
- ☐ Passengers asking you directions
- ☐ Passengers getting off the bus

Check your nearside external and internal mirrors before you move away, in case anyone is attempting to board or leave your vehicle. Always give passengers time to be seated safely before you move away.

5.13 Mark one answer DBC s2

Your bus is fitted with seat belts. When should passengers wear them?

- ☐ During journeys of more than 20 minutes
- ☐ Only when travelling in EU countries
- ☐ Only when travelling on motorways
- ☐ At all times

As the driver of a bus, you may be responsible for several passengers at any given time. If a situation occurs where you have to brake or steer harshly, your passengers could be thrown about the vehicle. In these circumstances, there's a significant risk of injury from fittings on board, such as luggage racks, handrails and poles. If seat belts are provided for passengers, they should wear them.

5.14 Mark one answer DBC s4

Which of these is a legal requirement for every bus?

- ☐ A fire extinguisher
- ☐ A current timetable
- ☐ A mobile phone or radio
- ☐ A working tachograph

Every bus must carry a fire extinguisher. Make sure that you know where it's located and how to use it, so that you're fully prepared in the event of a fire.

5.15 Mark one answer DBC s4

What must be clearly labelled on a bus?

- ☐ The location of the air vents
- ☐ The location of the first-aid equipment
- ☐ The location of the security cameras
- ☐ The location of the bus station

It's essential to know the location of first-aid equipment, the fuel cut-off device and the fire extinguisher on every bus that you drive. Take time to familiarise yourself with their location whenever you drive a different vehicle. When you take your PCV test, you'll be asked questions on safety.

5.16 Mark one answer DBC s1

A passenger is boarding your bus. What does it mean if they're carrying a white stick with a red ring painted on it?

- ☐ They have a learning difficulty
- ☐ They have poor vision and hearing
- ☐ They have a physical disability
- ☐ They have a speech problem

A person carrying a white stick or a long white cane may be visually impaired. If the stick has a red ring or red-and-white checks painted on it, they also have impaired hearing. Be prepared to help if they appear to need it, or if they ask for assistance.

5.17 Mark one answer DBC s1

What should a bus driver do to help passengers have a good journey?

- ☐ Keep to the timetable regardless of comfort
- ☐ Drive on before people are seated
- ☐ Look ahead when issuing tickets
- ☐ Be courteous and polite

Nobody likes to be late, but you shouldn't let the pressure of your timetable make you feel you have to rush people. Being pleasant to your passengers and showing them some common courtesies will encourage them to use your service again.

5.18 Mark one answer DBC s1

Why should you look at passengers when you're collecting fares from them?

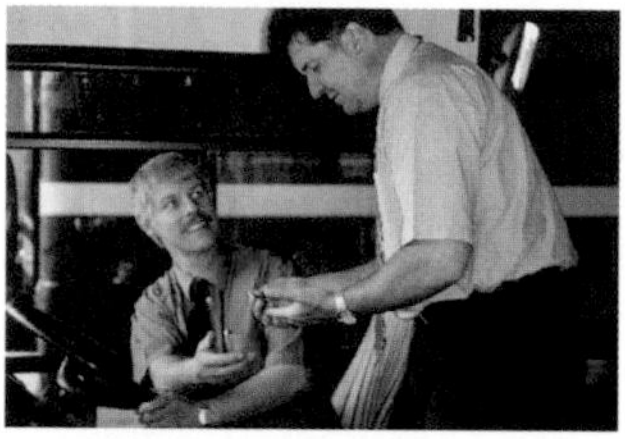

- ☐ To see if they're carrying bags
- ☐ To remove the threat of violence
- ☐ To be courteous and polite
- ☐ To show them you're in charge

You're a representative of your company, so showing passengers that their custom is appreciated will encourage them to travel with you again.

5.19 Mark one answer DBC s1

What should you do when dealing with passengers who have hearing difficulties?

- ☐ Shout as loudly as you can
- ☐ Ask another passenger to help them
- ☐ Hurry to get them to their seats
- ☐ Look at them when speaking to them

Passengers who are hard of hearing may want to lip read. Make sure that they're able to see your face clearly when you speak to them.

5.20 Mark one answer DBC s1

What's the purpose of a kneeling bus?

- ☐ To improve passenger comfort on bumpy roads
- ☐ To help with access under low bridges
- ☐ To allow the step height to be raised and lowered
- ☐ To give more clearance over speed ramps

A kneeling bus uses air suspension to lower the front entrance, giving easier access for passengers. This is especially helpful for people who have difficulty climbing steps, are disabled or use wheelchairs. Make sure you're properly trained to use this equipment.

5.21 Mark one answer DBC s1

Why would this sign be fitted to the front and rear of a bus?

- ☐ The bus may be carrying children
- ☐ Children must be accompanied by an adult
- ☐ The bus is carrying blind people
- ☐ The driver will help disabled people

If you're carrying children on your vehicle and it isn't on a scheduled route used by the general public, the bus must display this sign to the front and rear. When carrying children to and from school, it's likely that you'll have to make several stops in places other than recognised bus stops. Think carefully before you stop. Don't cause unnecessary inconvenience to other road users.

5.22 Mark one answer DBC s1

You're driving a bus displaying this sign. When can you use its hazard warning lights?

- ☐ When stopped at a pedestrian crossing
- ☐ When children are getting on or off the vehicle
- ☐ When approaching a school crossing patrol
- ☐ When there's a sign warning of a school ahead

You may be driving in the rush hour, when traffic is heavy, so when you stop you're permitted to show your hazard warning lights. This will tell other road users that children are getting on and off the bus. Look out for passing traffic and make sure that all your passengers get on and off safely.

5.23 🚌 **Mark one answer** **DBC s4**

Your bus has broken down at night. Why should you move your passengers to the front of the bus?

- ☐ To keep the bus stable
- ☐ To help you see clearly out of the back window
- ☐ To limit injuries in the event of a rear-end collision
- ☐ To keep them informed about the breakdown

Passenger safety should be your priority. You should take every precaution to ensure they're out of danger.

5.24 🚌 **Mark one answer** **DBC s4**

What should you do when you're driving a coach at night and you have passengers on board?

- ☐ Stop more often
- ☐ Switch off the radio
- ☐ Keep the interior lit
- ☐ Close the curtains

Passengers should be able to move about the vehicle in safety. You should make sure that all interior lights are in working order before you start your journey.

5.25 🚌 **Mark one answer** **DBC s4**

What should you do if there's a fire on the upper deck of your bus?

- ☐ Move all the passengers to the lower deck
- ☐ Get all the passengers off the bus
- ☐ Make sure passengers have their belongings
- ☐ Open all the windows on the upper deck

Fire can spread extremely quickly. Your priority is the safety of your passengers: you must get them off the bus straight away. If at all possible, disconnect electrical lines and cut off the fuel supply. Fire can destroy a vehicle very quickly.

5.26 Mark one answer DBC s1

What are kneeling buses designed to improve?

- ☐ Access for the driver
- ☐ Access to the luggage bay
- ☐ Access for passengers
- ☐ Access for maintenance

Some buses are equipped with air or hydraulic systems that allow the step level to be lowered. These are known as kneeling buses and have been designed to allow people to board more easily. Don't forget to raise the step again before moving off.

5.27 Mark one answer DBC s4

You're driving a bus carrying passengers at night. Why should you always switch on the interior lights?

- ☐ It will help you see the road ahead
- ☐ So that passengers can see to move around
- ☐ It will help passengers to see outside
- ☐ So that you can see your controls

Passengers need a properly lit area so they can move around safely.

5.28 Mark one answer DBC s1

Which passengers should you especially watch for as you move off?

- ☐ Those smoking in the lower saloon
- ☐ Those standing in the upper saloon
- ☐ Those avoiding paying the correct fare
- ☐ Those attempting to board the bus

Always check your nearside mirror before moving away. A passenger may be attempting to open the door to board the bus, or running to jump aboard an open platform. Also check for passengers trying to get off the bus as you move away.

5.29 Mark one answer DBC s1

What could you do to help a passenger who has difficulty walking?

- ☐ Drive quickly, so that the passenger has a shorter journey
- ☐ Wait until the passenger has sat down before moving away
- ☐ Make sure the passenger has a window seat
- ☐ Suggest that the passenger stands near the door

Try to wait until your passengers have sat down before you move off. This is even more important if they're elderly or have difficulty walking. Offer help when you think it might be needed – and remember, a smile goes a long way.

5.30 Mark one answer DBC s1

You're the driver of a bus displaying reflective yellow school-bus signs. When are you allowed to use hazard warning lights?

- ☐ When parked to take a rest period
- ☐ When children are boarding the bus
- ☐ When slowing down to approach a bus stop
- ☐ When slowing down to find a parking space

Buses carrying schoolchildren must display a distinctive yellow reflective sign on the front and rear, unless they're running a scheduled service for the general public. Buses displaying the sign are permitted to use hazard warning lights when they've stopped for schoolchildren to get on or off.

5.31 Mark one answer DBC s3

You're parking your coach at a coastal resort. What should you do if your passengers will still have access to the vehicle?

- ☐ Ask the passengers to take all of their belongings
- ☐ Park with the gear lever in neutral
- ☐ Ensure the storage lockers are left open
- ☐ Leave a responsible person on the coach

If passengers have access to your coach, it must never be left unattended. In particular, they mustn't be able to get into the cab area of the vehicle. You or another responsible person must always be there to supervise the coach.

5.32 Mark one answer DBC s2

What should you be especially aware of while you're driving a double-deck bus on a road with a steep camber?

- ☐ Lampposts
- ☐ Parking meters
- ☐ Parked cars
- ☐ Litter bins

When driving high-sided vehicles, such as double-deck buses, changes in the road camber can affect your vehicle. As the bus leans towards the kerb, there's a risk that shop awnings, trees and street furniture, such as lampposts, could strike the top deck.

5.33 Mark one answer DBC s3

What's the likely weight difference between an empty bus and a bus with 75 passengers on board?

- ☐ 5 tonnes
- ☐ 10 tonnes
- ☐ 15 tonnes
- ☐ 20 tonnes

Your vehicle will handle very differently when it's full, compared with when it's empty. Having 75 passengers on board would increase the weight of the bus by about 5 tonnes – and the passengers may also have luggage. It will take longer to build up speed and the vehicle will be more difficult to stop. Forward planning and controlled braking will be required to allow for this extra weight.

5.34 Mark one answer DBC s3

It's important to be able to work out the weight difference between a full bus and an empty one. How many passengers will add about 1 tonne to the vehicle's overall weight?

- ☐ 8
- ☐ 15
- ☐ 25
- ☐ 30

You should also make allowances for any luggage your passengers may bring on board. An average of two cases per passenger on a 50-seat coach will add about 1.5 tonnes to the overall weight of your bus or coach.

5.35 Mark one answer DBC s1

What must buses and coaches display when they're used for school contract work?

- ☐ School-bus signs
- ☐ Flashing amber beacons
- ☐ Triangular warning signs
- ☐ No overtaking' signs

Vehicles carrying schoolchildren display the 'school bus' sign to alert other drivers to the danger of children crossing the road. As the bus driver, you must make every effort to ensure the children's safety when boarding or getting off your bus, as well as during their journey with you.

5.36 Mark one answer DBC s2

What should you do if a passenger comments on exhaust smoke in the bus?

- ☐ Report it as soon as you return to the depot
- ☐ Stop and have the fault put right
- ☐ Avoid heavy revving of the engine when stationary
- ☐ Have the emissions checked at the next vehicle inspection

If exhaust fumes are getting into the vehicle, this could be a danger to your passengers. Stop as soon as it's safe to do so and have the fault put right before continuing.

5.37 Mark one answer DBC s2

What do you have to be aware of when driving a double-deck bus on a road that has a steep camber?

- ☐ 'Keep left' islands
- ☐ A smooth road surface
- ☐ Pedestrian crossings
- ☐ Overhanging trees

A road that has a steep camber will slope sharply towards the kerb, causing your bus to lean towards the pavement or verge. This brings a risk of collision with objects such as overhanging trees, lampposts and bus-stop roofs.

5.38 Mark one answer DBC s1

What does good passenger care include?

- ☐ Making criticisms of other road users
- ☐ Providing a commentary on the route
- ☐ Talking to passengers while driving
- ☐ Helping passengers who need directions

Always show consideration to passengers, whether they have special needs or not. Try to imagine what assistance you would like if you were in their position.

5.39 Mark one answer DBC s1

Your bus is fitted with lifts or ramps for less mobile passengers. Who should operate this equipment?

- ☐ Wheelchair attendants
- ☐ Fully trained people
- ☐ Bus-company employees
- ☐ Accompanying nurses

Make sure that you're fully trained in the safe use of lifts, ramps and securing devices. If you drive a vehicle fitted with this equipment, never let untrained people operate it.

5.40 Mark one answer DBC s1

What does a kneeling bus improve?

- ☐ Access for disabled people
- ☐ Stability when cornering
- ☐ Passenger comfort at higher speeds
- ☐ Access for the driver

Kneeling buses are equipped with air or hydraulic systems that allow the vehicle body to be lowered. This improves access for disabled and elderly passengers. Remember to return the mechanism to its travelling position before moving off.

5.41 Mark one answer DBC s2

Your double-deck bus breaks down on a busy road. Where should you ask your passengers to move to?

- ☐ The rear of the bus
- ☐ The top deck
- ☐ The lower deck
- ☐ The front of the bus

The greatest risk to a stationary bus is being hit from behind. Moving your passengers forward could reduce the risk of injury if such a collision occurred.

5.42 Mark one answer DBC s1

What should you watch for when you drive away from a bus stop?

- ☐ Passengers who avoid paying the correct fare
- ☐ Passengers smoking in the lower saloon
- ☐ Passengers leaving the bus
- ☐ Passengers using expired travel passes

Even though you've started to drive away, passengers may still try to get off the bus. You should also check your nearside mirror for any passengers trying to get on as you're moving away.

5.43 Mark one answer DBC s1

What must you do before moving off when you're driving a service bus?

- ☐ Finish issuing tickets to passengers
- ☐ Ensure your fuel tank is above a quarter full
- ☐ Take off your sunglasses
- ☐ Make sure all passengers have a seat

On regular services, traffic congestion can soon put you behind schedule. Nevertheless, you have a responsibility to your passengers at all times not to take shortcuts or jeopardise their safety; for example, by driving off while you're still issuing tickets.

5.44 Mark one answer DBC s4

Your bus breaks down on the motorway. What should you do if it isn't feasible to remove your passengers from the vehicle?

- ☐ Move passengers to the rear of the bus
- ☐ Move passengers to the front of the bus
- ☐ Move passengers to the middle of the bus
- ☐ Tell passengers to stay in the same seats

Stop as far to the left as possible on the hard shoulder. Make sure all your passengers have moved as far as possible to the front of the vehicle; this will reduce the risk of injury if a passing vehicle runs into the back of the bus. Use an emergency telephone to call for assistance.

5.45 Mark one answer DBC s1

What's a bus driver's first priority?

- ☐ The safety and comfort of passengers
- ☐ Keeping to the timetable
- ☐ Completing tachograph records correctly
- ☐ Making sure the destination is clearly marked

Bus drivers have responsibilities beyond those of other drivers. Your passengers have paid for a service, and they should arrive at their chosen destination safely. Customers are more likely to travel with you again if they receive a safe, courteous and comfortable service.

5.46 Mark one answer DBC s1

You're driving a one-person-operated bus. What should you do while issuing tickets at a bus stop?

- ☐ Signal right, so you're ready to move off
- ☐ Turn on your hazard warning lights
- ☐ Make sure you've cancelled any signal
- ☐ Signal left, to let others know you're not ready to move off

Giving wrong signals causes uncertainty and confusion to other road users. Only give signals that are relevant and timed to match your actions.

5.47 Mark one answer DBC s1

You're the driver of a one-person-operated double-deck bus. How do you stay aware of passengers on the top deck?

- ☐ By counting passengers up and down the staircase
- ☐ By frequent checks upstairs while stopped at bus stops
- ☐ By listening to passengers in the upstairs gangway when approaching bus stops
- ☐ By making full use of the internal mirror system

If you're driving a one-person-operated double-deck bus, you must make sure that you use your internal mirrors. You need to be able to see passengers who are about to descend the stairs. Consider their safety and avoid harsh braking and steering.

5.48 Mark one answer DBC s1

You're driving a double-deck bus. Why do you need to know if anyone is using the stairs?

- ☐ The upper deck is only to be used when the lower deck is full
- ☐ Tickets for the upper deck are more expensive
- ☐ Passengers using the stairs could fall if the bus brakes or swerves
- ☐ No-one under 18 years old may use the stairs

You must always be aware of anyone using the stairs. Be particularly careful to avoid any sudden movement, such as when braking or cornering, as this could cause passengers to stumble.

5.49 Mark one answer DBC s1

Where should you stop when you're stopping to collect passengers at a bus stop?

- ☐ Close to the kerb
- ☐ Away from the kerb
- ☐ After the bus stop
- ☐ Before the bus stop

If passengers aren't waiting at the bus stop, it can be tempting to stop near where they're standing. This may not be a safe place for your vehicle or for other road users. Be professional and be safe; stop in the correct position.

5.50 Mark one answer DBC s1

Where should you stop to make it easy for passengers leaving or joining your bus?

- ☐ Next to soft grass
- ☐ Alongside guard rails
- ☐ Near parked cars
- ☐ Close to the kerb

You should stop as close to the kerb as you can, so that passengers can reach the safety of the pavement without any difficulty.

5.51 Mark one answer DBC s3

A bus stop is blocked and you can't get into it. What's the most important thing you should do before opening the exit door?

- ☐ Try to get the bus stop cleared
- ☐ Carry on to the next bus stop
- ☐ Check for traffic on the left
- ☐ Check for traffic on the right

Make sure that you allow your passengers to get on or off the vehicle safely. If you're unable to stop close to the kerb, don't open the doors until you're sure it's safe. You must take care of your passengers at all times.

5.52 Mark one answer DBC s3

Several parked cars are blocking your bus stop. What should you do before allowing passengers to leave the bus?

- ☐ Move on to the next bus stop
- ☐ Check it's clear of traffic on the left
- ☐ Try to find the car owners
- ☐ Check it's clear of traffic on the right

Always check your nearside mirror before opening the door to allow passengers to alight. When you've stopped away from the kerb, it's important to warn the passengers to look out for cyclists and to expect a long step down onto the road. Be ready to offer help if they need it.

5.53 Mark one answer DBC s3

What should you do before allowing passengers off your bus?

- ☐ Collect their used tickets
- ☐ Activate an audible warning system
- ☐ Check mirrors before opening doors
- ☐ Ask if they have luggage to collect

Passenger safety is your main responsibility. Before you allow them to step down from the bus, you should always check your mirrors to make sure there's nothing approaching that could endanger them.

5.54 Mark one answer DBC s3

Passengers may be in a hurry to get off the bus as you approach a bus stop. What should you do to reduce any dangers?

- ☐ Insist that passengers stay seated until the bus stops
- ☐ Stop just before the bus stop and let passengers get off
- ☐ Let passengers on to the bus before letting passengers off
- ☐ Keep the passenger doors closed until the bus has stopped

Passengers may have left their seats early and may be standing up, waiting to get off. Don't brake harshly or open the doors until the vehicle has stopped.

5.55 Mark one answer DBC s4

When can you drive a bus at night without having the interior lights switched on?

- ☐ When the passengers want to sleep
- ☐ When most passengers want the lights off
- ☐ When there are no standing passengers
- ☐ When there are no passengers

Whenever passengers are on board, the interior lighting must be left on. It may be turned off when the bus is empty.

5.56 Mark one answer DBC s4

You're driving a bus in hot weather. Can the passenger door be left open to let in fresh air?

- ☐ Yes, this is normal practice
- ☐ No, unless all passengers are seated
- ☐ Yes, unless carrying schoolchildren
- ☐ No, this isn't allowed

Under no circumstances should you drive with your passenger door open. An open door invites people to make rash decisions to enter or leave the vehicle as you're about to move away; this can be extremely dangerous.

5.57 Mark one answer DBC s2

How would your bus benefit from an air suspension system?

- ☐ Fuel consumption will be reduced
- ☐ Passenger comfort will be increased
- ☐ It will allow you to drive faster
- ☐ Your brakes will be more effective

Air suspension will increase passenger comfort and can also help to reduce wear on road surfaces – which is why it's also known as 'road-friendly suspension'.

5.58 Mark one answer DBC s1

A disabled person is getting on your bus. What should you do if they're having problems, but say they can manage?

- ☐ Be prepared to move off
- ☐ Smile and offer to help them
- ☐ Ask them politely to hurry up
- ☐ Do nothing; you can't leave your seat

Disabled people like to keep their independence. However, if they're having problems, be prepared to offer help.

5.59 Mark one answer DBC s2

When may the built-in TV and video equipment be used on a coach?

- ☐ When the coach is moving slowly
- ☐ When it can't be seen by the driver
- ☐ During long motorway journeys
- ☐ During a coach tour

To prevent drivers from becoming distracted, TV or video equipment must be out of their sight. While passengers are being entertained, they can be sure the driver is concentrating on delivering them safely to their destination.

5.60 Mark one answer DBC s1

What must be fitted to new buses, to benefit people with disabilities?

- ☐ Internal mirror
- ☐ Radio telephone
- ☐ Automatic transmission
- ☐ Priority seating

All new buses and coaches must be easily accessible for disabled passengers. They must have ramps, priority seating and colour-contrasted fittings.

5.61 Mark one answer DBC s1

Who should benefit from priority seating?

- ☐ Those with heavy luggage
- ☐ Those with a weekly pass
- ☐ Those who get on first
- ☐ Those with disabilities

Since 2000, all new buses and coaches must be accessible to people with disabilities. These vehicles must have ramps, priority seating and colour-contrasted handrails and steps.

5.62 Mark one answer DBC s1

A passenger using a wheelchair wants to get on your bus. When may you need to deny them access?

- ☐ When the wheelchair won't fit in the luggage rack
- ☐ When the boarding device has failed to work
- ☐ When the passenger must remain in a wheelchair
- ☐ When the heater on the vehicle isn't working

It's unfortunate, but machinery will sometimes break down despite the best efforts to service it. Don't risk the health and safety of a person with a disability, or other passengers, by struggling to get them on without the proper equipment.

5.63 Mark one answer DBC s1

Your bus is accessible to wheelchair users. In what situation can you refuse to allow them access?

- ☐ When the designated space is occupied
- ☐ When you're running behind schedule
- ☐ When traffic behind you will be held up
- ☐ When you're shortly due to take a break

It's illegal for bus and coach operators to discriminate against people with disabilities. You must always make reasonable adjustments to the way that any service is provided. This means that you must allow wheelchair users access to the designated space – unless it's already been taken by another wheelchair user.

5.64 | **Mark one answer** | **DBC s1**

You're driving a school bus carrying children. When can you make physical contact with a passenger?

- ☐ In a medical emergency
- ☐ To prevent fare dodging
- ☐ When carrying luggage
- ☐ When fastening seat belts

All physical contact with children should be avoided except in specific circumstances such as genuine self-defence, a medical emergency, or to prevent a serious offence or threat to anyone's safety.

5.65 | **Mark one answer** | **DBC s3**

What should you wear when making a walk-round check of your coach?

- ☐ Ear protectors
- ☐ Protective goggles
- ☐ High-visibility vest
- ☐ Face mask

You should make yourself clearly visible when you're working outside your vehicle and close to moving traffic.

5.66 | **Mark one answer** | **DBC s1**

You're about to drive on a route where you'll be picking up wheelchair users. What must be checked before leaving the depot?

- ☐ The boarding device
- ☐ The radio
- ☐ The door seals
- ☐ The heater

It's a legal requirement that you should make reasonable adjustments to provide for people with disabilities. You should make sure that any ramp or lift is in place and in good working order before leaving the depot.

5.67 Mark one answer DBC s1

When can you refuse access to a wheelchair user?

- ☐ When the lift has stopped working
- ☐ When the wheelchair won't fold up
- ☐ When the heater has stopped working
- ☐ When the wheelchair is too heavy to lift manually

You won't have failed in your statutory duties if the boarding device breaks down and there's no other means of assisting the passenger.

5.68 Mark one answer DBC s1

Why may you be unable to allow a person in a wheelchair to enter your bus?

- ☐ Passengers are standing in a wheelchair space and can't move elsewhere
- ☐ You'll take time to load the wheelchair and fall behind schedule
- ☐ The wheelchair passenger will have to stand up
- ☐ You can't collapse the wheelchair

If a wheelchair space is occupied by standing passengers and luggage, you might not be able to allow a wheelchair user on the bus. This may occur on a bus that's almost full to capacity. You won't have failed in your statutory duties towards the wheelchair user.

5.69 Mark one answer DBC s1

You're at a bus stop. What should you do if passengers are standing on the open platform?

- ☐ Put the chain across the platform
- ☐ Ask them to hold on tight
- ☐ Move off but drive more slowly
- ☐ Wait until they're inside the bus

A bus driver must never allow their passengers to ride on an open platform. Anyone standing on an open platform is at risk should the driver have to brake harshly, swerve or suddenly change direction.

Section six
The road

In this section, you'll learn about

- different weather conditions
- parking at night
- using lanes
- dealing with gradients
- reducing risk.

The road

Different weather conditions

Wind

Strong winds can affect large vehicles, especially when driving on

- high-level bridges
- open and exposed stretches of road or motorway.

Look for places where a sudden gust of wind might catch your vehicle, such as when passing gateways or openings between buildings.

Give motorcyclists and cyclists plenty of room as you pass them. They could be affected by buffeting or **turbulence** caused by your vehicle. Remember that they may also move out to avoid potholes or drain covers in the road.

Definition

turbulence
Air motion caused by moving objects.

TIP

When driving around bends in very windy conditions, look ahead for debris in your path. In severe conditions, you may even need to be alert for trees that have blown down across the road.

A strong gust could blow you off course or into another lane. To guard against this, a lane may be closed to create a 'buffer' lane. This lane is kept free to prevent vehicles being blown into the path of other drivers. Some routes across high areas may be closed completely in severe winds.

Take care in windy weather, especially when driving

- a double-deck bus or coach
- a high-floor coach
- an empty vehicle.

DBC s4

If you're driving an unladen **curtain-sided vehicle** in high winds, tie back and secure the curtains. This reduces the effect of side winds on your vehicle. High-sided vehicles such as box vans are more at risk from side winds when they're empty.

DGV s4

Definition

curtain-sided vehicle

An LGV with sides made of removable material (usually plastic sheeting) that's attached to the frame to allow loading from the side. The curtains aren't part of the securing system.

Vehicles or trailers where the loading platform is enclosed by a curtain that offers weather protection and keeps the load hidden from view. The curtains aren't part of the securing system.

Rain

Other road users will have difficulty seeing when there's heavy rain and spray.

TIP

Take care when you change lanes because the spray from your vehicle can seriously reduce the visibility for other road users.

Make sure that your vehicle's spray suppression equipment isn't damaged or faulty.

If very heavy rain and excess spray causes visibility to drop below 100 metres, you're allowed to use your fog lights. However, you **MUST** switch them off when conditions improve again.

TIP To stop in wet weather, allow at least twice the distance you would leave in good, dry conditions.

Rain reduces your tyres' grip on the road surface. If your steering suddenly feels light, this could mean a film of water has built up between your tyres and the road surface. This is known as aquaplaning. Slow down gently until you feel the steering return to normal.

TIP Your brakes are less efficient when they're wet. After driving through deep water, such as floodwater or a ford, lightly apply your brakes to dry them.

Cold

In cold, icy conditions make sure that your mirrors and the whole of the windscreen are clear of any frost or ice before you attempt to move off.

TIP If your steering feels light, this could mean there's ice on the road.

Make your manoeuvres slowly and gently. Avoid any sudden

- braking
- steering
- acceleration.

TIP In icy conditions, your stopping distance may be as much as 10 times further than you would need in good, dry conditions.

Remember also that motorcyclists and cyclists are especially vulnerable on slippery roads.

Fog

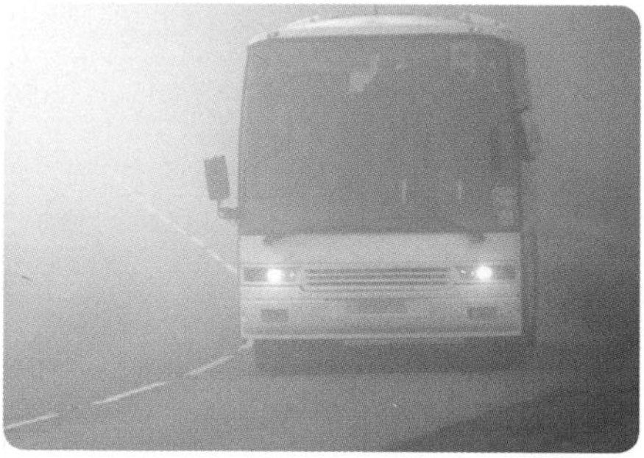

In fog, keep a safe distance from the vehicle in front. Large vehicles can briefly break up fog, making it seem thinner in places than it really is.

At night it's even more difficult to judge distance in fog. If you can see the rear lights of a vehicle in front, you're probably too close to stop in an emergency.

Use dipped headlights whenever visibility is reduced – you need to see and be seen.

DBC s4 **DGV** s4 **HC** r114, 121, 126, 227

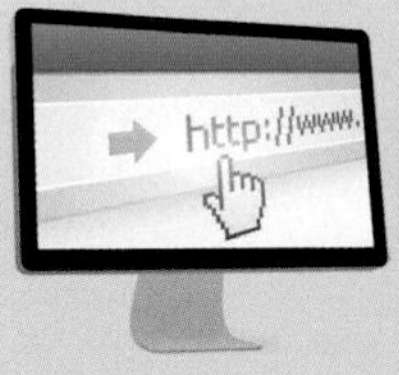

Find more information about how to drive safely in extreme weather conditions under the health and safety road topic at this link.

fta.co.uk/compliance-and-advice

Parking at night

At night, you **MUST NOT** park a large vehicle on a public road or in a lay-by without leaving the vehicle's parking lights switched on. You can only park without lights in an off-road parking area such as a coach park.

Vehicles less than 2500 kg laden weight are allowed to park in 30 mph zones without lights at night.

DBC s4 DGV s4 HC r248–250

Using lanes

Some bus lanes permit other designated vehicles (eg cycles or taxis) to use them during the times shown on the signs. Coaches may also be allowed to use a bus lane, unless it's designated for use only by local bus services.

DBC s3 HC r141

On motorways, you should use the left-hand lane unless you're overtaking. After overtaking another vehicle, check your mirrors and make sure that you're well clear of it before you move back to the left.

Any large vehicles fitted with speed limiters aren't allowed to use the right-hand lane of a motorway that has more than two lanes.

DBC s4 DGV s4 HC r264

Dealing with gradients

Travelling uphill or downhill will affect your vehicle's stopping distance. If you have to stop suddenly, try to do so in a straight line to reduce the risk of skidding.

DBC s2, 4 DGV s2, 4

If you're driving a large, heavy vehicle and are on a stretch of motorway with a long uphill gradient, you may slow down to the point where you could become a hazard to other road users. An extra lane (sometimes called a crawler or climber lane) may be provided. Use this lane to keep out of the way of faster traffic if your vehicle is heavy and slow.

DGV s4

Reducing risk

Before you leave off-road sites, take care to check your wheels for excess mud, especially in wet conditions. It's an offence to deposit mud on the road, so you may need to hose down your tyres and the underneath of the vehicle before you leave the site.

TIP

Check between double wheels for any large stones, bricks or other debris wedged between the tyres. These can fly out at speed and are a danger to other road users.

Engage the **differential lock** (diff-lock) when driving on off-road sites to assist traction. Remember to disengage it as soon as you're on a normal road surface again.

DGV **s2, 4**

Definition

differential lock

A feature that 'locks' both wheels on an axle so that they turn together, regardless of the traction available to either wheel.

Meeting the standards

You must be able to
be aware of what's around you (nearby and far away) at all times
drive at such a speed that you can always stop in the clear space ahead of you
make allowances for the extra vehicle length when towing a trailer, particularly • when turning • at junctions
make good use of mirrors and other aids to vision to identify and monitor other road users and hazards.

You must know and understand
how to read the road ahead and be prepared for the unexpected
the importance of keeping a safe separation distance in all weather and traffic conditions
which kinds of hazard to particularly look for in different environments, such as tractors on rural roads or deer on forest roads
how different types of mirror can make other road users appear to be nearer or further away than they actually are.

Notes

You can use this page to make your own notes or diagrams about the key points you need to remember.

Think about

- Is my vehicle at risk from high winds?
- Are there any exposed stretches of road, or high bridges, along my route?
- Do I know how to check my vehicle's spray suppression equipment?
- Can I feel when my vehicle aquaplanes?
- Do I carry a de-icing kit and keep it replenished?
- Do I know where I can park safely on my journey?
- Will my vehicle be slow when driving up a long uphill gradient?

Your notes

Things to discuss and practise with your instructor

These are just a few examples of what you could discuss and practise with your instructor. Read more about the road to come up with your own ideas.

Discuss with your instructor

- the diff-lock. What is it and when should you use it?
- how strong winds can affect your high-sided vehicle
- what to do when overtaking motorcyclists and cyclists in windy or rainy conditions
- your stopping distances in various weather conditions.

Practise with your instructor

- driving along unlit roads at night
- overtaking on a three-lane motorway
- driving around corners and straddling lanes.

6.1 Mark one answer DGV s4

What should you do when you overtake a cyclist on a two-way road?

- ☐ Pass closely, staying about 1 metre (3 feet 3 inches) from the kerb
- ☐ Go past quickly and move back to the left sharply
- ☐ Leave plenty of room and check your nearside mirror before returning to the left
- ☐ Give a right-turn signal after you've moved out to overtake

Only overtake when it's safe to do so, and don't use the size of your vehicle to intimidate approaching traffic. When overtaking a cyclist, leave plenty of room and, before returning to the left, check in your nearside mirror to make sure you've passed them safely.

6.2 Mark one answer DGV s4

Your lorry has a curtain-sided body. Your route will take you over several high-level bridges. Why should you check the weather forecast before starting your journey?

- ☐ Rain can make crossing bridges very difficult for large vehicles
- ☐ In hot weather, the bridges might be closed to heavy traffic
- ☐ You won't be able to cross the bridges if it's frosty
- ☐ High-level bridges may be closed to certain vehicles in windy weather

In exposed places, such as high-level bridges, viaducts or some motorways, high winds can cause the road to be closed to certain vehicles. If you're driving a high-sided vehicle and high winds are forecast, plan an alternative route in case your intended route is closed.

6.3 Mark one answer DGV s4

What should you do when you want to join a motorway where traffic is flowing freely?

- ☐ Keep to a low speed until you see a gap on the motorway
- ☐ Build up your speed on the slip road before joining the motorway
- ☐ Stop at the start of the slip road and move off when you see a gap
- ☐ Use the hard shoulder, if necessary, to build up speed

Normally, you should avoid having to stop before joining a motorway. Use the slip road to build up your speed so that you can emerge safely. If it's extremely busy, there may be rare occasions when you have to stop and then filter into the traffic. This is most likely to happen where traffic is queuing; for example, at roadworks.

6.4 Mark one answer DGV s4

What should you do before driving your high-sided lorry in windy weather?

- ☐ Check your wind deflector
- ☐ Check your spray-suppression equipment
- ☐ Plan your route in advance
- ☐ Only half-load your lorry

In high winds, some exposed roads, bridges and viaducts are closed to high-sided vehicles. Plan your route with care and try to avoid these areas. Also plan an alternative route in case of road closure.

6.5 Mark one answer DGV s4

You're overtaking another lorry. What should you do if you start to lose speed due to an uphill gradient?

- ☐ Continue at the same speed and in the same position
- ☐ Try to force the vehicle you were overtaking to drop back
- ☐ Try to force the vehicle you were overtaking to speed up
- ☐ Ease off and drop behind the vehicle you were trying to overtake

If you try to overtake another vehicle and realise that you're unable to complete the manoeuvre, ease off the accelerator and drop back behind the vehicle. If the vehicle you're trying to overtake is large, then it will take a long time to pass. You should assess whether you have the time and power to complete the manoeuvre before you try to overtake.

6.6 Mark one answer DGV s2

How could you reduce the problems caused by a strong side wind when driving an empty curtain-sided vehicle on an exposed bridge?

- ☐ Tie one curtain open and lock open the rear doors
- ☐ Leave both curtains closed
- ☐ Tie both curtains open
- ☐ Tie open the curtain on the side the wind's blowing from

If you're driving an empty curtain-sided vehicle, you can help to lower the resistance to side wind if you tie the curtains open. The air will then be able to flow across the flat bed of the vehicle. This will reduce any loss of control, particularly when crossing exposed viaducts or bridges.

6.7 Mark one answer DGV s2

When is a high-sided vehicle most affected by side wind?

- ☐ When it's stationary
- ☐ When it's travelling loaded
- ☐ When it's being reversed
- ☐ When it's travelling empty

Take care if you're driving an empty high-sided vehicle when it's windy. Watch for places where the conditions could suddenly change, such as a gap between buildings or when passing under a bridge. Reduce your speed and stay alert for other road users who are also affected by these weather conditions.

6.8 Mark one answer DGV s2

You're driving an empty curtain-sided vehicle. Why might you consider tying the curtains open?

- ☐ To use less fuel
- ☐ It's a legal requirement
- ☐ To prevent the curtains from tearing
- ☐ To reduce the effect of side wind

Closed curtains on large empty vehicles can hold the wind. Strong side wind can blow a semi-trailer off course and, in severe cases, can blow the whole vehicle over onto its side. To reduce the risk of this happening, you should tie the curtains open.

6.9 Mark one answer DGV s2

When will a box van be most affected by side wind?

- ☐ When travelling empty
- ☐ When stationary
- ☐ When travelling fully laden
- ☐ When reversing

High-sided vehicles such as box vans are affected by side wind, particularly when they're empty. Drivers should slow down and stay alert for other road users affected by the windy conditions.

6.10 Mark one answer DGV s2

You're driving a lorry towards a high bridge on a windy day. What should you expect?

- ☐ Changed weight limits
- ☐ No restrictions for lorries
- ☐ Minimum speed limits
- ☐ Lane closures

In windy weather, exposed stretches of road may have lane closures or there may be diversions for some types of vehicle. Consider this when you're planning your journey; delays will add to your driving time.

6.11 Mark one answer DGV s2

Which of these vehicles is most at risk from strong side wind?

- ☐ A laden lorry with box body
- ☐ An unladen lorry with box body
- ☐ An unladen lorry with platform body
- ☐ A laden lorry with platform body

An unladen vehicle that also has high sides will be vulnerable to the effects of strong side wind. An evenly distributed load will help to keep the vehicle more stable, but the higher the vehicle is, the more it will be affected. Be aware that such a vehicle may become unstable or difficult to control in windy conditions.

6.12 Mark one answer DGV s4

In strong winds, an overtaking lorry can cause additional air turbulence that can affect other road users. Which vehicle is most at risk from this turbulence?

- ☐ A car
- ☐ A delivery van
- ☐ A motorcycle
- ☐ A coach

Vehicles such as motorcycles, bicycles and caravans could be blown off course by strong winds. You should be aware that your vehicle could cause additional buffeting when you overtake, so allow them plenty of room.

6.13 Mark one answer DGV s2

Which vehicle is most at risk in windy conditions?

- ☐ High-sided lorry
- ☐ Saloon car
- ☐ Tractor unit
- ☐ Single-deck bus

High-sided vehicles are severely affected by side wind because of the large surface area of the bodywork. The risk of loss of control is increased if the vehicle is unladen.

6.14 Mark one answer DGV s2

When are air deflectors most effective?

- ☐ When there's a side wind
- ☐ When there's a headwind
- ☐ When reversing
- ☐ When there's a strong tailwind

Air deflectors reduce wind resistance by streamlining the vehicle. They're most effective when there's a headwind. The streamlining also reduces the vehicle's fuel consumption.

6.15 Mark one answer DGV s4

You're driving your lorry on the motorway. What should you do when visibility is reduced by heavy rain and spray?

- ☐ Maintain a constant speed
- ☐ Use main-beam headlights
- ☐ Double your dry-weather separation distance
- ☐ Stay in the left-hand lane

Rain makes the road surface slippery and reduces tyre grip. Doubling your separation distance should allow you the extra time and space you'll need to slow down and stop safely.

6.16 Mark one answer DGV s4

What should you do before driving your lorry away from a wet construction site at the side of a motorway?

- ☐ Turn on your amber beacon
- ☐ Drain the air tanks
- ☐ Hose down the wheels
- ☐ Set your tachograph to 'other work'

Before leaving a construction site, check your vehicle for mud and for debris wedged between double rear wheels. It's important that you prevent any mud or debris from being deposited on the road, where it could be a danger to other road users.

6.17 Mark one answer DGV s4

You're driving a lorry from a wet construction site onto a motorway. Why must you take extra precautions before leaving the site?

- ☐ Your view from the cab will be obscured by plant machinery
- ☐ It's an offence to emerge from a works site onto a motorway
- ☐ Your lorry's spray-suppression equipment will be inoperative
- ☐ It's an offence to deposit mud on a road

You should take precautions against dropping mud on the road, because it's a hazard to other road users. It's also an offence for which you could be prosecuted.

6.18 Mark one answer DGV s2

What ground conditions are suitable for a tipper lorry to tip its load?

- ☐ Soft and level
- ☐ Soft and downhill
- ☐ Solid and uphill
- ☐ Solid and level

When discharging a load from a tipper vehicle, the centre of gravity is raised as the tipper body rises. To avoid the vehicle rolling over, it's vital that it's positioned on a level, solid surface before the tipper mechanism is engaged.

6.19 Mark one answer DGV s4

You're driving a lorry of more than 7.5 tonnes maximum authorised mass along a motorway. When can you use the right-hand lane to overtake?

- ☐ When the motorway has two lanes
- ☐ When the motorway has three lanes
- ☐ When there's a 50 mph speed limit
- ☐ When there's a 40 mph speed limit

You may use the right-hand lane to overtake on a two-lane motorway. Lorries weighing more than 7.5 tonnes, vehicles that are required to be fitted with speed limiters, and vehicles towing trailers aren't allowed to use the right-hand lane of a motorway with three or more lanes, except in special circumstances.

6.20 Mark one answer DGV s4

You're driving on a motorway. You look in your mirrors and see smoke coming from your rear tyres. What should you do?

- ☐ Reduce speed for the rest of your journey
- ☐ Drive on the hard shoulder until the smoke stops
- ☐ Stop as soon as it's safe to do so
- ☐ Ignore it; this is normal when driving at speed

A tyre may burst or shred if it overheats. This could result in serious loss of control. Stop on the hard shoulder as soon as it's safe to do so and call for help.

6.21 Mark one answer DGV s4

You're on a wet, level road, driving at 50 mph. What's the minimum time gap you should leave between your vehicle and the one in front?

- ☐ One second
- ☐ Two seconds
- ☐ Three seconds
- ☐ Four seconds

The usual advice is to leave a minimum time gap of two seconds, but on wet roads you should leave at least double that time. Always take the road and weather conditions into account when you're driving.

6.22 Mark one answer DGV s4

You're driving an articulated lorry on a three-lane motorway. When can you drive in the right-hand lane?

- ☐ When overtaking a slow-moving car in the middle lane
- ☐ When the escort vehicle of an oversized load signals you to pass
- ☐ If no speed limiter is fitted to your lorry
- ☐ If your lorry is unladen

Articulated lorries aren't allowed to drive in the right-hand lane of motorways with three or more lanes. The only exceptions are when an escort vehicle gives you a signal, when road signs indicate, or when directed by a police officer.

6.23 Mark one answer DGV s4

You're driving on a motorway and it's raining heavily. What should you do when spray is causing poor visibility?

- ☐ Use dipped headlights
- ☐ Use sidelights only
- ☐ Remove spray-suppression equipment
- ☐ Use the two-second rule

Whenever visibility is reduced, use dipped headlights to make your vehicle easier to see. During and after heavy rain, you should also reduce speed and increase your distance from the vehicle in front.

6.24 Mark one answer DBC s2, DGV s2

What might happen if using the brakes continuously causes them to overheat?

- ☐ The brakes will work better
- ☐ The brakes will be less effective
- ☐ The brake pedal will become stiff
- ☐ The air pressure will increase

Whenever you brake, the brakes convert the moving energy of your vehicle into heat. Continuous use of the brakes, particularly from high speed or on long steep descents, can cause the shoes and drums to overheat. This will make them less effective and, in some cases, they may not work at all.

6.25 Mark one answer DBC s4, DGV s4

What do you need to be careful of when you're driving in high winds?

- ☐ Fallen trees
- ☐ Poor visibility
- ☐ A risk of grounding
- ☐ Steep gradients

In severe weather conditions, you should be aware of the dangers of fallen trees or damaged branches that may fall on your vehicle. Be prepared in case a fallen tree is lying across the road just around a bend.

6.26 Mark one answer DBC s4, DGV s4

What do you need to be aware of when you drive over an exposed suspension bridge?

- ☐ The swaying of the bridge
- ☐ The width of the lanes
- ☐ The effect of strong side wind
- ☐ The overhanging bridge cables

On exposed bridges, the wind can be particularly strong. If the bridge has two lanes, one may be closed in windy conditions to act as a buffer lane. In severe conditions, these bridges are often closed to high-sided vehicles, caravans and motorcycles.

6.27 Mark one answer DBC s4, DGV s4

You're driving through the night. What could it mean if your steering feels lighter than normal?

- ☐ Your tyres are losing pressure
- ☐ There's a leak in the power-steering pump
- ☐ There's ice or frost on the road
- ☐ There isn't enough tread on your tyres

When you're in a warm cab, you may not notice the drop in outside temperature. One of the first signs of an icy road is a lack of 'feel' in the steering. If this happens, slow down and increase the distance between your vehicle and the one in front.

6.28 Mark one answer DBC s4, DGV s4

You're in the centre lane of a motorway, overtaking another vehicle. What should you do before returning to the left-hand lane?

- ☐ Signal left, then check your mirror
- ☐ Wait for the other driver to flash their headlights
- ☐ Check ahead and then your nearside mirror
- ☐ Check behind for fast traffic in the right-hand lane

Never return to the left-hand lane until you're sure that you're well clear of the vehicle you've just overtaken. Don't cut in too soon and don't rely on the other driver flashing you in. Make your own judgement about when it's safe to move into the left-hand lane.

6.29 Mark one answer DBC s6, DGV s6

What should you do as you approach a zebra crossing?

- ☐ Stop before the zigzag lines
- ☐ Wave pedestrians across the road
- ☐ Sound the horn and flash headlights
- ☐ Be prepared to stop in good time

Always keep your speed down as you approach a zebra crossing. This will allow you to stop safely if necessary. Look well ahead for anyone about to reach the crossing or waiting to cross.

6.30 Mark one answer DBC s4, DGV s4

You're driving a high-sided vehicle on a motorway. You should anticipate the effects of side wind. When is this most likely to happen?

- ☐ When passing signs
- ☐ On exposed sections
- ☐ In contraflow systems
- ☐ In service areas

Motorways often have exposed sections of road, placing high-sided vehicles at risk in side wind. When there are extremes of weather, listen to and watch out for weather forecasts. That way, you can plan your journey for the safest time, or make a decision on whether you should travel at all.

6.31 Mark one answer — DBC s4, DGV s4

You're driving a high-sided vehicle in very windy conditions. Where should you avoid travelling?

- ☐ Across suspension bridges
- ☐ Up steep hills
- ☐ Along country lanes
- ☐ Through road tunnels

In windy weather, try to avoid routes where you would have to drive across suspension bridges and other exposed stretches of road. Plan a route that takes you across lower ground and where you'll be less exposed.

6.32 Mark one answer — DBC s4, DGV s4

You're driving in heavy rain. What should you do if your steering suddenly feels very light?

- ☐ Brake very sharply
- ☐ Steer towards the centre line
- ☐ Ease off the accelerator
- ☐ Increase your speed

In wet conditions, water can build up between your tyres and the road surface. This is known as aquaplaning and it will cause your vehicle to slide. Try to regain control by easing off the accelerator to gradually lose speed. Don't brake or steer until your tyres are gripping the road again.

6.33 Mark one answer — DBC s4, DGV s4

Why should you be especially aware of parked cars when you're driving at night?

- ☐ There are more of them
- ☐ They're more likely to move off
- ☐ The drivers may be drunk
- ☐ They can park without lights

Vehicles below 2500 kg laden weight may park without lights on roads with a speed limit no greater than 30 mph. In dark streets, you may not be able to see them clearly.

6.34 Mark one answer — DBC s4, DGV s4

You're waiting at a T-junction. What should you do if you see a vehicle is coming from the right, with its left indicator flashing?

- ☐ Move out and accelerate hard
- ☐ Wait until the vehicle starts to turn in
- ☐ Pull out before the vehicle reaches the junction
- ☐ Move out slowly

Try to anticipate the actions of other road users. When you're waiting at a junction, don't emerge until you're sure of their intentions; their signals might be misleading.

6.35 Mark one answer DBC s4, DGV s4

You're driving along a motorway and see this sign. What does it mean?

- ☐ There's a steep downhill section ahead
- ☐ Only lorries and buses are allowed to use that lane
- ☐ Vehicles fitted with speed limiters must use that lane
- ☐ There's a long uphill gradient ahead

Where a motorway has a long or steep gradient, slow-moving large vehicles might get in the way of other traffic. To help the traffic flow, an extra lane – called a crawler lane – is sometimes found on these gradients. This helps the traffic to flow by providing the slower heavy vehicles with a dedicated lane on the left.

6.36 Mark one answer HC r152

You're about to pass this car. What's the main hazard you should be aware of?

- ☐ Sunlight reflecting off the car windscreen
- ☐ The driver's door may suddenly open
- ☐ Its alarm may be activated
- ☐ The narrow pavement on the right

You don't know what the driver of this car will do. Be prepared for any hazard. The driver may not have seen you. Planning ahead will allow you to be in the correct gear and at the correct speed to cope if this hazard develops.

6.37 Mark one answer DBC s4, DGV s4

What should you do when you park your vehicle on a two-way road at night?

- ☐ Use the parking lights
- ☐ Switch off all lights
- ☐ Put a traffic cone behind your vehicle
- ☐ Park underneath a street lamp

Large vehicles are required by law to use parking lights when parked at night, unless in a designated off-road parking area. Also, unless you're in a one-way street, you must only park on the left-hand side. It's generally much better to park off-road if you can, particularly if you're going to be stationary for a long period of time.

6.38 Mark one answer DBC s2, DGV s2

When would weight transference be most likely to occur?

- ☐ When you're reversing
- ☐ When you're loading
- ☐ When you're overtaking
- ☐ When you're cornering

You must take extra care when your vehicle is carrying a load. When cornering, weight will be transferred away from the direction in which you're turning. Also, when you're braking, weight can be transferred in several different directions.

6.39 Mark one answer DBC s2, DGV s2

How will gravity affect a laden vehicle going downhill?

- ☐ The vehicle will use more fuel
- ☐ Stopping distances will be decreased
- ☐ More braking effort will be required
- ☐ The load will be more stable

Always plan ahead and take note of gradient warning signs. On downhill gradients, you can help control your vehicle's speed and avoid brake fade by selecting a lower gear in good time.

6.40 Mark one answer DBC s2, DGV s2

When do long vehicles sometimes need to straddle lanes?

- ☐ When turning at tight junctions
- ☐ When travelling empty
- ☐ When driving on motorways
- ☐ When coming to contraflow systems

When driving a long vehicle around corners, it's sometimes necessary to adopt a different road position to avoid mounting the kerb or colliding with street furniture such as lampposts or traffic signs. Other road users may not understand what you intend to do next. Watch them carefully and always signal in good time.

6.41 Mark one answer DBC s4, DGV s4

What should you do when you drive in windy weather?

- ☐ Maintain a constant speed in exposed areas
- ☐ Anticipate how the weather may affect other road users
- ☐ Always keep to your intended route
- ☐ Stay close to smaller or vulnerable vehicles as you overtake them

Cyclists, motorcyclists, caravans and high-sided vehicles can be very vulnerable in high winds. Plan well ahead and consider the difficulties they may be having. Remember, some routes may be closed to certain vehicles and you may need to use an alternative route.

6.42 Mark one answer DBC s4, DGV s4

High-sided vehicles can be affected by side wind. On which type of road is this most likely?

- ☐ Narrow country lanes
- ☐ Open roads
- ☐ Slip roads
- ☐ Motorway underpasses

You should take account of the weather forecast when you plan your route. You're most likely to be subjected to side wind on exposed routes; for example, on long bridges, open roads and exposed stretches of motorway.

6.43 Mark one answer DBC s4, DGV s4

Where would you anticipate problems when you're driving a high-sided vehicle on a windy day?

- ☐ In road tunnels
- ☐ On high-level roads
- ☐ In dead ground
- ☐ On ring roads

Listen to the weather forecast and plan your route to avoid exposed or high-level roads if windy weather is expected.

6.44 Mark one answer DBC s4, DGV s4

Which vehicle is least likely to be affected by high wind?

- ☐ 
- ☐ 
- ☐ 
- ☐ 

The car is the most stable of the vehicles shown. However, all vehicles are affected by side wind and buffeting from larger vehicles. Be aware that your speed and road position could cause unnecessary turbulence when passing smaller vehicles.

6.45 Mark one answer DBC s4, DGV s4

What's a buffer lane?

- ☐ A safety lane for vehicles blown off course
- ☐ A lane for overtaking safely
- ☐ A lane to park in until the wind drops
- ☐ The only lane to be used in high winds

During high winds, one of the lanes on a high bridge might be closed to traffic to create a buffer lane. The buffer lane is kept free to prevent vehicles from being blown into the path of other road users in the next lane. The closure of this lane may cause traffic congestion and delay.

6.46 Mark one answer DBC s4, DGV s4

Where would you expect to find buffer lanes?

- ☐ In built-up areas
- ☐ On high-level bridges
- ☐ On country roads
- ☐ In roadworks

Buffer lanes are often found in places that are frequently subjected to high winds. Examples include high-level bridges, high-level roads, exposed viaducts and exposed stretches of motorway. Listen to the weather forecast, which should tell you of any need to re-plan your route.

6.47 Mark one answer DBC s4, DGV s4

When is a buffer lane most likely to be in use?

- ☐ When it's windy
- ☐ When it's raining
- ☐ When it's foggy
- ☐ When it's icy

A buffer lane is used when the wind begins to cause a risk to high-sided vehicles. At other times, it will be a normal lane. Don't use the buffer lane unless your vehicle has been blown off course into the lane, or you need to use it to avoid an incident or collision.

6.48 Mark one answer DBC s4, DGV s4

What's the best way to control your vehicle when driving on snow?

- ☐ Keep the engine revs high and spin the wheels
- ☐ Drive in your vehicle's lowest gear
- ☐ Keep the engine revs high and slip the clutch
- ☐ Drive slowly, in a higher gear than normal

Driving slowly will give you better control of your vehicle. By using a high gear, you reduce the chance of your wheels spinning and skidding on the snow.

6.49 Mark one answer DBC s4, DGV s4

You're driving a large vehicle in gusty conditions. Which vehicle is most likely to be affected by buffeting when you overtake it?

- ☐ A motorcycle
- ☐ A flat-bed lorry
- ☐ A car
- ☐ A loaded tanker

The lighter the vehicle, the more likely it is to be affected by the buffeting from your vehicle in windy weather. Give motorcyclists and cyclists extra room when you overtake them, as your vehicle's draught can easily upset their control and balance.

6.50 Mark one answer DBC s4, DGV s4

It's very windy and you're overtaking a motorcyclist on a motorway. What should you be aware of that may cause them problems?

- ☐ Your exhaust smoke
- ☐ Your engine noise
- ☐ The buffeting effect
- ☐ The tyre noise

Give them plenty of room. Watch carefully in your nearside mirror, as the high winds and the buffeting effect from your vehicle may blow them off course.

6.51 Mark one answer DBC s4, DGV s4

Why should you be aware of motorcyclists on a motorway when there are gusty winds?

- ☐ They may be blown into your path
- ☐ They may leave at the next exit
- ☐ They may suddenly stop on the hard shoulder
- ☐ They may position to turn right

On motorways, there are often exposed stretches of road where vulnerable vehicles can be affected by side wind. A gust of wind can even blow a motorcyclist out of their lane. Try to anticipate how the wind may affect others and give them plenty of room.

6.52 Mark one answer DBC s4, DGV s4

Which road users are most at risk from the turbulence caused by your vehicle?

- ☐ Lorry drivers
- ☐ Coach drivers
- ☐ Tractor drivers
- ☐ Car drivers towing caravans

Remember that the buffeting caused by large vehicles can affect other road users, including

- cars towing caravans
- motorcyclists.

6.53 Mark one answer DBC s4, DGV s4

Which road users are most likely to be affected by the turbulence created by large vehicles travelling at speed?

- ☐ Drivers of low-loaders
- ☐ Drivers towing caravans
- ☐ Drivers of tankers
- ☐ Drivers of emergency vehicles

Caravans are relatively light but have a large surface area. This makes them vulnerable to the turbulence created by large vehicles travelling at speed. The blast of air can destabilise the caravan, causing it to snake, and in extreme cases they can be blown over.

6.54 Mark one answer DBC s4, DGV s4

You're overtaking a motorcycle in windy conditions. Why should you check your nearside mirror?

- ☐ To check your road position
- ☐ To see if the rider is in control of their motorcycle
- ☐ To see if other vehicles have been affected
- ☐ To check that it's properly adjusted

You must check the nearside mirror during and after overtaking a motorcyclist. Make sure that the rider is still in control and that your vehicle's draught hasn't caused them to lose their balance.

6.55 Mark one answer DBC s4, DGV s4

Which vehicles are most likely to be affected by high winds?

- ☐ Slow-moving vehicles
- ☐ Track-laying vehicles
- ☐ Front-wheel-drive vehicles
- ☐ Curtain-sided vehicles

In very windy conditions, certain types of vehicle can be blown off course. These include box vans, curtain-siders, caravans and motorcycles. Make allowances for how the wind might affect these vehicles, especially if you're overtaking.

6.56 Mark one answer DBC s4, DGV s4

You're driving a high-sided vehicle on a motorway. Where should you be especially aware of the effects of side wind?

- ☐ In cuttings
- ☐ Through tunnels
- ☐ Across viaducts
- ☐ Passing large signs

On motorways, be especially careful when driving over bridges and viaducts. The road here is exposed and in windy weather there's a risk of strong side wind, which may blow you off course or even blow your vehicle over. In severe windy weather, these routes are often closed to drivers of high-sided vehicles.

6.57 Mark one answer DBC s4, DGV s4

You're on a motorway and the surface is still wet after rain. Why should you take care when you're overtaking?

- ☐ Wet roads may cause more buffeting
- ☐ Other vehicles will have their lights on
- ☐ Vehicles may be parked on the hard shoulder
- ☐ The road may still be slippery

Rain, combined with oil, rubber particles, dirt and debris on the road surface, reduces the tyres' grip on the road. It can be especially slippery when it rains after a long dry spell.

6.58 Mark one answer DBC s4, DGV s4

It's raining. Why would a motorcyclist steer around drain covers on a bend?

- ☐ To reduce the risk of skidding
- ☐ To avoid puncturing the tyres
- ☐ To improve their view around the bend
- ☐ To avoid splashing pedestrians

Drain covers usually have a metal surface. When wet, these covers are slippery and can cause a motorcycle to skid; be prepared for riders to avoid them.

6.59 Mark one answer DBC s4, DGV s4

What's the minimum distance you should allow for braking when it's raining heavily?

- ☐ The normal distance
- ☐ Twice the normal distance
- ☐ Three times the normal distance
- ☐ Five times the normal distance

If you're travelling in heavy rain, you should be aware that it could take twice as long for you to stop as it would in dry weather. Increase your distance from the vehicle in front to allow for the longer stopping distance.

6.60 Mark one answer DBC s4, DGV s4

You're driving on a motorway. What should you do if your view ahead is restricted by heavy spray?

- ☐ Move into the lane on the right
- ☐ Use the two-second rule
- ☐ Switch on your dipped headlights
- ☐ Leave the motorway at the next exit

If you're travelling on a motorway and your view ahead is poor, then you should reduce your speed. Leave at least four seconds between your vehicle and the one in front, and make sure that others can see you by using your dipped headlights.

6.61 Mark one answer DBC s4, DGV s4

What's the greatest risk to your safety when you're overtaking on a motorway in very wet weather?

- ☐ Your engine may get flooded
- ☐ Your braking distance may be reduced
- ☐ Your steering may become heavy
- ☐ Your tyres may lose grip

In very wet conditions, tyres may be unable to clear all the water from their path. This can result in a layer of water building up between the road and the tyre – a situation known as aquaplaning. To regain control, ease off the accelerator and, as the vehicle slows, the tyres should regain their grip. It's safer to reduce the risk of aquaplaning occurring by driving more slowly in very wet weather.

6.62 Mark one answer DBC s4, DGV s4

What should you be aware of when you're overtaking on a motorway while it's raining heavily?

- ☐ Manhole covers will be slippery
- ☐ Visibility will be reduced by spray from traffic
- ☐ Your braking distances will be reduced
- ☐ There'll be distracting bright reflections

Other road users can be affected by the spray from large vehicles, severely reducing their vision. Always take care when you need to change lanes: signal in good time and watch your mirrors carefully before pulling out in these conditions.

6.63 Mark one answer DBC s4, DGV s4

Why can visibility be worse when driving at high speed in wet weather?

- ☐ Drivers bunch together when it's raining
- ☐ Headlights will dazzle you more easily
- ☐ Windscreen wipers obstruct your view
- ☐ More spray will be thrown up

Higher speeds will cause more spray to be thrown up by vehicles' tyres – particularly the tyres of larger vehicles and trailers. Always use dipped headlights in poor visibility caused by rain and spray.

6.64 Mark one answer DBC s4, DGV s4

What should you do if you intend to overtake a lorry that's throwing up spray?

- ☐ Get much closer before moving out
- ☐ Wait until the other driver gives a left signal
- ☐ Move out earlier than normal
- ☐ Wait for the lorry to slow down on a hill

If you wish to overtake a vehicle that's throwing up spray, move out earlier than normal. This will help to limit the reduction in vision that the rear spray will cause before you pass.

6.65 Mark one answer DBC s4, DGV s4

You're driving in wet weather. What effect will faulty spray-suppression equipment have on other drivers?

- ☐ Their vision will be increased
- ☐ Their vision will be reduced
- ☐ They'll be able to overtake more easily
- ☐ They'll need to use intermittent wipers

Spray-suppression equipment reduces the amount of spray thrown up to other road users. Check the equipment regularly to make sure it's secure and undamaged.

6.66 Mark one answer DBC s4, DGV s4

Why are brushes fitted to this vehicle's wheel arch?

- ☐ To clear mud from the tyres on building sites
- ☐ To remove objects from the tyre tread
- ☐ To stop snow building up behind the wheel
- ☐ To reduce spray on wet roads

The brushes are part of a spray-suppression system. They reduce the amount of water thrown up at the sides and rear of the vehicle. Check them regularly for security. If they become worn, make sure they're replaced.

6.67 Mark one answer DBC s4, DGV s4

When should you check your vehicle's spray-suppression equipment?

- ☐ Only when you'll be using a motorway
- ☐ Before setting out on every journey
- ☐ Only as part of a pre-winter check
- ☐ Once per year before the MOT test

You should always check all your spray-suppression equipment before a journey. Don't ignore it just because it's dry when you set out. The weather can change on the way.

6.68 Mark one answer DBC s4, DGV s4

What component fitted to your vehicle will help following drivers in wet weather?

- ☐ Spray reducers
- ☐ Side-panel skirts
- ☐ Wind deflectors
- ☐ Catalytic converter

Spray reducers or spray-suppression equipment are attachments fitted around the wheel arches of large vehicles. They effectively trap spray thrown up from the wheels, which would otherwise reduce visibility for other road users.

6.69 Mark one answer DBC s4, DGV s4

This vehicle is fitted with spray-suppression equipment. In what weather conditions will it be most useful?

- [ ] Rain
- [ ] Snow
- [ ] Wind
- [ ] Fog

The spray-suppression equipment helps to prevent water from being thrown up and causing reduced visibility, for you and other drivers. Driving at a reduced speed in wet conditions will also help to reduce the amount of water thrown up by your vehicle's tyres.

6.70 Mark one answer DBC s4, DGV s4

How should you drive through flood water?

- [ ] Slowly, in a low gear, with engine speed high
- [ ] Slowly, in a high gear, with engine speed low
- [ ] As quickly as possible to cause the least delay
- [ ] At normal speed when you have spray reducers fitted

It's important to use a low gear and keep the engine speed high to prevent water from entering the engine through the exhaust system. In these circumstances, it may be necessary to slip the clutch to keep the road speed as low as possible.

6.71 Mark one answer DBC s4, DGV s4

You're on a motorway, approaching a working snow plough. Why should you stay behind it?

- [ ] It's illegal to overtake snow ploughs
- [ ] Snow ploughs are left-hand-drive only
- [ ] Your speed could cause snow to drift behind
- [ ] There may be deep snow ahead

There's no way of knowing how deep the snow could be ahead. If you overtake and then become stuck in the snow, you may cause problems for the snow plough, particularly if you block the road.

6.72 Mark one answer DBC s4, DGV s4

You're driving in heavy rain. Why do you need to increase your distance from the vehicle in front?

- ☐ To prevent rain from entering the vehicle's braking system
- ☐ To let other motorists move into the gap
- ☐ To allow for an increased stopping distance
- ☐ To reduce the risk of water spraying into the filters

Extra care is needed when driving in heavy rain. Reduce your speed and switch on dipped headlights. Tyres will have less grip on a wet road and it will take up to twice the distance to stop when compared with good dry conditions.

6.73 Mark one answer DBC s4, DGV s4

You're driving a large vehicle in heavy rain on a motorway. You slow down to reduce the risk of aquaplaning. For what other reason should you slow down?

- ☐ To force other drivers to act properly and slow down
- ☐ To reduce the amount of spray thrown up
- ☐ To prevent water from entering the braking system
- ☐ To stop the electrics getting wet

Driving at high speed on a wet motorway can throw up dense spray, making it very difficult for other drivers to see ahead. Driving more slowly reduces the amount of spray and this makes it safer for other drivers.

6.74 Mark one answer HC r114, 226

You're driving on a motorway in heavy rain. When are you allowed to use high-intensity rear fog lights?

- ☐ When visibility is more than 100 metres (328 feet)
- ☐ Only when the national speed limit applies
- ☐ Only when you're being followed closely by other traffic
- ☐ When visibility is reduced to 100 metres (328 feet) or less

Heavy rain and the resulting spray from large vehicles travelling at speed can seriously affect visibility on the motorway. Reduce your speed and use your headlights. Don't use high-intensity rear lights unless visibility is reduced to 100 metres (328 feet) or less.

6.75 Mark one answer — DBC s4, DGV s4

You're driving on a motorway in heavy rain. What could cause your steering to become unresponsive?

- ☐ Water reducing the tyres' grip on the road
- ☐ Tyres becoming hotter in the bad weather
- ☐ Braking gently and in good time
- ☐ Water entering the braking system

If a build-up of water prevents the tyres from gripping the road, the steering can become unresponsive. If this happens, ease off the accelerator and slow down gradually.

6.76 Mark one answer — DBC s2, DGV s2

You're driving downhill. How will this affect your vehicle?

- ☐ It will need more engine power
- ☐ It will take longer to stop
- ☐ It will increase fuel consumption
- ☐ It will be easier to change direction

When driving downhill, gravity will cause the vehicle to increase speed. More braking effort will be required, and stopping distances will increase.

6.77 Mark one answer — DBC s4, DGV s1

The road is wet. Why should you slow down as you approach this pedestrian?

- ☐ Because there are no road markings
- ☐ To avoid splashing them
- ☐ Because they have priority
- ☐ To encourage them to cross

When the roads are wet, have consideration for pedestrians walking or standing near the kerb.

6.78 Mark one answer — DBC s1, 4, DGV s1

Why is tailgating dangerous?

- ☐ You're likely to be distracted
- ☐ Your engine will overheat
- ☐ Your view to the rear is reduced
- ☐ Your view ahead is reduced

Tailgating is very dangerous because there isn't enough room to stop if an emergency arises. It's also intimidating for the driver that you're following.

6.79 Mark one answer DBC s4, DGV s4

When may you use your vehicle's hazard warning lights while it's moving?

- ☐ When you've just overtaken another vehicle
- ☐ When you need to reverse for some distance
- ☐ When you're on a motorway and traffic ahead slows suddenly
- ☐ When one of your lights has failed

While moving, hazard warning lights may only be used on a motorway or unrestricted dual carriageway to warn drivers behind of a need to slow down, due to a hazard ahead.

6.80 Mark one answer DBC s4, DGV s4

You're driving in town and see these lights flashing. What would you expect to see ahead?

- ☐ Contraflow system
- ☐ Uneven road surface
- ☐ Children crossing the road
- ☐ Roadworks ahead

These lights warn that children are likely to be crossing the road on their way to and from school. Slow down and watch out for them.

6.81 Mark one answer DBC s4, DGV s4

Where are these lights found?

- ☐ On approach to a level crossing
- ☐ Near a fire station
- ☐ On approach to a motorway
- ☐ Near a school

Flashing amber lights are found near schools. They warn you that children are likely to be crossing the road on their way to and from school. Drive slowly until you're clear of the area.

6.82 Mark one answer DBC s4, DGV s4

What problem may you have when driving at night?

- ☐ It's more difficult to keep to speed limits
- ☐ There'll be an increase in traffic
- ☐ You'll have reduced visibility
- ☐ It takes longer to stop

You must be able to stop safely in the distance that you can see to be clear ahead. This will be the distance illuminated by your headlights or by street lights and will generally be less than you can see in daylight.

6.83 Mark one answer DBC s4, DGV s4

Why should you take extra care when overtaking at night?

- ☐ Every driver will normally be tired
- ☐ Large vehicles are subject to a 10% speed reduction
- ☐ Speed and distance are harder to judge
- ☐ Most towns aren't adequately lit

In the darkness, it will be difficult to assess the road ahead – especially if there are bends or hills, which may prevent you from seeing oncoming vehicles.

6.84 Mark one answer DBC s4, DGV s4, HC r132

You're driving along a motorway in thick fog at night. The reflective studs are red on your left and white on your right. Which lane are you in?

- ☐ The right-hand lane
- ☐ The hard shoulder
- ☐ The left-hand lane
- ☐ The middle lane

The red studs on your left indicate the edge of the hard shoulder, while the white studs are used between the lanes. Using more than one reference point will help you to be certain of your position.

6.85 Mark one answer DBC s4, DGV s4

You're driving in the left-hand lane on a three-lane motorway. What should you check for before moving into the middle lane to overtake a slower vehicle?

- ☐ Traffic behind in the right-hand lane, returning to the middle lane
- ☐ Traffic that's intending to leave at the next exit
- ☐ Traffic ahead moving into the right-hand lane
- ☐ Traffic behind that's trying to pass on your left

Traffic behind that's returning from the right-hand lane is a hazard to be aware of before you move into the middle lane to overtake. Checking your blind spot can help you see these vehicles before you commit to changing lanes.

6.86 Mark one answer DBC s4, DGV s4

What should you do before changing lanes to overtake on a motorway?

- ☐ Check your mirrors carefully
- ☐ Change to a lower gear
- ☐ Look over your left shoulder
- ☐ Increase your speed gently

Check your mirrors to make sure that it's safe to overtake. Remember that traffic coming up behind will be travelling at a faster speed.

6.87 Mark one answer DBC s4, DGV s4

Which lane should you use when you're driving on a clear motorway, at the speed limit for your vehicle?

- ☐ Any lane
- ☐ The middle lane
- ☐ The right-hand lane
- ☐ The left-hand lane

You should remain in the left-hand lane unless you're overtaking other vehicles. Don't be inconsiderate and stay in the centre lane or other overtaking lanes when you've finished overtaking.

6.88 Mark one answer DBC s4, DGV s4

You're driving in the left-hand lane of a motorway. What should you do when you see a large vehicle merging from a slip road ahead?

- ☐ Try to race and get ahead of it
- ☐ Leave the other vehicle to adjust its speed
- ☐ Stay at the maximum speed allowed for your vehicle
- ☐ Be ready to adjust your speed

It's not always possible for drivers in the left-hand lane to change lanes to let joining traffic merge. However, you should be ready to adjust your speed to help the joining driver merge safely.

6.89 Mark one answer DBC s4, DGV s4

You're driving at the speed limit in the left-hand lane of a motorway. What should you do if you see lorries merging from a slip road ahead?

- ☐ Try to accelerate past them
- ☐ Move to the next lane if it's safe
- ☐ Brake heavily and let them merge
- ☐ Expect them to stop and let you pass

Only change lanes if doing so won't inconvenience another road user. Merging drivers have a responsibility to give way, if necessary, and not force their way onto the main carriageway. Planning ahead will ensure you're prepared to deal with these situations.

6.90 Mark one answer DBC s4, DGV s4

Why should you take care before moving into the centre lane of a three-lane motorway?

- ☐ The centre lane is narrower than the left-hand lane
- ☐ The bridge height clearance will be less in the centre lane
- ☐ Traffic in the centre lane may be travelling much faster than you
- ☐ The reflective studs may be missing in the centre lane

Always indicate your intentions in good time; this will allow others time to alter their speed and course if necessary. Always check your blind spots before changing lanes.

6.91 Mark one answer DBC s4, DGV s4

When may vehicles over 7.5 tonnes maximum authorised mass (MAM) use the right-hand lane of a motorway to overtake?

- ☐ When the motorway has three lanes
- ☐ When vehicles are stopped on the hard shoulder
- ☐ When the motorway has two lanes
- ☐ When other vehicles are turning right

Goods and passenger vehicles with a MAM of more than 7.5 tonnes, those required to be fitted with a speed limiter, and vehicles towing trailers must not use the right-hand lane on a motorway with more than two lanes, unless there are exceptional circumstances.

6.92 Mark one answer DBC s4, DGV s4

You've just overtaken another vehicle on a motorway. What should you avoid when moving back to the left?

- ☐ Cutting in
- ☐ Increasing your speed
- ☐ Changing gear
- ☐ Signalling

After overtaking, check your nearside mirror carefully before moving back to the left. Don't rely on the other driver flashing their headlights, because only you can make the decision that it's safe.

6.93 Mark one answer DBC s4, DGV s4

What's a crawler lane for?

- ☐ To enable other traffic to overtake on the nearside
- ☐ To enable large vehicles to park out of the way
- ☐ To enable slow-moving traffic to move further over to the left on uphill gradients
- ☐ To enable emergency vehicles to get quickly to the scene of an incident

On a motorway where there's a long uphill gradient, there may be a crawler lane. This helps the traffic to flow by making an extra lane available on the left for slower, heavy vehicles.

6.94 Mark one answer DBC s4, DGV s4

What does this motorway sign mean?

- ☐ You're approaching a long downhill slope
- ☐ You're approaching a long uphill slope
- ☐ You're approaching a 'lorries only' lane
- ☐ You're approaching a service area

The term 'crawler lane' doesn't mean the lane is only for extremely slow vehicles. It's advising you of an extra lane on the left. Crawler lanes are usually built on sections of road where the length of the gradient is such that some large vehicles will be slowed to the point where they become a hazard for other road users.

6.95 Mark one answer — DBC s4, DGV s4

When should you use a crawler lane?

- ☐ When letting faster traffic overtake you
- ☐ When turning right from a major road
- ☐ When parking to have a rest
- ☐ When slowing down for a motorway exit

Many vehicles are very powerful and can maintain speed even when climbing a gradient. Even if your vehicle is capable of maintaining speed, you can still use the crawler lane to make it easier for other road users to overtake safely.

6.96 Mark one answer — HC r167

What should you do before overtaking on a single carriageway road?

- ☐ Flash your headlights at oncoming traffic
- ☐ Look ahead for road junctions
- ☐ Drive very close to the vehicle in front
- ☐ Make a final check in your left-hand mirror

Before overtaking, you need to make sure it's safe to carry out the manoeuvre. As part of your check, look well ahead for road junctions. You shouldn't overtake as you approach a road junction, because if a vehicle emerges, or has to wait to turn, you could risk a collision.

6.97 Mark one answer — HC r166–167

When shouldn't you overtake?

- ☐ When the speed limit has been reduced
- ☐ When you're approaching motorway slip roads
- ☐ When you would have to break the speed limit
- ☐ When your view of the road ahead is clear

Never commit yourself to overtaking unless you're absolutely certain that you have the time and space to complete the manoeuvre safely. Don't take any risks that will endanger other road users.

6.98 Mark one answer — DBC s4, DGV s4

You've just overtaken another large vehicle. How do you know when it's safe to move back to the nearside lane?

- ☐ By waiting for the driver you've overtaken to flash their headlights
- ☐ By checking your nearside mirror
- ☐ By using your hazard warning lights as a signal
- ☐ By moving to the nearside, hoping the other vehicle will slow down

While overtaking, you'll need to judge carefully when to return to the nearside lane. Check your left-hand (nearside) mirror to make sure the rear of your vehicle is well clear. You should allow for the length of both vehicles and take into account the speed difference. Don't cut in too soon on the vehicle you've overtaken, and don't rely on signals from other drivers; they may be signalling to someone else.

6.99 Mark one answer DBC s4, DGV s4

You're driving on a dual carriageway and intend to overtake the vehicle ahead. What should you do if there's a car approaching quickly behind you in the right-hand lane?

- ☐ Stay behind the slower vehicle
- ☐ Signal right and move out
- ☐ Move up closer to the slower vehicle
- ☐ Signal left while waiting for the car to pass

You should allow the faster-moving vehicle to overtake safely. If you pull out to overtake at this time, you'll cause the car behind to brake sharply. Don't use the size of your vehicle to intimidate. Be professional and show consideration to other road users.

6.100 Mark one answer DBC s4, DGV s4

What should you do after overtaking on a dual carriageway?

- ☐ Move back to the left when it's safe to do so
- ☐ Indicate left, then right
- ☐ Wait for the other driver to flash their headlights
- ☐ Switch your rear lights on and off

Signal if necessary and move back to the left as soon as it's safe to do so. Don't drive for long distances in the centre or right-hand lane unnecessarily.

6.101 Mark one answer DBC s6, DGV s6

You're driving a long vehicle and want to turn right at a roundabout ahead. How should you signal if you need to occupy the left-hand lane?

- ☐ Signal left on approach
- ☐ Signal right on approach
- ☐ Give no signal on approach
- ☐ Give a right signal after entering the roundabout

There are times when, due to the size of your vehicle, you'll have to take up part of another lane. If you need to do this, make sure that you use effective observation all around. Be aware that other road users might not understand the reasons for your position on the road. You should signal your intentions and take up your position in good time.

6.102 Mark one answer DBC s6, DGV s6

When do drivers of long vehicles need to straddle lanes?

- ☐ When they need to avoid braking sharply
- ☐ When driving on motorways
- ☐ When they need to avoid mounting the kerb
- ☐ When coming to contraflow systems

When you have to straddle the lanes at small roundabouts or junctions, always signal in good time. Be alert for inexperienced road users who may, without realising, put themselves into a dangerous position.

6.103 Mark one answer DBC s4, DGV s4, HC r288

What should you do as you approach these roadworks?

- ☐ Check your mirrors
- ☐ Select a higher gear
- ☐ Flash your headlights
- ☐ Use your endurance brake

Before changing direction, you should check your mirrors for any vehicles that may be overtaking. An overtaking driver or rider might not have seen the narrowing road ahead. You should let them get past before you move across to the right.

6.104 Mark one answer DBC s4, DGV s4

What should you do when you're overtaking a motorcyclist in very windy conditions?

- ☐ Allow extra room
- ☐ Overtake slowly
- ☐ Sound your horn as you pass
- ☐ Keep close as you pass

Cyclists and motorcyclists may become unbalanced by your vehicle passing too close. Plan ahead and give them plenty of room. Remember that your large vehicle can cause a vacuum, which will suck the rider towards it. The noise and close proximity of your vehicle can also startle them, and could cause them to swerve suddenly.

6.105 Mark one answer DBC s4, DGV s4

You're driving behind two cyclists. They're approaching a roundabout in the left-hand lane. What should you expect them to do?

- ☐ Go in any direction
- ☐ Turn left
- ☐ Turn right
- ☐ Go straight ahead

When following cyclists onto a roundabout, be aware that they might not take the exit you expect them to. Cyclists approaching in the left-hand lane may be turning right. They may not have been able to get into the correct lane due to heavy traffic, so give them room.

6.106 Mark one answer DBC s6, DGV s6

What should you do before you start to overtake a lorry?

- ☐ Look well ahead for uphill gradients
- ☐ Check your position in the left-hand mirror
- ☐ Change quickly to a higher gear
- ☐ Close right up before pulling out

Plan your overtaking carefully. The weight of your vehicle, combined with the restrictions imposed by a speed limiter, is likely to make your vehicle lose speed on uphill gradients. You should take this into consideration before attempting to overtake another moving vehicle.

6.107 Mark one answer DBC s4, DGV s4

You're in the left-hand lane on a three-lane motorway. Why should you check for any vehicles in the right-hand lane before you overtake?

- ☐ They may be moving faster than you
- ☐ They may move back to the middle lane as you move out
- ☐ They may cut in sharply behind you
- ☐ They may accelerate briskly in front of you

Vehicles overtaking in the right-hand lane may return to the centre lane when they've finished their manoeuvre. You should look for this before starting to pull out. Don't rely on the size of your vehicle to claim right of way.

6.108 Mark one answer DBC s4, DGV s4

What does it indicate if you see white lights on a vehicle at night?

- ☐ It's moving away from you
- ☐ It's stationary and facing away from you
- ☐ It's ahead of you and braking
- ☐ It's facing towards you

At night, you can tell whether a vehicle is facing towards or away from you by the colour of its lights. Additionally, its brake lights or indicators can give you information about its direction and intentions. This information helps you to make safe driving decisions.

6.109 Mark one answer DBC s4, DGV s4

You're driving at night. Why should you ensure that all your lights are clean and working correctly?

- ☐ To allow following drivers to use dipped headlights
- ☐ To prevent the battery from overcharging
- ☐ So that other road users can see the size of your vehicle
- ☐ So that the intensity of street lighting can be reduced

As well as headlights and tail lights, large vehicles have additional lighting to help other road users see the size of the vehicle. All lights should be clean and working properly. Driving at night when your vehicle is poorly lit is dangerous for you and other road users.

6.110 Mark one answer — DBC s4, DGV s4, HC r114, 226

You're driving in fog. When should high-intensity rear fog lights be used?

- ☐ When visibility falls below 100 metres (328 feet)
- ☐ When visibility falls below 200 metres (656 feet)
- ☐ When visibility falls below 300 metres (984 feet)
- ☐ When visibility falls below 400 metres (1312 feet)

In dense fog, large vehicles are no more visible than any other. If visibility falls below 100 metres (328 feet), use your fog lights to help other drivers see you.

6.111 Mark one answer — HC r114, 226

You're driving in fog that's starting to clear. When should you switch off your high-intensity rear fog lights?

- ☐ When visibility is more than 10 metres (32 feet)
- ☐ When visibility is more than 50 metres (164 feet)
- ☐ When visibility is more than 75 metres (246 feet)
- ☐ When visibility is more than 100 metres (328 feet)

Switch off your fog lights when fog starts to clear and visibility improves to more than 100 metres (328 feet). Don't leave your fog lights on when the fog has cleared, because they can cause unnecessary glare or distraction for other drivers.

6.112 Mark one answer — DBC s4, DGV s4, HC r126

In fast traffic, when should you leave a two-second gap between your vehicle and the one in front?

- ☐ When the roads are dry
- ☐ When it's raining and the roads are wet
- ☐ When the roads are icy
- ☐ When you're driving in fog

In good dry conditions, you should leave a two-second gap between your vehicle and the one ahead. When the weather is bad, you'll have to keep a greater distance, because it will take you longer to stop. In wet weather it will take up to twice the distance; in icy weather this could increase to ten times as far.

6.113 Mark one answer DBC s4, DGV s4

What should you do if you see a vehicle too close behind when you're driving in traffic on the motorway?

- ☐ Increase your distance from the vehicle in front
- ☐ Touch the brake pedal sharply to show your brake lights
- ☐ Briskly accelerate away from the vehicle behind
- ☐ Switch your rear fog lamps on and off

The faster the traffic is moving, the greater the distance that needs to be kept between vehicles to maintain safety margins. If you find another vehicle driving too close behind you, gradually reduce your speed to increase your distance from the vehicle in front. You'll then be able to brake more gently when you need to, reducing the likelihood of the vehicle behind running into the back of your vehicle.

6.114 Mark one answer DBC s4, DGV s4

What should you do if an overtaking vehicle pulls in too close in front of you?

- ☐ Slow down
- ☐ Stay close behind
- ☐ Overtake when you can
- ☐ Flash your headlights

Always maintain a safe following distance from the vehicle in front. If another vehicle pulls into the gap, slow down to re-establish a safe following distance.

Following too close is dangerous and places you and other road users in unnecessary danger; your view of the road ahead will be seriously reduced, and you won't be able to see or plan effectively.

6.115 Mark one answer DBC s4, DGV s4, HC r126

You're driving at 50 mph (80 km/h) on a dry, level road. What gap should you leave between your vehicle and the one in front?

- ☐ One vehicle length
- ☐ Two vehicle lengths
- ☐ At least a one-second gap
- ☐ At least a two-second gap

'Tailgating' – travelling too close to the vehicle in front – is very dangerous, because it doesn't give you enough distance to stop safely.

Leave at least a two-second gap in good conditions, and increase this to four seconds in wet weather.

6.116 Mark one answer DBC s4, DGV s4

You're following a large vehicle. How can you improve your view ahead?

- ☐ Drop further back
- ☐ Move over to the right
- ☐ Move over to the left
- ☐ Overtake as soon as you can

Staying well behind the vehicle in front will improve your view of the road ahead. This will give you more time and room to react to hazards. Use the two-second rule to make sure you're not too close. Don't move out to the right, as this could put you into the path of an oncoming vehicle or obstruct smaller vehicles that wish to overtake.

6.117 Mark one answer DBC s4, DGV s4

What must you be certain of before you overtake at night?

- ☐ You can see well ahead
- ☐ The road's well lit
- ☐ There's an overtaking lane
- ☐ You're outside a built-up area

It's more difficult to judge speed and distance accurately in the dark. Darkness can also hide bends and dips in the road. Plan ahead, and be sure that you have time and space to complete your overtaking safely.

6.118 Mark one answer DBC s4, DGV s4, HC r126

How much longer will your stopping distance be in snow than in dry weather?

- ☐ Twice as long
- ☐ Four times as long
- ☐ Five times as long
- ☐ Ten times as long

In icy or snowy weather, your stopping distance can increase by up to ten times. Because snowy weather increases the distance needed to stop, you must look further ahead and leave an increased safety margin.

6.119 Mark one answer DBC s4, DGV s4

How should you slow down or stop when you're driving on an icy road?

- ☐ Brake gently and in good time
- ☐ Always use the retarder
- ☐ Change down through the gears
- ☐ Use the parking brake in a rapid on-and-off movement

When you have to slow down or stop, you should always avoid harsh, late braking. This is particularly important on icy or slippery roads, where you need to brake gently and in good time to make sure you keep control of your vehicle.

6.120 Mark one answer DBC s4, DGV s4

How should you use the brakes in icy weather?

- ☐ Suddenly
- ☐ By pumping the brake pedal
- ☐ Gently
- ☐ After using the gears first

In poor conditions, all braking should be controlled, in good time and when travelling in a straight line. Try to avoid braking and turning at the same time. Look well ahead to assess and plan your actions.

6.121 Mark one answer DBC s4, DGV s4, HC r126

When should you use the two-second rule?

- ☐ To keep a safe distance from the vehicle in front when it's raining
- ☐ To keep a safe distance from the vehicle in front when conditions are good
- ☐ To keep a safe distance from the vehicle in front when it's icy
- ☐ To keep a safe distance from the vehicle in front in queuing traffic

Use the two-second rule when conditions are good and you're driving on roads carrying faster-moving traffic. Choose a fixed reference point, and when the vehicle ahead passes that point say to yourself 'Only a fool breaks the two-second rule.' If you reach the object before you finish saying this, you're too close and you need to drop back and check again.

6.122 Mark one answer DBC s4, DGV s4

Why is it dangerous to follow a large goods vehicle too closely?

- ☐ Your field of vision will be seriously reduced
- ☐ Your cab will fill with fumes
- ☐ Your engine will overheat
- ☐ Your brakes will overheat

Staying back will increase your view of the road ahead. This will help you to see any hazards that might occur and will allow you more time to react. It will also allow the driver in front to see you in their mirrors.

6.123 Mark one answer DBC s4, DGV s4, HC r126

What time gap should you leave when you're following a vehicle on a wet road?

- ☐ At least one second
- ☐ At least two seconds
- ☐ At least three seconds
- ☐ At least four seconds

Wet roads will increase the time it takes you to stop. Four seconds is the minimum gap you should leave on wet roads. This is double the gap you should normally keep from the vehicle in front on good, dry roads.

6.124 Mark one answer DBC s4, DGV s4

The entrances to roundabouts are often slippery when they're wet. How should you deal with this hazard?

- ☐ Stay in the left-hand lane
- ☐ Brake in good time
- ☐ Use the handbrake
- ☐ Stop before emerging

The entrances to roundabouts are often worn. This can make the surface slippery, especially when it's wet. You should plan your approach and brake in good time.

6.125 Mark one answer DBC s2, 4, DGV s4

You're driving in wet conditions. What should you try to do before braking?

- ☐ Make sure that the gear lever is in neutral
- ☐ Make sure that all spray-suppression equipment is working
- ☐ Make sure that there's no mist on your rear-view mirrors
- ☐ Make sure that your vehicle is travelling in a straight line

If you need to brake when the road surface is wet, do so while your vehicle is travelling in a straight line. This will reduce the risk of skidding. As a professional driver, you should routinely brake in good time, so that you don't brake and change direction at the same time.

6.126 Mark one answer DBC s4, DGV s4

You're driving a fully laden vehicle approaching an uphill gradient. What will happen if you start to overtake?

- ☐ You'll be able to get past more quickly
- ☐ The weight of the load will help you go faster
- ☐ You'll take longer to get past
- ☐ You'll need to use a high gear

Always plan well ahead before overtaking. Take into consideration the size of your vehicle and whether there's a gradient ahead. The load on your vehicle will slow you down on an uphill gradient. It may be better to wait until you're on the level before trying to overtake.

6.127 Mark one answer DBC s4, DGV s4

You're on a motorway and there's a strong side wind. Which vehicle is particularly at risk?

- ☐ A motorcycle
- ☐ A taxi
- ☐ A sports car
- ☐ A road tanker

Motorcycles can be very vulnerable to strong side wind. Allow extra room when overtaking them in windy weather. Check your nearside mirror while alongside and before moving back to the left.

6.128 Mark one answer DBC s4, DGV s4

You're driving at a speed of 50 mph (80 km/h) in good, dry conditions. What distance should you stay behind the vehicle in front?

- ☐ At least 23 metres (75 feet)
- ☐ At least 36 metres (118 feet)
- ☐ At least 53 metres (175 feet)
- ☐ At least 73 metres (240 feet)

You should always leave a safety margin between your vehicle and the one in front. This gap will give you a better view of the road ahead. It will also allow you time to react if the traffic in front changes speed or direction.

6.129 Mark one answer DBC s4, DGV s4

You're driving on a motorway after it's been raining heavily. What should you do if spray is being thrown up, causing poor visibility?

- ☐ Keep a two-second gap
- ☐ Reduce your separation distance
- ☐ Remove spray-suppression equipment
- ☐ Use dipped headlights

When visibility is poor or the road is wet, leave a greater separation distance – at least a four-second time gap from the vehicle in front. You should also reduce speed and use dipped headlights.

6.130 Mark one answer DBC s4

You'll be driving a high-floor coach. You'll be going over several high bridges. Why should you check the weather forecast before starting your journey?

- ☐ Rain can make crossing bridges very difficult for buses and coaches
- ☐ In hot weather, the bridges might be closed to heavy traffic
- ☐ You won't be able to climb the bridges if it's frosty
- ☐ Some roads may be closed to certain vehicles in high winds

Some exposed places are frequently subject to high winds. These include high-level bridges and roads, viaducts and some motorways. Check the weather forecast and plan an alternative route. Remember, you're responsible for your passengers' safety. Don't ignore warnings of severe winds.

6.131 Mark one answer DBC s4

It's raining heavily. What effect could this have when you overtake a cyclist?

- ☐ Spray could affect their control
- ☐ Exhaust fumes are more toxic when it's raining
- ☐ Noise from your engine will be louder
- ☐ Buffeting will be reduced

When passing other road users – especially motorcyclists and cyclists – in wet weather, the spray from your vehicle could affect their control. Leave plenty of room as you pass, and look in your left-hand mirror to check that they're still in control.

6.132 Mark one answer DBC s4

How is a strong wind likely to affect your bus?

- ☐ Steering will be easier
- ☐ Your stopping distances will be reduced
- ☐ Your stopping distances will be greater
- ☐ Steering will be more difficult

Be alert for places where the road is shielded from the wind. When you leave this shelter, the wind may suddenly blow you off course, making steering more difficult.

6.133 Mark one answer DBC s4

What will affect a double-deck bus more than a single-deck bus?

- ☐ Strong wind
- ☐ Heavy rain
- ☐ Thick fog
- ☐ Dense spray

If you're driving a double-deck bus, you'll feel the effect of the wind more than you would on a single-deck vehicle. This is because the vehicle's height increases the surface area that can be affected by the wind.

6.134 Mark one answer DBC s4

You're driving a bus on a windy day. What should you do when you're overtaking a motorcyclist?

- ☐ Pass close to shield them from the wind
- ☐ Move back early to protect them from buffeting
- ☐ Pass wide in case they're blown off course
- ☐ Signal left so they know you're pulling in

In strong wind, motorcyclists can be blown off course. Allow them plenty of room when you're overtaking, in case they're blown into your path.

6.135 Mark one answer DBC s4

In which type of weather is spray-suppression equipment particularly useful?

- ☐ Rain
- ☐ Ice
- ☐ Fog
- ☐ Wind

If there's heavy rain, the spray-suppression equipment fitted to your vehicle will protect other road users from loss of vision when following or passing your vehicle. Check the wheel arches to make sure the fitments haven't worked loose and no parts have broken off. Well-maintained equipment will force the spray back down onto the road, instead of out to the rear and the sides of the vehicle.

6.136 Mark one answer DBC s2

You're driving a long bus. When would you need to straddle lanes?

- ☐ Only when joining a bus lane
- ☐ On all bends and corners
- ☐ On the approach to all roundabouts
- ☐ To avoid mounting the kerb

There may be times when you have to take up part of another lane in order to make a turn or manoeuvre. Use your mirrors and check for other traffic that might try to move into the gap you've left to make your turn. Signal and take up your position in good time.

6.137 Mark one answer DBC s4

What must bus and coach drivers do when they park their vehicle on the road at night?

- ☐ They must park under street lights
- ☐ They must park within 25 metres (82 feet) of a street light
- ☐ They must leave all the interior lights switched on
- ☐ They must leave their parking lights switched on

All buses and coaches and most minibuses – depending on their weight – must have their parking lights switched on when they're parked on the road at night. Make sure that all your lights are clean and working properly.

6.138 Mark one answer DBC s4

You're driving a bus at night on a road without street lighting. Why could it be dangerous to overtake?

- ☐ There may be unseen dips or bends in the road
- ☐ You may dazzle other drivers
- ☐ It's harder to concentrate
- ☐ It's harder to keep control in the dark

On unlit roads, it's more difficult to see bends, junctions or dips ahead. This could prevent you from seeing oncoming traffic, pedestrians or cyclists. Unless the road is well lit or you're on a dual carriageway, ask yourself whether overtaking is absolutely necessary.

6.139 Mark one answer DBC s4

Why do some buses have marker lights along their sides?

- ☐ To make them easier to overtake
- ☐ To help the driver when reversing
- ☐ To help any passengers getting on or off
- ☐ To make them easier to see at junctions

Newer buses and coaches have marker lights along the side; these ensure that they're visible as they emerge at junctions, roundabouts and in other situations. If your vehicle has marker lights fitted, they must all be clean and working.

6.140 Mark one answer DBC s3

You're driving a bus. Where should you be aware of the risk of grounding?

- ☐ On a hump bridge
- ☐ In a crawler lane
- ☐ On a left-hand bend
- ☐ In a narrow street

You must be aware of the risk of grounding; for example, on a hump bridge. Look ahead for advance warning signs. If you decide to divert to avoid the hazard, make sure your new route doesn't also have a restriction that would be impassable for your vehicle.

Section seven

Incidents, accidents and emergencies

In this section, you'll learn about

- breakdowns
- what to do at the scene of an incident
- dealing with a vehicle fire
- reporting an incident
- safety in tunnels.

Incidents, accidents and emergencies

Breakdowns

If your vehicle breaks down and you have a warning triangle, place it at least 45 metres (147 feet) behind the vehicle.

Don't try to place a warning triangle when you're on a

- motorway
- dual carriageway
- hard shoulder
- slip road.

DBC s4 DGV s4 HC r274, 277

If your vehicle breaks down on a motorway

- pull over onto the hard shoulder
- turn on your hazard warning lights
- go to the nearest emergency telephone and call for assistance.

While using the emergency telephone, face the oncoming traffic and look out for any vehicles veering off the road and onto the hard shoulder.

DBC s4 DGV s4 HC r274, 277

Watch the Think! 'Van of Elvises' video to find what to do if you break down on a motorway.

youtube.com/thinkuk

If you're driving a bus with passengers and your vehicle breaks down, pull over as far to the left as possible. Move your passengers as far forward in the vehicle as you can, in case another vehicle runs into the back of yours.

DBC **s4**

What to do at the scene of an incident

If you're first to arrive at the scene of an incident where people are seriously injured, your first priorities are to guard against further collisions and prevent a fire.

Do

- warn other traffic
- switch off all engines and fuel supplies
- call the emergency services or ensure that someone else calls them
- move uninjured people clear of the scene.

Don't

- move people trapped in vehicles unless they're in danger.

If the incident involves a vehicle carrying dangerous goods

- tell the emergency services
- give details of any information on the hazard labels shown on the vehicle(s).

DBC **s4** **DGV** **s4**

If you're trained to give first aid, remember **DR ABC**

- Check for **D**anger.
- Check the casualty's **R**esponse.
- Then check their **A**irway, **B**reathing and **C**irculation, and give CPR if necessary.

TIP

If you don't know how to give first aid, think about taking some training – you could save a life.

DBC s4 DGV s4 HC r283, p131–133

If you arrive at the scene of an incident

- reassure your passengers
- confirm whether the emergency services have been called, if they're not already at the scene.

DBC s4

Dealing with a vehicle fire

If your vehicle catches fire or you suspect it has

- stop as quickly and safely as possible
- get any passengers off the vehicle as quickly as possible and lead them to a safe place
- telephone 999 or get someone else to do so immediately
- tackle the source with a suitable fire extinguisher if it's safe to do so without putting your own life in danger.

DBC s4 DGV s4

Find checklists for fire prevention at these links.

- **iru.org/resources/iru-library/iru-checklist-against-fire-for-buses-and-coaches**
- **iru.org/resources/iru-library/checklist-against-fire-for-trucks**

If you're driving a vehicle carrying dangerous goods, you must have the appropriate firefighting equipment on board and be trained in its use.

DGV s2, 3, 4 **HC** r284–285

Reporting an incident

The headroom under bridges in the UK is at least 5 metres (16 feet 6 inches) unless otherwise marked.

TIP If the bridge is arched, the overhead clearance is normally only between the limits marked on the bridge. You may need to use the centre of the road when passing under the arch.

If your vehicle collides with a railway bridge, report the collision to the railway authority immediately to avoid the chance of a serious incident.

- Call the number shown on the bridge identification plate.
- If there's no plate, call 999.

DBC s3 **DGV** s2

If you're involved in an incident that either damages property or causes injury you **MUST** stop, even if you can't find the owner of the property.

You're obliged to give your details to anyone who has **reasonable grounds** to ask for them. If you don't do this, you must report the collision to the police within 24 hours. (In Northern Ireland, you must report it immediately.)

DBC s3 **DGV** s2 **HC** r281–287

Definition

reasonable grounds
A fair reason.

If the police stop you, you **MUST** be able to show your

- driving licence
- insurance certificate
- vehicle test certificate (MOT).

If you can't produce them at the time, you may be asked to take them to a police station within seven days.

DBC s4 **DGV** s4

Find good-practice guides to the prevention of bridge strikes at this website.

www.gov.uk

Safety in tunnels

Before driving into a tunnel

- remove your sunglasses
- switch on your dipped headlights
- tune your radio to a local station to get advance warning of any incident, congestion or roadworks.

While driving in a tunnel, make sure that you keep a safe distance from other traffic. Obey any messages on variable message signs, road signs or signals from tunnel officials.

If your vehicle breaks down in a tunnel

- switch on your hazard warning lights
- take any passengers or co-workers to the nearest exit point
- call for help from an emergency telephone.

If your vehicle catches fire in a tunnel but it can still be driven safely, drive it out of the tunnel.

If the vehicle can't be driven safely

- pull over to the side of the road
- switch off the engine and fuel supply
- switch on the hazard warning lights
- lead any passengers to the nearest exit point
- without putting yourself in danger, try to put out the fire
- if that's not possible, move to the tunnel's nearest emergency exit and call for help from the emergency phone.

DBC s4 **DGV** s4 **HC** r126

Meeting the standards

At the scene of an incident, you must be able to

stop and park your vehicle in a safe place, if necessary

make sure that warning is given to other road users

give help to others if you can

where possible, record information about what you saw or the scene that you found. It may be helpful to take photographs and draw sketch plans.

You must know and understand

where possible, how to keep control of the vehicle if it breaks down

how and when to use a warning triangle or hazard warning lights

what the law says about stopping if you're involved in an incident that causes damage or injury to

- any other person
- another vehicle
- an animal
- someone's property

how to contact the emergency services and how important it is to give them accurate information.

Notes

You can use this page to make your own notes or diagrams about the key points you need to remember.

Think about

- Does your vehicle carry a warning triangle?
- Are you confident you know what to do at the scene of an incident?
- Is your first aid training up to date or could you be trained to give first aid?
- If you carry dangerous goods, are you trained to use the firefighting equipment you must carry?
- Are you aware of any low bridges on your journey?
- Do you have all the documentation you may be asked to show at the side of the road?

Your notes

Things to discuss and practise with your instructor

These are just a few examples of what you could discuss and practise with your instructor. Some are for PCV, some are for LGV and some are for both. Read more about incidents, accidents and emergencies to come up with your own ideas.

Discuss with your instructor

- what to do with your passengers at the scene of an accident
- what to do with a passenger if they
  - become unconscious
  - stop breathing
  - are suffering from shock
- what to do if your bus catches fire on a normal road and in a tunnel.

- what training you must take before you can carry toxic substances.

- what you should do if you arrive at the scene of a crash and find
  - someone bleeding badly with nothing embedded in their wound
  - someone with a burn
  - someone who isn't breathing normally
  - an injured motorcyclist wearing a helmet
- what it means if you see a driver displaying a 'help' pennant.

Practise with your instructor

Hopefully, you won't have the opportunity to practise what to do in the event of an incident or emergency during your lesson. Instead, practise with your instructor

- learning the rules relating to
  - breakdowns on all roads, including motorways
  - obstructions
  - incidents, eg warning signs and flashing lights
  - passing and being involved in a crash
  - incidents involving dangerous goods
  - which documents you'll need to produce if you're involved in a crash
  - incidents in tunnels.

7.1 Mark one answer DGV s4

What should you do when there's a fire in your vehicle's engine compartment?

- ☐ Open all windows
- ☐ Try to remove the load
- ☐ Flag down a passing motorist
- ☐ Cut off the fuel supply

If you suspect a fire, try to isolate the source. If possible, disconnect electrical leads and cut off the fuel supply.

7.2 Mark one answer DGV s2, 3

What training must you take before driving a lorry loaded with toxic substances?

- ☐ How to use fire-fighting equipment
- ☐ How to operate a forklift truck
- ☐ How to operate a lorry-mounted crane
- ☐ How to use breathing apparatus

Vehicles carrying dangerous goods and other materials that may pose a hazard are subject to detailed emergency procedures. These procedures must be followed. Appropriate fire-fighting equipment must be available and the driver must be trained in its use.

7.3 Mark one answer DBC s3, DGV s2

You're approaching a bridge that has no height restriction on it. What's the minimum height under the bridge?

- ☐ 3.6 metres (11 feet 10 inches)
- ☐ 4.4 metres (14 feet 5 inches)
- ☐ 4.8 metres (16 feet)
- ☐ 5 metres (16 feet 6 inches)

The headroom under bridges in the UK is at least 5 metres (16 feet 6 inches), unless otherwise stated. Where the overhead clearance is arched, this headroom is normally only between the limits marked.

7.4 Mark one answer DBC s4, DGV s4

What must you do if you're involved in a traffic incident?

- ☐ Drive on for help
- ☐ Inform the police within seven days
- ☐ Stop at the scene
- ☐ Drive to the nearest police station

If your vehicle is involved in a traffic incident, you must stop. Follow the guidance set out in The Highway Code if there's injury or damage to any other person, vehicle or property.

7.5 Mark one answer NI EXEMPT DBC s4, DGV s4

Your vehicle is involved in an incident where someone is injured. You don't produce your insurance certificate at the time. How soon must you report the incident to the police?

- ☐ 24 hours
- ☐ 48 hours
- ☐ 72 hours
- ☐ Seven days

If you don't give your name and address at the time of an incident, report it to the police as soon as you can, or in any case within 24 hours. If any other person is injured and you don't produce your insurance certificate to the police at the time of the incident, or to any other person who has reasonable grounds to request it, you must also

- report the incident to the police as soon as possible, or in any case within 24 hours
- produce your insurance certificate to the police either when reporting the incident or within seven days (five days in Northern Ireland) at any police station.

7.6 Mark one answer DBC s4, DGV s4

You're at the scene of an incident. What does it mean if there's a plain orange rectangle displayed on one of the vehicles?

- ☐ It's carrying dangerous goods
- ☐ It's carrying a first-aid kit
- ☐ It's carrying medical supplies
- ☐ It's unladen

Vehicles that carry dangerous goods have badges displayed on the side and rear. The badges are orange and show the type of material that's being carried. Make a note of this and report it to the emergency services when you contact them.

7.7 Mark one answer DBC s4, DGV s4

You arrive at an incident where someone is suffering from severe burns. What should you do to help?

- ☐ Douse the burns with cool water
- ☐ Remove anything stuck to the burns
- ☐ Burst blisters that form on the burns
- ☐ Apply ointment to the burns

Clean, cool water will cool the burning tissue and help prevent further damage. It will also reduce swelling, minimise shock and alleviate pain.

7.8 Mark one answer — DBC s4, DGV s4

You're waiting for the emergency services to arrive at the scene of an incident. A pedestrian's leg is bleeding heavily, but there's nothing embedded in the wound. What could you do to help?

- ☐ Apply firm pressure to the wound
- ☐ Dab the wound to stop the bleeding
- ☐ Put the casualty into the recovery position
- ☐ Wrap an ice pack near the wound

Before helping the pedestrian, you should protect yourself from contact with blood. If there's nothing embedded in the wound, you can stem the flow of blood by applying pressure to the wound area. As soon as practical, fasten a pad to the wound with a bandage or cloth.

7.9 Mark one answer — DBC s4, DGV s4

At the scene of an incident, an injured person is lying in a busy road. What's the first thing you should do to help?

- ☐ Treat the person for shock
- ☐ Warn other traffic
- ☐ Place them in the recovery position
- ☐ Make sure the injured person is kept warm

You should warn other road users that there has been a incident. You can do this by displaying an advance warning triangle or by switching on hazard warning lights or other lights. However you choose to warn people of the incident, make sure it doesn't put you or others at risk.

7.10 Mark one answer — DBC s4, DGV s4

You're the first person to arrive at an incident where people are badly injured. What's the first thing you should do?

- ☐ Switch on your own hazard warning lights
- ☐ Contact the local radio station to tell them about the incident
- ☐ Try and get people who are injured to drink something
- ☐ Move the people who are injured clear of their vehicles

If you're the first to arrive at an incident, further collisions and fire are the immediate concerns. Switch on your hazard warning lights to warn other traffic. Stop any running vehicle engines to reduce the risk of fire. Don't assume that someone else has called the emergency services – do it yourself. Don't move casualties trapped in vehicles unless they're in danger.

7.11 Mark one answer — DBC s4, DGV s4

You arrive at an incident where a motorcyclist is lying in the road, unconscious and bleeding. The area is safe and the emergency services have been called. What's the first priority when checking the rider?

- ☐ Remove their helmet
- ☐ Stop the bleeding
- ☐ Check their airway is open
- ☐ Move them off the road

At a traffic incident, the danger of further collisions and fire needs to be dealt with and the emergency services must be called. When helping a casualty, follow the DR ABC procedure, as this has been designed by medical experts to give the best outcome while waiting for the emergency services to arrive. Don't remove a motorcyclist's helmet unless it's essential to do so.

7.12 Mark one answer — DBC s4, DGV s4

You arrive at the scene of an incident and someone is unconscious. What's the first thing you should do to help them?

- ☐ Take the numbers of the vehicles involved
- ☐ Try to get them to drink water
- ☐ Check that they're breathing
- ☐ Look for any witnesses

The priority when dealing with an unconscious person is to make sure their airway is open and they can breathe. Stay with the casualty and make sure someone rings for an ambulance.

7.13 Mark one answer — DBC s4, DGV s4

You've stopped at the scene of an incident where there are casualties. What should you do to help while waiting for the emergency services to arrive?

- ☐ Keep injured people warm and comfortable
- ☐ Clear debris from the road
- ☐ Keep injured people on the move by walking them around
- ☐ Give injured people a warm drink

If you stop to give help at an incident and there are casualties, don't move injured people unless there's further danger. Keep them warm, comfortable and calm. Don't give them anything to drink.

7.14 Mark one answer DBC s4, DGV s4

You arrive at the scene of an incident where someone is injured. What should be given urgent priority?

- ☐ Deal with the danger from further collisions or fire
- ☐ Get the casualty a warm drink
- ☐ Take details of any witnesses
- ☐ Take the numbers of the vehicles involved

Make the area safe and call the emergency services first. If you're able to give first aid, follow the DR ABC procedure. If you feel you aren't capable of carrying out first aid, consider taking some training. It could save a life.

7.15 Mark one answer DBC s4, DGV s4

You're at the scene of an incident. How can you help someone suffering from shock?

- ☐ Reassure them confidently
- ☐ Offer them a cigarette
- ☐ Keep them moving
- ☐ Give them a warm drink

Immediate treatment can help to reduce the effects of shock. Reassure the casualty and try to keep them warm and comfortable. Only move a casualty if it's necessary for their safety.

7.16 Mark one answer — DBC s4, DGV s4

What should you do if you're the first person to arrive at the scene of an incident?

- ☐ Warn other traffic to prevent further collisions
- ☐ Take photographs of the scene
- ☐ Offer a calming cigarette to those involved
- ☐ Ask everyone who isn't injured to leave

Warn other traffic of the incident, without putting yourself or others at risk. Once the danger of further collisions is dealt with, call the emergency services. While you're waiting for them to arrive, keeping casualties or witnesses calm is important, but never offer a cigarette because of the risk of fire.

7.17 Mark one answer — DBC s4, DGV s4

There's been an incident. What can you do to help a driver who's suffering from shock?

- ☐ Give them a drink
- ☐ Offer them a cigarette
- ☐ Ask who caused the collision
- ☐ Reassure them confidently

You can help someone suffering from shock by talking to them and reassuring them confidently. Don't leave them alone or give them anything to eat or drink.

7.18 Mark one answer — DBC s4, DGV s4

You arrive at the scene of an incident where a motorcyclist is lying in the middle of the road and unconscious. What's the first thing you should do?

- ☐ Move the rider out of the road
- ☐ Warn other traffic
- ☐ Clear the road of debris
- ☐ Give the rider reassurance

At the scene of an incident, the first priority is to prevent any further collisions by warning traffic. You can warn other traffic by switching on hazard warning lights or displaying an advance warning triangle, or by any other appropriate means.

7.19 Mark one answer DBC s4, DGV s4

You're trying to resuscitate a casualty who isn't breathing. For how long should you continue compressions if they haven't started breathing again?

- ☐ Give up if you think they're dead
- ☐ Keep trying for up to two minutes
- ☐ Until medical help arrives and takes over
- ☐ Keep trying for up to four minutes

If you're trying to resuscitate someone, it's important that you don't give up on them. Make sure that someone has called the emergency services and continue trying to resuscitate the casualty until a medical professional takes over.

7.20 Mark one answer DBC s4, DGV s4

When should you stop trying to resuscitate a casualty?

- ☐ When you think they're dead
- ☐ When they can breathe without help
- ☐ When they've turned blue
- ☐ When you think the ambulance is coming

If you're trying to resuscitate a casualty, don't give up until a medical professional takes over. Never assume that someone's dead. People who aren't breathing can survive longer than you may imagine when given CPR.

7.21 Mark one answer DBC s4, DGV s4

You arrive at the scene of an incident. A casualty has an injured arm that's bleeding heavily. What could you do to help while waiting for the emergency services to arrive?

- ☐ Remove clothing around the wound
- ☐ Dab the wound
- ☐ Get them a drink
- ☐ Try to stem the flow of blood

You should first protect yourself from exposure to blood – many first-aid kits contain surgical gloves. To stem the flow of blood, you should apply firm pressure over the wound. If there's an object in the wound, you'll need to build up padding on either side and make sure you don't press on the object.

7.22 Mark one answer DBC s4, DGV s4

You arrive at the scene of an incident. The emergency services have already been called. A pedestrian is bleeding heavily from a leg wound, but their leg isn't broken. What could you do to help them?

- ☐ Dab the wound to stop the bleeding
- ☐ Give them some painkiller tablets
- ☐ Apply firm pressure to the wound
- ☐ Fetch them a warm drink

You should first protect yourself from exposure to blood – many first-aid kits contain surgical gloves. To stem the flow of blood, you should apply firm pressure over the wound. If there's an object stuck in the wound, you should leave it in place and build up padding on either side, making sure you don't press on the object.

7.23 Mark one answer DBC s4, DGV s4

At an incident, a casualty is unconscious but still breathing. Why would you move them?

- ☐ Because an ambulance is on the way
- ☐ Because bystanders advise you to
- ☐ Because there's further danger
- ☐ Because they look uncomfortable

Moving a casualty could cause further injury, so it's important that this is only done if there's danger to them where they are.

7.24 Mark one answer DBC s4, DGV s4

At an incident, you suspect a casualty has back injuries. What's the best way to help them?

- ☐ Offer them a drink
- ☐ Leave them where they are
- ☐ Raise their legs
- ☐ Offer them a cigarette

If you attempt to move a casualty who has a spinal injury, it could cause further damage. Make sure an ambulance has been called for and talk to the casualty to keep them calm.

7.25 Mark one answer DBC s4, DGV s4

The area around an incident is safe. What's the best way to help an injured person?

- ☐ Get them out of their vehicle
- ☐ Give them a drink
- ☐ Give them something to eat
- ☐ Keep them in the vehicle

Don't move a casualty who's in a safe place, as this could worsen any injury they may have. Only move a casualty if they're in danger – for example, from passing traffic. You may also need to move them if you need to give lifesaving first aid, such as opening their airway.

7.26 Mark one answer DBC s4, DGV s4

You're stopped by a police officer for a suspected motoring offence. Which document will you always be asked to produce?

- ☐ Your driving licence
- ☐ Your vehicle registration document
- ☐ Your vehicle's MOT certificate
- ☐ Your theory test certificate

If you can't produce your driving licence when asked to do so by a police officer, you may be asked to produce it at a police station within a certain time.

7.27 Mark one answer DBC s4, DGV s4

At the scene of an incident, a person has become hysterical. How should you calm them down?

- ☐ By leaving them to quietly recover
- ☐ By engaging them in banter
- ☐ By giving them a hot drink
- ☐ By talking to them quietly and firmly

Someone who has been involved in or witnessed a crash will be in no mood for humour. The best way to calm them is by remaining calm yourself, and talking quietly and firmly to them.

7.28 Mark one answer DBC s4, DGV s4

You're at an incident. Why may it be harmful to move a casualty?

- ☐ You could damage your back
- ☐ You could get blood on your hands
- ☐ You could be accused of assault
- ☐ You could cause more injury

Be especially careful about moving casualties at the scene of an incident. Inexperienced handling of a casualty could cause more injury, or even prove to be fatal. Only move casualties if they're in danger; for example, from fire.

7.29 Mark one answer DBC s4, DGV s4

At the scene of a crash, a casualty is in contact with live electrical cables. What should you use to break the electrical connection?

- ☐ A metal pole
- ☐ Woollen gloves
- ☐ A damp piece of cloth
- ☐ A length of wood

At the scene of an incident, if there are any casualties who are in contact with live electricity, don't touch them with your hands – even if you're wearing gloves. You should use a dry, non-conducting item, such as a wooden sweeping brush or plastic spade. You mustn't give first aid until you're sure the electrical contact has been broken, otherwise you'll be putting yourself in danger.

7.30 Mark one answer DBC s4, DGV s4

Where should you stand when using a motorway emergency telephone?

- ☐ In front of the barrier
- ☐ Facing the oncoming traffic
- ☐ With your back to the traffic
- ☐ Looking towards the grass verge

Motorway phones are free and easily located. You should face the oncoming traffic while using them, so that you can see approaching vehicles.

7.31 Mark one answer DBC s4, DGV s4

Which type of fire extinguisher shouldn't be used on flammable liquids?

- ☐ Water (red)
- ☐ Foam (cream)
- ☐ Dry powder (blue)
- ☐ Carbon dioxide (black)

It's very important to know the correct fire extinguisher to use. They'll carry a coloured label or badge telling you their contents. Make sure you learn which extinguisher is which, as using the wrong one could be dangerous and may spread the fire further. For example, you should never use a water extinguisher on burning oil.

7.32 Mark one answer DBC s4, DGV s4

Your vehicle breaks down on a motorway. What should you do if you can't get it fully onto the hard shoulder?

- ☐ Stand at the edge of the carriageway to warn others
- ☐ Place a warning triangle in the lane behind your vehicle
- ☐ Wear a bright jacket and stand in the lane behind your vehicle
- ☐ Call for help using the nearest emergency telephone

Standing on or near the edge of the carriageway is very dangerous. Your priority is to alert the emergency services by using an emergency roadside telephone. This connects to an operator, who'll get help out to you as soon as possible. In most cases, this is preferable to using a mobile phone, as it will show the operator your precise location.

7.33 Mark one answer DBC s4, DGV s4

Your engine catches fire. What should you do before attempting to put the fire out?

- ☐ Shut off the fuel supply
- ☐ Open the engine housing wide
- ☐ Drive to the nearest fire station
- ☐ Empty the air tanks

An engine fire is serious. If the fire breaches the fuel lines, it can easily spread to the fuel tank. If that happens, both the vehicle and its cargo will probably be lost. Therefore, your priority is to shut off the fuel supply.

7.34 Mark one answer DBC s4, DGV s4

What should you do before driving into a tunnel?

- ☐ Switch off your radio
- ☐ Remove your sunglasses
- ☐ Close your sunroof
- ☐ Switch on your windscreen wipers

Before driving into a tunnel, you should take off your sunglasses. If you don't, you'll find that your vision is restricted – even in a tunnel that appears to be well lit.

7.35 Mark one answer DBC s4, DGV s4

What should you do when you drive into a tunnel?

- ☐ Use parking lights
- ☐ Use front spotlights
- ☐ Use dipped headlights
- ☐ Use rear fog lights

Before entering a tunnel, you should switch on your dipped headlights; these help you to see and be seen. In many tunnels, using headlights is a legal requirement. You should also remove your sunglasses before you drive into a tunnel.

7.36 Mark one answer DBC s4, DGV s4

What should you do before entering a tunnel?

- ☐ Put on your sunglasses
- ☐ Check your tyre pressures
- ☐ Change to a lower gear
- ☐ Tune your radio to a local channel

On the approach to many tunnels, a board will indicate a local channel or radio frequency that you should tune into. This should give a warning of any incident or congestion in the tunnel ahead. Severe loss of life has occurred in tunnel fires. Getting advance warning of any problems ahead will help you to take appropriate action in good time.

7.37	Mark one answer	DBC s4, DGV s4

You're driving through a congested tunnel. What should you do if you have to stop?

- ☐ Stop very close to the vehicle in front to save space
- ☐ Ignore any message signs as they're never up to date
- ☐ Keep a safe distance from the vehicle in front
- ☐ Put your headlights on full beam

It's important to keep a safe distance from the vehicle in front at all times. This still applies in congested tunnels, even if you're moving very slowly or have stopped. If the vehicle in front breaks down, you may need room to manoeuvre past it.

7.38	Mark one answer	DBC s4, DGV s4

You're driving through a tunnel. What should you do if your vehicle breaks down?

- ☐ Switch on hazard warning lights
- ☐ Remain in your vehicle
- ☐ Wait for the police to find you
- ☐ Rely on CCTV cameras seeing you

If your vehicle breaks down in a tunnel, it could present a danger to other traffic. First, switch on your hazard warning lights. If there are passengers in your vehicle, take them to the nearest exit point. You should then call for help from an emergency telephone. Don't rely on being found by the police or being seen by a CCTV camera. The longer a vehicle stays in an exposed position, the more danger it poses to other traffic.

7.39	Mark one answer	DBC s4, DGV s4

Your vehicle catches fire while you're driving through a tunnel. What should you do if the vehicle can still be driven?

- ☐ Leave it where it is, with the engine running
- ☐ Stop, then walk to an emergency telephone point
- ☐ Park it away from the carriageway
- ☐ Drive it out of the tunnel if you can do so

If it's possible, and you can do so without causing further danger, it's safer to drive a vehicle that's on fire out of a tunnel. The greatest danger in a tunnel fire is smoke and suffocation.

7.40 Mark one answer DBC s4, DGV s4

What should you do when you're driving through a tunnel in heavy traffic?

- ☐ Stay close to the vehicle in front to reduce congestion
- ☐ Switch off your dipped headlights to reduce dazzle
- ☐ Closely follow the tail lights of other vehicles
- ☐ Follow the instructions given by variable message signs

In congested tunnels, you should always obey the instructions given by variable message signs and tunnel officials. Listen out for radio announcements and observe the road signs and signals. Keep your distance from the vehicle in front.

7.41 Mark one answer DBC s4, DGV s4

A vehicle has rolled over and caught fire. The driver's hands and arms have been burned. How could you help the driver?

- ☐ Douse the burns with cool water
- ☐ Try to put out the fire
- ☐ Remove smouldering clothing
- ☐ Remove anything sticking to the burns

Don't remove anything sticking to a burn. You may cause further damage and introduce infection into the wound. If you can, douse the burn with clean, cool water for at least 10 minutes.

7.42 Mark one answer DBC s4, DGV s4

At an incident, a casualty has stopped breathing. How could you help them?

- ☐ By opening their airway
- ☐ By tilting their head as far forward as possible
- ☐ By raising their legs
- ☐ By giving them something to drink

To help a casualty who isn't breathing, you should open their airway by placing your fingers under their chin and lifting it forward. This will open their airway but if there are no signs of breathing you should then start CPR.

7.43 Mark one answer NI EXEMPT DBC s4, DGV s4

You've been involved in an incident and damaged some property. What should you do if nobody else is present and your vehicle is still roadworthy?

- ☐ Stop, then report the incident to the police within 24 hours
- ☐ Leave the scene. Don't report the incident if there are no witnesses
- ☐ Stop, then report the incident to the police after 48 hours
- ☐ Leave the scene. Don't report the incident if there were no injuries

If you're involved in an incident that causes damage or injury to a person or property, you must stop. You must also give your details to anyone who has reasonable grounds to ask for them. If you don't do this, you must inform the police as soon as possible and in any case within 24 hours.

7.44 Mark one answer DBC s4, DGV s4

You've had an incident while driving through a tunnel. You aren't injured but your vehicle can't be driven. What's the first thing you should do in this situation?

- ☐ Rely on other drivers to phone the police
- ☐ Switch off the engine and switch on hazard warning lights
- ☐ Take the names of witnesses and other drivers
- ☐ Sweep up any debris that's in the road

If you're involved in an incident in a tunnel, be aware of the danger this can cause to other traffic. You should immediately switch on your hazard warning lights and switch off your engine.

7.45 Mark one answer DBC s4, DGV s4

While driving through a tunnel, your vehicle catches fire. What's the first thing you should do if the vehicle can't be driven out of the tunnel?

- ☐ Wait for the police; tunnels are regularly patrolled
- ☐ Stay with your vehicle; you'll be seen by the CCTV cameras
- ☐ Stop at the side of the road and switch off the engine
- ☐ Don't put out the fire; wait for the emergency services

If your vehicle catches fire in a tunnel and it can't be driven, stop on the side of the road, as far away from passing traffic as possible. Turn off the engine and warn other traffic by switching on your hazard warning lights.

7.46 Mark one answer — DBC s4, DGV s4

You're at the scene of an incident. What would suggest someone is suffering from shock?

- ☐ A flushed complexion
- ☐ Being alert
- ☐ Rapid shallow breathing
- ☐ Being very hungry

The effects of shock may not be immediately obvious. Try to recognise the warning signs early; prompt treatment can help to deal with it. Symptoms include a rapid pulse, pale grey skin, sweating and rapid shallow breathing.

7.47 Mark one answer — DBC s4, DGV s4

What's the first thing you should try to do if your vehicle catches fire while you're driving through a tunnel?

- ☐ Drive out of the tunnel
- ☐ Stop at the side of the tunnel
- ☐ Stop and put out the fire
- ☐ Stop and leave the vehicle immediately

If your vehicle catches fire in a tunnel, you should first try to drive it out – but only if you can do so safely.

7.48 Mark one answer — DBC s4, DGV s4

You arrive at the scene of an incident. How should you deal with someone who is hysterical?

- ☐ Talk to them quietly and firmly
- ☐ Let them wander off to calm down
- ☐ Restrain them physically
- ☐ Shout loudly to get their attention

A person who's hysterical may do things that you wouldn't expect. Try to keep the distressed person calm and safe until the emergency services arrive.

7.49 Mark one answer — DBC s4, DGV s4

You're at the scene of a traffic incident. What are the symptoms of someone suffering from shock?

- ☐ Rapid pulse and sweating
- ☐ Flushed complexion and deep breathing
- ☐ Slow pulse and dry skin
- ☐ Muscle spasms and an itchy rash

Prompt treatment can help to deal with shock. Warning signs to look for include rapid pulse, sweating, pale grey skin, and rapid shallow breathing. While waiting for help to arrive, keep the casualty warm and comfortable, and speak reassuringly to them.

7.50 — Mark one answer — DBC s4, DGV s4

You're at the scene of a collision. People are injured. What should be done when the risk of further collisions has been dealt with?

- ☐ Pull people who are hurt out of their vehicles
- ☐ Clear a parking area for the emergency services
- ☐ Make sure the emergency services have been called
- ☐ Get the casualties to drink something sweet

Make sure you or someone else has called the emergency services. Check the casualties and, if you feel able to give first aid, do so. Alternatively, ask any bystanders if they can help.

7.51 — Mark one answer — DBC s4

You're driving a coach carrying elderly people. You arrive at the scene of an incident. What should you do if the emergency services have already arrived?

- ☐ Ask your passengers to find out what's happening
- ☐ Don't tell your passengers anything in case you upset them
- ☐ Leave your passengers on the bus and see what's happening
- ☐ Tell your passengers what's happening without upsetting them

At the scene of an incident, you should tell your passengers what has happened, while reassuring them that they're not in danger. Explain to them that the emergency services are dealing with it and they should remain on the coach.

7.52 — Mark one answer — DBC s4

What should you do if your bus has been involved in an incident with other vehicles?

- ☐ Switch off your headlights
- ☐ Switch off the fuel supply
- ☐ Switch off the air conditioning
- ☐ Switch off the destination signs

The risk of fire will be reduced if all fuel supplies are switched off. However, don't put yourself or your passengers at any risk. If there's a fire, call the fire service.

7.53 Mark one answer — NI EXEMPT — DBC s4

What must you do if your bus has hit a parked vehicle and the owner can't be found?

- ☐ Report the incident to the police within seven days
- ☐ Report the incident to the owner as soon as possible
- ☐ Report the incident to the owner within seven days
- ☐ Report the incident to the police within 24 hours

If you damage a parked vehicle and the owner isn't around, you must report it to the police within 24 hours. This applies to any property you might have damaged. In Northern Ireland, all incidents must be reported to the police immediately.

7.54 Mark one answer — DBC s4

How should you care for a passenger who's in shock?

- ☐ Give them liquids
- ☐ Keep them moving
- ☐ Encourage them to sleep
- ☐ Keep them warm

After an incident, passengers who aren't obviously injured may be suffering from shock. Reassure them and keep them warm. Don't give them anything to eat or drink. Call the emergency services and keep yourself and any injured passengers calm until professional help arrives.

7.55 Mark one answer — DBC s4

A passenger on your bus has stopped breathing. You help them by giving CPR. When should you stop doing this?

- ☐ When they can breathe on their own
- ☐ When you think the passenger has died
- ☐ When their skin has turned blue
- ☐ When you think the ambulance is coming

You should only stop resuscitation when the passenger can breathe on their own or a medical professional takes over.

7.56 Mark one answer DBC s4

Your bus is involved in an incident. What should you do if you have a passenger who's unconscious but still breathing?

- ☐ Get medical help
- ☐ Check their pulse
- ☐ Give them liquid
- ☐ Lie them on their back

If one of your passengers is unconscious but breathing, get medical help immediately. Only move them if there's a danger of further injury and you're moving them to a place of safety while you wait for help to arrive.

7.57 Mark one answer DBC s4

Your bus has stopped at the scene of an incident. Why should you consider moving your passengers to the front of the bus?

- ☐ To improve weight distribution
- ☐ To reduce the risk of injury from a rear-end collision
- ☐ To be closer to emergency exits
- ☐ To be witnesses to the incident

If there's a risk that another vehicle could run into the back of your bus, you should move your passengers forward to a safer position.

7.58 Mark one answer NI EXEMPT DBC s3

Your bus hits a low railway bridge. Nobody is injured. When and to whom should you report the incident?

- ☐ Immediately, to your employer
- ☐ Within 24 hours, to the railway authority
- ☐ Within seven days, to the police
- ☐ Immediately, to the railway authority

If your bus hits a railway bridge, you must inform the railway authority immediately, as the impact may have damaged the bridge and made it unsafe. You should find the telephone number on or near the bridge. It's very important to do this as soon as possible, so that all rail traffic is stopped. You must also inform the police immediately, or in any case within 24 hours.

7.59 Mark one answer DBC s4

While driving through a tunnel, your coach catches fire. What should you do if the vehicle can't be driven any further?

- ☐ Move the passengers to the rear of the coach
- ☐ Get the passengers off and keep them together
- ☐ Make sure the passengers stay in their seats
- ☐ Move the passengers to the front of the coach

If your coach catches fire and it can't go any further, stop at the side of the road, switch the engine off and switch on your hazard warning lights. Your immediate responsibility is to your passengers. Get them off the coach, keep them together and take them to the nearest emergency exit.

7.60 Mark one answer DBC s4

You're carrying a bus full of passengers. What must you do if you see smoke coming from the engine compartment?

- ☐ Move the passengers to the front of the bus
- ☐ Open the engine cover to investigate
- ☐ Drive to the bus station for a replacement bus
- ☐ Get everyone off the bus quickly and safely

Smoke from the engine compartment is very serious, as fire within a vehicle can spread very quickly. Your priority is to make sure that all passengers get off the bus as quickly and safely as possible.

7.61 Mark one answer DBC s4

Your bus catches fire in a tunnel and can't be driven any further. What should you do when you've stopped safely at the side of the road?

- ☐ Move your passengers to the nearest tunnel exit
- ☐ Use the extinguisher to put out the fire
- ☐ Keep your passengers together in front of the vehicle
- ☐ Stand behind your vehicle and warn other drivers

If your vehicle catches fire and you can drive it out of the tunnel, you should do so. If you can't do this, stop at the side of the road and switch off the engine. Move your passengers calmly but quickly to the nearest exit.

Section eight
Vehicle condition

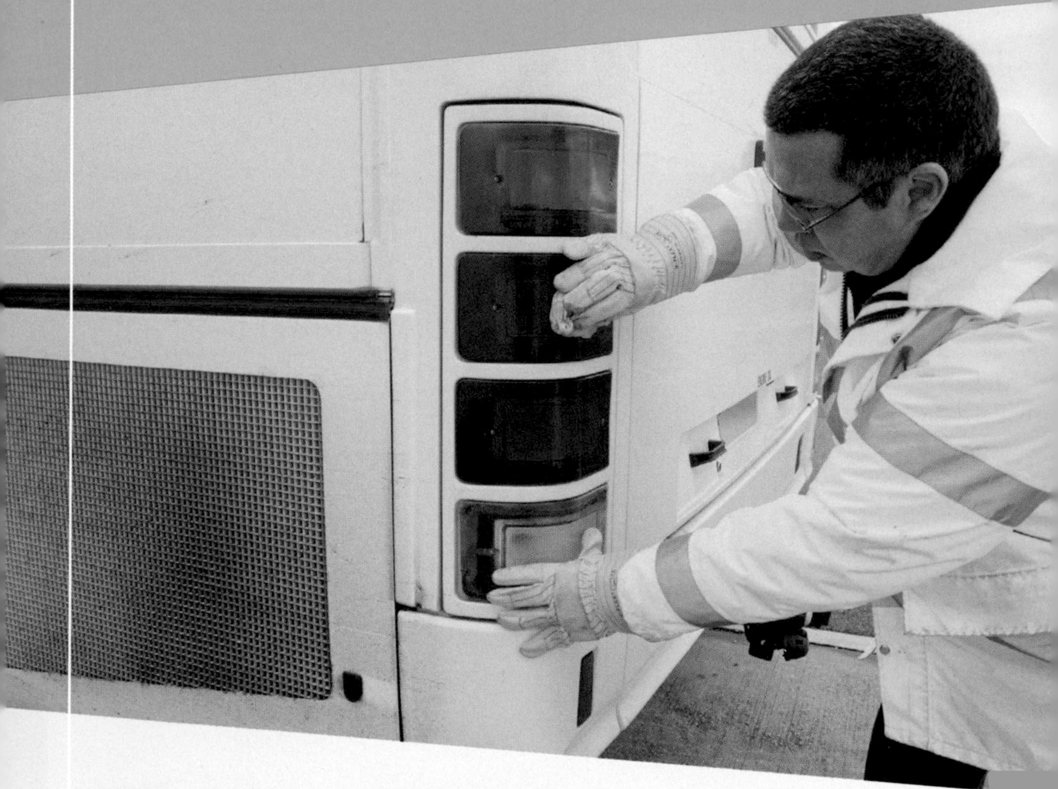

In this section, you'll learn about

- wheels and tyres
- vehicle maintenance and minor repairs
- cold weather
- warning lights and buzzers
- trailer coupling.

Vehicle condition

Wheels and tyres

All the tyres on your vehicle and on any trailer must be in good condition. Regularly check the tread depth of your tyres. Also look carefully for any serious faults, such as bulges, exposed ply, splits or cuts longer than one inch. Any of these, or too little tread, mean the tyre is illegal.

The tyres should be suitable for the load that you're carrying.

TIP

Tyres have codes on the sidewalls indicating the maximum load and speed that the tyre is designed for.

Keep the tyres correctly inflated. This will help prevent failure and also improve your fuel consumption.

- Under-inflated tyres will increase wear at the outer edge of the tread area.
- Over-inflated tyres will distort the tread and increase wear in the centre of the tread area.

If double-wheel tyres are touching or rubbing, this could indicate either under-inflation or overloading of your vehicle. Overloading the axles can cause tyres to fail.

TIP To help prevent your tyres overheating when driving at higher speeds for long distances, make sure that they're inflated to the correct pressure.

Energy-saving tyres have reduced **rolling resistance** and can improve your fuel economy.

Definition

rolling resistance

The force resisting the motion when a wheel rolls on a surface.

Turning the steering onto **full lock** when manoeuvring can cause the tyres to 'scrub' on the road surface, wearing down the tread.

DBC s2, 4 DGV s2, 3, 4

Definition

full lock

When the steering wheel is turned as far as it will go to either the left or right.

Some coaches are designed to have different-sized wheels on the front and the rear axles. Never mix different-sized wheels on the same axle.

DBC s2

Check the tyres on three-axle vehicles frequently, as punctures and faults can be more difficult to detect – especially on the inside wheels and the inner walls of double-wheel set-ups. While checking the tyres, also check for damaged or missing wheel nuts.

Try to turn around tight corners at a very low speed, especially when driving a three-axle double-deck vehicle. This will minimise any scrubbing effect on the rear tyres.

DBC s2

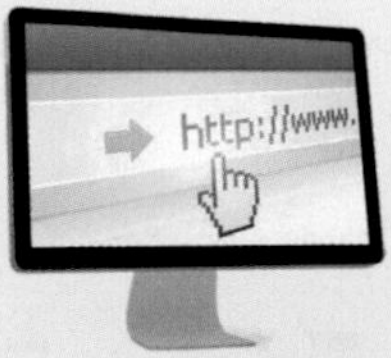

Read more about tyre safety at this website.

tyresafe.org/tyre-safety/truck-tyre-safety

Vehicle maintenance and minor repairs

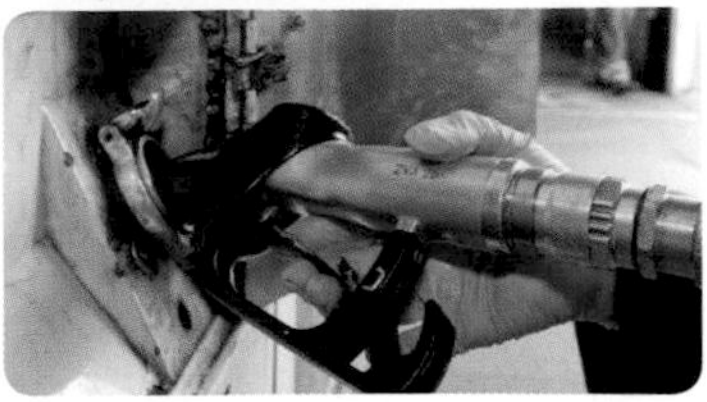

Wear gloves during regular and routine tasks such as checking the engine oil level and refuelling. This will help keep you clean and prevent infections or skin conditions such as eczema or dermatitis.

DBC s2 DGV s3

Whenever you drive your vehicle on the road, you're responsible for its condition. Before any journey, carry out a walk-round check. Your walk-round check should include items that need to be checked from inside the vehicle, such as warning lights and devices, brakes, and seat belts.

Have any faults rectified before you use the vehicle.

It's a good idea to carry spare bulbs and fuses with you in case you need them during a journey.

DBC s2, 3 DGV s3, 4

If your power-assisted steering fails when you're travelling, your vehicle will be difficult or almost impossible to steer. Stop as soon as possible, in a safe place, and seek assistance.

DBC s2 DGV s2

You **MUST NOT** knowingly use a faulty vehicle. The Driver and Vehicle Standards Agency (DVSA) carries out spot checks of vehicle condition. If a DVSA officer finds serious defects, the vehicle will be prohibited from further use and the traffic commissioners will be notified.

DBC s3 DGV s5

Read DVSA's guide to maintaining roadworthiness at this website.

safedrivingforlife.info/sites/default/files/PDF/guide-to-maintaining-roadworthiness.pdf

Cold weather

In cold weather, check every day that the air tanks on the brake systems are drained of moisture. On modern vehicles this is automated but on older vehicles you'll need to do it manually. Draining is necessary because the system takes in and compresses air. Air contains moisture and in cold weather this can turn to ice, which can block valves. This may result in loss of air pressure or even a total system failure.

DBC s2 **DGV** s2

At very low temperatures, diesel fuel can freeze, changing from a liquid to a waxy solid. Most diesel fuel contains some anti-waxing additives but in very low temperatures you may need to add further additive. Waxy diesel fuel leads to blockages throughout the fuel system, but especially in the fuel pipes and injectors. It may also cause increased wear in the injector pump.

DBC s2 **DGV** s2

To help start cold engines, use the pre-heating devices in your engine cylinders. These work by warming up the combustion chambers and this leads to reliable engine starting. When starting your vehicle's engine in cold conditions, you should be guided by the manufacturer's recommended procedure.

DBC s2 **DGV** s2

Warning lights and buzzers

All large vehicles have warning lights to alert the driver to faults such as low engine oil pressure. You should be familiar with the meaning of these lights and you shouldn't drive a vehicle until any fault indicated has been rectified.

Both a warning light and a buzzer give the warning for low air pressure. These warnings will be activated when you start the vehicle for the first time each day, but should stop when the air pressure reaches its operating level. Don't drive the vehicle while these warnings are on.

Some anti-lock brake system warning lights don't come on with the ignition and on these vehicles there should be a 'check' switch on the dashboard instead. Use this before you drive the vehicle.

DBC s2 DGV s2

Trailer coupling

Fifth-wheel couplings need routine maintenance every month or every 10 000 km. This should include a check for damage or defects, including to the **kingpin locking mechanism**.

On drawbar units, the eyelet coupling should be inspected for damage or wear. Drawbar and fifth-wheel couplings will need regular lubrication with special heavy-duty grease.

When coupling a trailer with a fifth-wheel coupling, make sure that you secure the kingpin release handle with the **dog-clip**.

Definition

fifth-wheel coupling

A coupling mechanism between a tractor unit and a trailer.

kingpin locking mechanism

The mechanism that locks the pin of the trailer to the fifth wheel of the tractor unit.

dog-clip

The clip that prevents unintentional movement of the kingpin release handle.

Before uncoupling a trailer, you must

- select a place with safe and level ground
- apply the trailer parking brake
- lower the landing gear and stow the handle away safely.

DGV s2

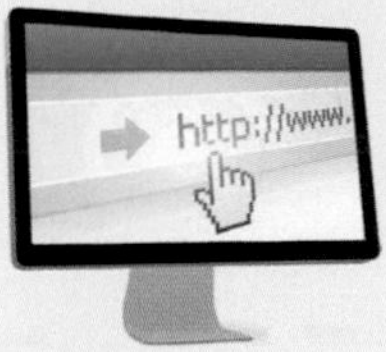

Register at the Society of Operations Engineers (SOE) website for free technical guides to roadworthiness and coupling/uncoupling.

soe.org.uk/resources/technical-guides/

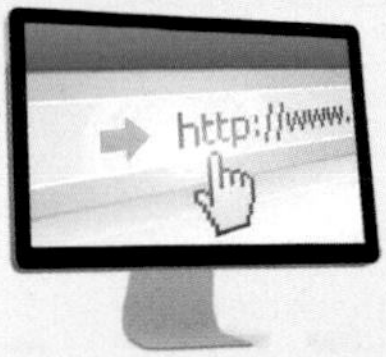

Find more information about coupling and uncoupling at this link.

hse.gov.uk/workplacetransport/information/coupling.htm

Meeting the standards

You must be able to
familiarise yourself with the vehicle if it's the first time you've driven it
conduct first-use walk-round and pre-journey checks and set up the vehicle correctly
categorise and report any vehicle defects and know what action to take; for example • remove the vehicle from service for safety-critical defects • report defects that aren't safety critical and continue to use the vehicle.

You must know and understand
which first-use walk-round and pre-journey checks are needed and what adjustments to make
how to check which sort of fuel your vehicle uses
how to check whether your vehicle uses a fuel additive, which sort it uses and how to keep it topped up.

Notes

You can use this page to make your own notes or diagrams about the key points you need to remember.

Think about

- Do I understand the load and speed codes on my tyres?
- Do I know my vehicle has the right tyres?
- Do I keep and use gloves for routine tasks and maintenance?
- Do I know what to look for during a walk-round check of my vehicle?
- Do my air tanks drain automatically or do I know how to do this manually?
- Do I know how to maintain and inspect my trailer coupling?

Your notes

Things to discuss and practise with your instructor

These are just a few examples of what you could discuss and practise with your instructor. Some are for PCV, some are for LGV and some are for both. Read more about vehicle condition to come up with your own ideas.

Discuss with your instructor

- how to use full steering lock when driving a three-axle double-deck bus
- why you have to be careful when driving a bus with different wheel sizes
- when and why a coach driver should wear gloves.

- what you should do to your wheels when leaving a building site, and why
- what the fifth wheel coupling is and how often it should be inspected.

- how diesel reacts to extremes of temperature
- Driver and Vehicle Standards Agency (DVSA) spot checks.

Practise with your instructor

- coupling and uncoupling a trailer.

- checking the condition of your tyres and wheels
- recognising the warning lights on your instrument panel
- checking the oil in your vehicle.

8.1 Mark one answer DGV s4

What's the main reason for cleaning your wheels and tyres when you leave a building site?

- ☐ To keep the tyres in good condition
- ☐ To prevent the tyres from damaging the road surface
- ☐ To reduce air leakage from the tyre valves
- ☐ To stop mud dropping onto the road

If your wheels leave mud on the road, you must arrange for it to be cleared. A slippery, muddy surface could cause danger to other road users.

8.2 Mark one answer DGV s2

Your vehicle has double rear wheels. Why should you check them before leaving a building site?

- ☐ To make sure the diff-lock is engaged
- ☐ To make sure the diff-lock is disengaged
- ☐ To check that the load-sensing valve is operating correctly
- ☐ To make sure bricks or debris aren't wedged between the wheels

While driving on a building site, bricks, mud or other debris can become lodged between a lorry's double rear wheels. When you leave the site and increase your road speed, anything lodged between the tyres could be thrown out. This flying debris can be a hazard for drivers of following vehicles and can create dangerous road conditions, especially for motorcyclists.

8.3 Mark one answer DGV s4

What should you check before you drive off a muddy building site onto the road?

- ☐ The diff-lock is disengaged
- ☐ The diff-lock is engaged
- ☐ The steering lock is engaged
- ☐ The twist lock is disengaged

Attempting to drive at normal speeds with the diff-lock engaged is dangerous; it will severely affect your steering control. The diff-lock is designed to be used in slippery conditions, at low speed, when maximum traction is required.

8.4 Mark one answer DGV s2

You're driving a lorry along a motorway. You notice that you're losing tread from one of your tyres. What should you do?

- ☐ Stop on the hard shoulder and phone for assistance
- ☐ Stop on the hard shoulder and change the wheel
- ☐ Continue driving to the next service station
- ☐ Continue driving and leave by the next exit

Continuous high speeds on the motorway can cause the tyres to become hot and to shred. If you notice this happening, you must stop on the hard shoulder as soon as it's safe to do so. It's dangerous to drive a defective vehicle.

8.5 Mark one answer DGV s2, 4

What's the minimum depth of tread required over three-quarters of the breadth of a lorry tyre?

- ☐ 1 mm
- ☐ 1.5 mm
- ☐ 2.5 mm
- ☐ 5 mm

Tyres are your lorry's only contact with the road. It's essential that this contact gives you the grip you need to control your vehicle at all times. For that reason, goods vehicles must have a tread depth of at least 1 mm across three-quarters of the breadth of the tread and in a continuous band around the entire circumference.

8.6 Mark one answer DGV s6

What's the final thing you should do after recoupling a trailer?

- ☐ Connect the brake lines
- ☐ Release the trailer parking brake
- ☐ Connect the electrical lines
- ☐ Raise the trailer legs

It's important to work methodically when uncoupling or recoupling a tractor unit and trailer. After recoupling, check that all connections, systems and lights are working correctly.

8.7 Mark one answer DGV s2

When should your lorry coupling system (fifth wheel) be checked and lubricated?

- ☐ Only prior to an MOT test
- ☐ Every 6 months
- ☐ Yearly
- ☐ Regularly

Maintenance of the fifth wheel should be carried out regularly, ideally monthly or every 10 000 km.

8.8 Mark one answer DGV s2

You're driving a new articulated lorry that's fully laden. You notice that the steering feels heavy. What's the most likely reason?

- ☐ The road is icy
- ☐ The power steering is faulty
- ☐ The lorry is new
- ☐ The load on the trailer has shifted

Heavy steering could mean there's a fault with the power steering. It should be investigated as soon as possible. Many large vehicles can become undriveable if the power steering fails completely.

8.9 Mark one answer DGV s2

When should wheel nuts be checked?

- ☐ After driving down a steep hill
- ☐ Soon after initial tightening
- ☐ Just after driving on a motorway
- ☐ After unloading

Always have wheel nuts checked soon after their initial tightening. They should be tightened to the torque specified by the vehicle manufacturer.

8.10 Mark one answer DGV s6

What would you secure with a dog clip?

- ☐ The kingpin release handle
- ☐ The air lines
- ☐ The parking brake
- ☐ The diff-lock

When recoupling, you must connect the dog clip to secure the kingpin release handle.

8.11 Mark one answer DGV s2

How frequently should the components of a fifth-wheel coupling be inspected?

- ☐ Daily
- ☐ Weekly
- ☐ Monthly
- ☐ Yearly

A fifth wheel must be maintained properly. It requires regular lubrication and inspection. This should be carried out monthly or every 10 000 km – whichever comes first.

8.12 Mark one answer DGV s2

What should you check for wear or damage on a drawbar unit?

- ☐ Dog clip
- ☐ Fifth wheel
- ☐ Kingpin release handle
- ☐ Eyelet coupling

The eyelet coupling on drawbar units should be checked regularly for damage or wear. It should be kept lubricated with heavy-duty grease.

8.13 Mark one answer DGV s2

What's the fifth-wheel coupling used for?

- ☐ To connect the tractor unit to the trailer
- ☐ To support the trailer when it's detached
- ☐ To prevent the trailer from jack-knifing
- ☐ To attach air lines to the trailer

The fifth-wheel coupling is a device to connect the tractor unit to the trailer. It allows articulation between the tractor and trailer, and it should be regularly maintained.

8.14 Mark one answer DGV s2

How often should the components of the fifth-wheel coupling be inspected?

- ☐ Every 3000 miles (5000 km)
- ☐ Every 6000 miles (10 000 km)
- ☐ Every 9000 miles (15 000 km)
- ☐ Every 12 000 miles (20 000 km)

Fifth-wheel couplings need regular inspection and lubrication. This should be carried out either every 6000 miles (10 000 km) or monthly – whichever is the sooner.

8.15 Mark one answer DGV s2

What happens to diesel fuel when it gets hot?

- ☐ It expands
- ☐ It liquefies
- ☐ It shrinks
- ☐ It waxes

When a full diesel fuel tank heats up, the fuel expands and can be forced out of the breather vent and onto the road. This wastes fuel and makes the road surface very dangerous for other road users.

8.16 Mark one answer DBC s2, DGV s2

You're checking your vehicle's tyres before starting a long motorway journey. What check should be made on each tyre?

- ☐ Air pressure
- ☐ Tracking
- ☐ Valve clearance
- ☐ Tyre profile

All tyres must be properly inflated and in good condition. Tyre failure on a motorway can have serious consequences. This can be avoided by checking your tyres at the start of each journey.

8.17 Mark one answer DBC s2, 3, DGV s2, 3

What should you do if you notice that two wheel nuts are missing from one of the wheels on your vehicle?

- ☐ Continue your journey
- ☐ Drive to the nearest tyre depot
- ☐ Use a nut from another wheel
- ☐ Park and phone for assistance

If you notice any missing wheel nuts, park and phone for assistance. It's essential that all wheel fixings are in place and tightened to the torque specified by the manufacturer. Checks should be made before you start any journey, but make a visual check every time you start up again after a rest stop.

8.18 Mark one answer DBC s2, DGV s2, 4

What would be illegal?

- ☐ Using different makes of tyre on the same axle
- ☐ Using a tyre with a recut tread
- ☐ Using a tyre with a tread depth of 1.3 mm
- ☐ Using a tyre with exposed ply or cord

You should check your tyres as part of your routine check of the vehicle. A damaged tyre – such as one with exposed ply – can have a major effect on your vehicle's handling and will make it a danger to you and other road users.

8.19 Mark one answer DBC s2, DGV s2

In very cold weather, moisture can freeze in your vehicle's air storage tanks. What can you do to help prevent this from happening?

- ☐ Cover the air tanks with a blanket
- ☐ Drain the tanks daily
- ☐ Use the brakes frequently
- ☐ Pump the brakes

In very cold weather, moisture in the storage tanks can freeze. Ice can form in the pipes and this will result in loss of pressure, which could cause brake failure. Make sure that you drain the tanks daily as part of a routine. Most modern vehicles are fitted with automatic bleed valves. Check that they're working properly and that air-drying systems are effective.

8.20 Mark one answer DBC s2, DGV s2

What does this warning light on the instrument panel mean?

- [ ] Low oil pressure
- [ ] Battery discharge
- [ ] Braking-system fault
- [ ] Door open

You should be familiar with all warning lights and buzzers fitted to your vehicle. If you're driving the vehicle for the first time, make sure that you know the function of each. If the brake warning light shows a fault in the system, stop as soon as it's safe to do so. Report the fault, and don't continue until it's been corrected.

8.21 Mark one answer DBC s2, DGV s2

You're driving along a motorway. What should you do if the air-pressure warning device starts to operate?

- [ ] Stop immediately in the lane you're in
- [ ] Continue slowly to the next service area
- [ ] Stop on the hard shoulder as soon as possible
- [ ] Leave the motorway at the next exit

If the air-pressure warning activates, you should have enough air to allow you to stop safely on the hard shoulder. Don't delay stopping, as further loss of air may cause the brakes to lock on. Switch on the hazard warning lights. Use the nearest emergency telephone to call for assistance.

8.22 Mark one answer DBC s4, DGV s4

Your vehicle has broken down at night on a two-way road. How should you try to leave the vehicle?

- [ ] On the left of the road
- [ ] Partly on the pavement
- [ ] On a grass verge
- [ ] On the right of the road

If your vehicle breaks down, try to stop on the left, facing in the same direction as the flow of traffic. This will help other drivers to see you, as your red reflectors will show to the rear. Don't stop on the pavement, as the weight of the vehicle may damage paving stones and underground services.

8.23 Mark one answer DBC s2, DGV s6

What should you do if your power-assisted steering fails suddenly?

- ☐ Continue driving to the nearest repair centre
- ☐ Return to the depot
- ☐ Continue your journey at a slower speed
- ☐ Park and get help

Faulty power steering will make your steering wheel very difficult or almost impossible to turn. Continuing to drive could cause danger to you and other road users. You should park safely and get help.

8.24 Mark one answer DBC s2, DGV s6

Your steering suddenly becomes heavy to turn. What could make this happen?

- ☐ An uneven road surface
- ☐ A loss of air-brake pressure
- ☐ A faulty parking brake
- ☐ A fault with the power-assisted steering

If the steering becomes heavy, the power-assisted steering may have failed. It's also possible that your vehicle has a puncture or the load might have shifted. You should stop safely, investigate the cause and call for help if necessary.

8.25 Mark one answer DBC s2, DGV s2

What should you do if the brake pedal becomes hard to press?

- ☐ Continue to drive and report it at the end of the day
- ☐ Pump the brake pedal continuously
- ☐ Drain the air tanks and then continue
- ☐ Park and phone for help

As soon as you detect a fault on your vehicle, you must take action. Where faulty brakes are concerned, you need to seek help immediately. Always report minor faults as soon as you detect them. Minor faults can become major ones if they aren't repaired quickly.

8.26 Mark one answer DBC s2, DGV s2

Why do air tanks on brake systems need to be drained?

- ☐ To remove excess coolant
- ☐ To remove rain water that's seeped into the system
- ☐ To remove any oil leaks that collect here
- ☐ To remove moisture drawn in from the atmosphere

Moisture in the air condenses as the air pressure increases. This moisture can be transmitted around the braking system and is especially dangerous in cold weather. It can lead to ice forming and blocking the valves and pipes.

8.27 Mark one answer — DBC s2, DGV s2

You're checking your vehicle. What should you do if you discover an air leak in the braking system?

- ☐ Drive slowly to the nearest garage
- ☐ Check the leak from time to time on your journey
- ☐ Leave the vehicle parked and report the fault immediately
- ☐ Start your journey and report the fault on your return

Under no circumstances should you attempt to move or drive a vehicle with an air leak in the braking system. Report the fault immediately, or arrange to have it repaired. Place a warning sign in the cab, telling other drivers not to drive the vehicle and why.

8.28 Mark one answer — DBC s2, DGV s2

A loud buzzer sounds in your vehicle. What's this most likely to indicate?

- ☐ Low oil pressure
- ☐ Low air pressure
- ☐ Low tyre pressure
- ☐ Low fuel level

Warning buzzers are linked to many systems on modern vehicles, including the air brakes. A warning light on the dashboard may help you identify the system that's caused the problem. Stop in a safe place until the fault has been identified and put right. Get professional help if necessary.

8.29 Mark one answer — DBC s2, DGV s2

What proportion of the width of a tyre must have at least the legal minimum depth of tread?

- ☐ One-quarter
- ☐ One-half
- ☐ Five-eighths
- ☐ Three-quarters

Keeping your vehicle's tyres in good condition will give them a good grip on the road and will improve the vehicle's overall stability. Each tyre must have at least the minimum legal depth of tread across three-quarters of its width. Renew tyres well before they wear to less than that.

8.30 Mark one answer — DBC s4, DGV s4

You discover that one of your rear brake-light bulbs has failed. How soon should it be replaced?

- ☐ Immediately
- ☐ On return to your base
- ☐ Within five days
- ☐ At the next service

Bulbs should be replaced as soon as you're aware that they've failed. Carry a stock of all the various bulbs used on your vehicle, so you can repair a fault without delay.

8.31 Mark one answer DBC s2, DGV s2

What can you add to diesel fuel to prevent it from becoming less effective at low temperatures?

- ☐ Anti-freeze
- ☐ Anti-waxing additives
- ☐ Petrol
- ☐ Paraffin

In extremely cold weather, you'll have to use diesel fuel with anti-waxing additives to stop the fuel lines from freezing up. During the winter months, these additives are usually put in by the fuel companies.

8.32 Mark one answer DBC s2, DGV s3

You're about to start a long journey midway through the day. What should you do if you notice that the sidelights work but the headlights are faulty?

- ☐ Don't drive until they're repaired
- ☐ Drive only until the light begins to fade
- ☐ Avoid driving on motorways after dark
- ☐ Drive only if the weather is good

To comply with the law, all lights must be in good working order, even in daylight when they're not being used. Before you set out, make sure that everything is working and get any faulty lights fixed. You may need to use your headlights or other lights if you're delayed or find yourself driving in conditions of reduced visibility.

8.33 Mark one answer DBC s2, DGV s2

Before starting a journey, you want to check your brake-system warning lights. What can you do when these aren't operated by the ignition switch?

- ☐ Look for a check switch on the dashboard
- ☐ Get someone behind to check your brake lights
- ☐ Check them at the end of your journey
- ☐ Pump the brake pedal a number of times

A warning-lights check is sometimes performed automatically when the ignition is switched on. However, you may need to do this manually by operating a separate check switch. Never start a journey without carrying out this check. If there's a problem, have it repaired before you set off.

8.34 Mark one answer DBC s2, 4, DGV s2

On motorways, you're usually driving at higher speeds for long distances. What effect can this have on your tyres?

- ☐ They'll be more likely to overheat and disintegrate
- ☐ They'll have lower rolling resistance
- ☐ They'll need less air pressure
- ☐ They'll become very slippery

Driving at higher speeds and for longer periods, such as on motorways, can cause your tyres to overheat and disintegrate. You should make a point of checking them when you stop for a break.

8.35 Mark one answer DBC s2, DGV s2

You notice that one of your tyres has a bulge in the side wall. What will happen if you drive the vehicle?

- ☐ Your tachograph reading won't be accurate
- ☐ Your speedometer will give an incorrect reading
- ☐ The vehicle will become unstable on corners
- ☐ You'll break the law and risk prosecution

It's a legal requirement that your tyres have at least the minimum permitted depth of tread and are in good condition before you start any journey. Make sure that you inspect them before setting off, and at regular intervals.

8.36 Mark one answer DBC s2, DGV s4

What's the purpose of a preheating device?

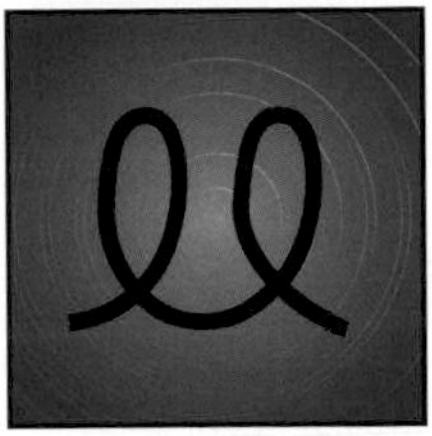

- ☐ To heat the cab
- ☐ To heat the gearbox
- ☐ To heat the combustion chambers
- ☐ To heat the seat

Most diesel engines have glow plugs to preheat the combustion chambers, so helping the engine to start when it's cold. When the ignition is switched on, the preheat warning lamp will show and the glow-plugs will rapidly heat up.

In cold weather, the starter should only be operated when the glow plugs reach operating temperature and the warning light goes out.

8.37 Mark one answer DBC s2, DGV s2

Where does a high-pressure fuel injector deliver fuel?

- ☐ Into the carburettors
- ☐ Into the combustion chamber
- ☐ Into the manifold
- ☐ Into the crankcase

Most diesel engines use a high-pressure fuel-injector system that will deliver pressurised fuel directly into the combustion chamber of the engine. This system is known as a direct-injection engine.

8.38 Mark one answer DBC s4, DGV s2

Why is it important to avoid overfilling the engine with oil?

- ☐ It could leave an oil stain on the road
- ☐ It will increase the amount of exhaust gases
- ☐ It could increase pressure in the engine and cause damage
- ☐ It will damage the exhaust system

Too much oil in an engine can be just as bad as too little. Overfilling an engine can create excess pressure, cause oil leakage through seals and result in expensive damage.

8.39 Mark one answer DBC s2, DGV s2

Why should you use an approved coolant solution in your engine's cooling system?

- ☐ To protect the coolant from freezing
- ☐ For easier starting from cold
- ☐ To prevent the air tank from freezing
- ☐ For effective cab heating

Coolant solution is also known as anti-freeze. It provides protection from freezing and also contains a corrosion inhibitor that prolongs the life of the cooling system. Remember to check the coolant level regularly.

8.40 Mark one answer DBC s2, DGV s2

What's it advisable to do when you replace a tubeless tyre?

- ☐ Fit the same valve
- ☐ Replace the valve
- ☐ Have the valve checked
- ☐ Clean the valve

Always get expert advice when dealing with tyres. It's good practice to have a new valve fitted when replacing a tubeless tyre. Good garages and specialist tyre services know the regulations.

8.41 Mark one answer DBC s2, DGV s2

When should you check the oil level in your engine?

- ☐ When the engine is running
- ☐ When the engine is cold
- ☐ When the engine is warm
- ☐ When the engine is hot

Oil is vital for lubrication of the engine. It should be checked regularly and topped up as necessary. You should always check your engine's oil level when the engine is cold and your vehicle is parked on the level.

8.42 Mark one answer DBC s2, DGV s2

When should anti-freeze be used in the cooling system?

- ☐ In winter only
- ☐ All year round
- ☐ In summer only
- ☐ When starting from cold

Today, all water-cooled engines use a mixture of water and anti-freeze to make up the coolant. As well as helping to keep the engine at its correct operating temperature, the anti-freeze acts as a corrosion inhibitor to prolong the life of the cooling system.

8.43 Mark one answer DBC s2, DGV s2

What does this warning light on the instrument panel mean?

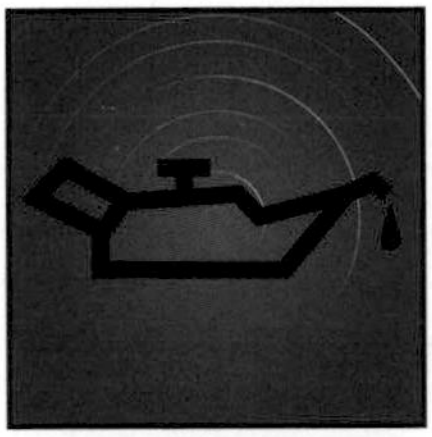

- ☐ Low fuel pressure
- ☐ Low oil pressure
- ☐ Low water pressure
- ☐ Low air pressure

You should be familiar with all the warning lights fitted to your vehicle. The oil warning light indicates low oil pressure or lack of oil. If it lights up, report the fault and don't continue until it's been corrected.

8.44 Mark one answer DBC s2, DGV s2

What does it mean if the ignition warning light comes on while you're driving?

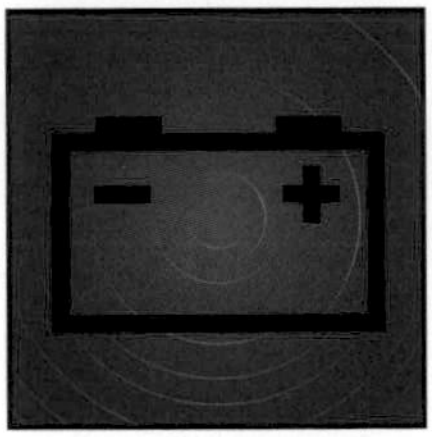

- ☐ The oil pressure is low
- ☐ There's an electrical fault
- ☐ The air pressure is low
- ☐ There's a hydraulic fault

The ignition light should go out when the engine has started. If it fails to go out or comes on while you're driving, it means that there's an electrical fault of some kind.

8.45 Mark one answer DBC s2, DGV s2

What do the codes shown on the side walls of bus and lorry tyres refer to?

- ☐ Tread pattern
- ☐ Minimum temperature
- ☐ Maximum load
- ☐ Running pressure

Tyres have codes shown on the wall of the tyre. These refer to the maximum load and speed capability of the tyre.

8.46 Mark one answer DBC s2, DGV s3

What will happen if you follow a regular vehicle maintenance schedule?

- ☐ It will reduce breakdowns
- ☐ It will increase fuel bills
- ☐ It will allow heavier loads to be carried
- ☐ It will reduce insurance premiums

You can reduce the chance of a breakdown with routine maintenance and regular checks on your vehicle.

8.47 Mark one answer DBC s2, DGV s2

What should you do before changing a wheel on your vehicle?

- ☐ Leave the parking brake off
- ☐ Use wheel chocks if available
- ☐ Dismantle the wheel and tyre
- ☐ Get someone to check the other tyres

Always make sure your vehicle is secure and level before a wheel is changed. Apply the parking brake and use chocks if they're available. When the replacement wheel is fitted, all wheel nuts/studs should be retightened to the correct torque.

On a motorway, never attempt to change a wheel yourself. Use the emergency telephones to call for assistance.

8.48 Mark one answer DBC s3, DGV s2

You've had to change a wheel on your vehicle. When should the wheel nuts be checked again?

- ☐ At the next service interval
- ☐ When they're cold
- ☐ When they're hot
- ☐ Shortly afterwards

When refitting a wheel, the nuts should be tightened to the correct torque. They should be rechecked after about 30 minutes if the vehicle hasn't moved, or after 40 to 80 km (25 to 50 miles) of driving.

8.49 Mark one answer DBC s2, DGV s2

How can vehicle breakdowns be reduced?

- ☐ By driving slowly
- ☐ By regular servicing
- ☐ By regular cleaning
- ☐ By avoiding bad weather

Following the vehicle manufacturer's guidelines for service intervals will enable worn components to be replaced before they fail. This will help prevent costly breakdowns.

8.50 Mark one answer DBC s2, DGV s2

Why should your engine oil be changed at the recommended intervals?

- ☐ To reduce friction and wear
- ☐ For better steering control
- ☐ To prevent oil leaks
- ☐ To improve clutch wear

The oil and filter will get dirty in normal use. If you follow the service-interval guidance, your engine will remain protected and properly lubricated, making it less likely to break down.

8.51 Mark one answer DBC s2, DGV s2

Where should you park your vehicle before checking the engine oil level?

- ☐ On sloping ground
- ☐ On a steep gradient
- ☐ On flat ground
- ☐ On a downhill slope

If you check the engine oil level when your vehicle is standing on a slope, the reading on the dipstick won't be accurate. Park it on flat ground, so the dipstick shows the true level.

8.52 Mark one answer DBC s2, DGV s2

What could happen if you overfill your engine with oil?

- ☐ The handling may improve
- ☐ Exhaust emissions could be reduced
- ☐ Some gaskets might be damaged
- ☐ You could wait longer before the next service

If you overfill the engine with oil, you could cause the engine to build up too much pressure when it's running. This could cause damage to oil seals and gaskets.

8.53 Mark one answer DBC s2, DGV s6

You've stopped on a firm, level surface. What's the first thing you must do before you uncouple the trailer?

- ☐ Lower the trailer legs to the ground
- ☐ Apply the parking brake
- ☐ Release the brake air lines
- ☐ Uncouple the electrical lines

Before leaving the cab, it's very important to secure the vehicle by applying the parking brake. After leaving the cab, apply the trailer parking brake.

8.54 Mark one answer DGV s6

You're uncoupling a trailer. What must you do before disconnecting any of the air lines?

- ☐ Drain the air tanks
- ☐ Apply the trailer parking brake
- ☐ Lower the landing gear
- ☐ Disconnect the electrical line

Whenever you uncouple a trailer, you must work through the uncoupling process methodically. Start by making sure that the brakes are applied on both the vehicle and the trailer.

8.55 Mark one answer DBC s2, 3, DGV s2

When should you check the wheel nuts on your vehicle?

- ☐ Before any journey
- ☐ Only before long trips
- ☐ Every 1000 miles (1600 km)
- ☐ Annually

Always have a walk round your vehicle and visually check the wheel nuts whenever you take a break. When a wheel is replaced, it's important to have the wheel nuts rechecked shortly after their initial tightening. Wheel nuts must always be tightened to the torque specified by the manufacturer.

8.56 Mark one answer DBC s2, DGV s2

What should you do if thick black smoke is coming from the exhaust of your vehicle?

- ☐ Continue on to the nearest garage
- ☐ Return to your depot and report the problem
- ☐ Stop in a safe place and get help
- ☐ Drive slowly with your hazard warning lights on

Causing excessive smoke is an offence and could contribute to causing an incident or collision. You could also end up creating serious traffic problems if your vehicle breaks down in a difficult location.

8.57 Mark one answer — DBC s2, DGV s2

You hit the kerb at speed. What part of your vehicle should you check for damage?

- ☐ Exhaust
- ☐ Brakes
- ☐ Lights
- ☐ Tyres

Hitting the kerb at speed can split the tyre or put the steering and suspension geometry out of alignment. Have it checked as soon as possible. A sudden puncture in a front tyre could result in a loss of steering control and a serious crash.

8.58 Mark one answer — DBC s2, DGV s2

What's most likely to cause a burst tyre?

- ☐ Frequent gear changing in varying conditions
- ☐ Running tyres under-inflated
- ☐ Always operating in cool weather
- ☐ Mixing tyres with different tread depth

Tyres can become very hot during sustained high-speed driving. If they're under-inflated or damaged, there's a risk they'll burst or disintegrate. Check for excessive heat when you stop for a break.

8.59 Mark one answer — DBC s2, DGV s2

When can 'selective' or 'block' gear changing be used?

- ☐ To change down in gear only
- ☐ To change up in gear only
- ☐ To change gear to a low speed only
- ☐ To change either up or down in gear

'Block' gear changing means changing up or down to the most appropriate gear without going through each gear in turn. It reduces the number of gear changes, and this can improve fuel consumption.

8.60 Mark one answer — DBC s2, DGV s2

When would you use 'kickdown' on a vehicle that has automatic transmission?

- ☐ To give quicker acceleration
- ☐ To apply the emergency brakes
- ☐ To stop more smoothly
- ☐ To go down a steep hill

Depending on road speed, pressing the accelerator pedal firmly to the floor will activate a switch that allows the gearbox to select a lower gear for improved acceleration.

8.61 Mark one answer DBC s2, DGV s6

What's the most likely cause if your steering starts to feel heavy?

- ☐ Faulty power steering
- ☐ An icy road
- ☐ A burst rear tyre
- ☐ A wet road

Modern large vehicles are fitted with power-assisted steering. If the steering feels heavy, then there's likely to be a fault, which should be investigated and repaired as soon as possible.

8.62 Mark one answer DBC s2, DGV s2

Your vehicle suffers a tyre blow-out. How could this create a hazard for other road users?

- ☐ Scattered debris
- ☐ Skid marks
- ☐ Suspension failure
- ☐ Axle damage

When a tyre explodes, fragments are thrown over a wide area. This can create a serious hazard for other drivers. You should regularly check your tyres for cuts, damage and pressure. Frequent checks and proper maintenance can help prevent a blow-out.

8.63 Mark one answer NI EXEMPT DBC s3, DGV s3

The Driver and Vehicle Standards Agency (DVSA) and the police carry out spot checks for faulty vehicles. What will happen to the vehicle if serious defects are found?

- ☐ It will be impounded until a new driver is found
- ☐ It will be restricted to 30 mph for the remainder of the journey
- ☐ It will be prohibited from further use until the defects are rectified
- ☐ It will be ordered back to the depot to unload goods or passengers

DVSA or the police can order an immediate prohibition. The vehicle may be immobilised and you won't be able to drive it until the faults have been rectified. Details are notified to the traffic commissioner. Never use a vehicle that you know is faulty.

8.64 Mark one answer NI EXEMPT DBC s3, DGV s3

At a Driver and Vehicle Standards Agency (DVSA) roadside check, your vehicle is found to have serious defects and you may no longer use it. Who will DVSA share this information with?

- ☐ The Driver and Vehicle Licensing Agency
- ☐ The Traffic Commissioner
- ☐ The Road Transport Industry Training Body
- ☐ The Bus, Coach and Commercial Vehicle Council

DVSA and the police carry out frequent roadside checks of commercial vehicles. Where serious defects are found, the vehicle is taken out of use until they're put right. Details of the prohibition are shared with the traffic commissioner.

8.65 Mark one answer DBC s2, DGV s2

What could happen if there isn't enough oil in your engine?

- ☐ It may run faster
- ☐ It may break down
- ☐ It may use less fuel
- ☐ It may produce more power

You should check your oil level regularly and pay attention to any oil warning devices. A damaged engine can be very expensive to mend. It could also cause sudden and unexpected loss of control, putting you and others in danger. If your engine seizes, you should make every effort to stop in a safe place.

8.66 Mark one answer DBC s2, DGV s2

While driving, your engine-oil warning light comes on. Why could it be dangerous to continue driving?

- ☐ The brakes will fail
- ☐ The engine may be damaged
- ☐ You'll need to have the vehicle serviced
- ☐ You'll need to replace the carburettor

Lack of oil could cause the engine to seize, and this could cause sudden and unexpected loss of control. You could break down in a dangerous position and put yourself and other road users at risk.

8.67 Mark one answer DBC s2, DGV s2

Bus and lorry tyres have codes on their side walls. What do these codes refer to?

- ☐ Running pressure
- ☐ Speed capability
- ☐ Minimum temperature
- ☐ Tread depth

Codes are shown on the wall of the tyre. These refer to the maximum load and speed capability of the tyre.

8.68 Mark one answer — DBC s2, DGV s2

What's the advantage of energy-saving tyres?

- ☐ Greater tread depth
- ☐ Reduced rolling resistance
- ☐ Less tread depth
- ☐ Increased rolling resistance

When changing or fitting new tyres, consider buying energy-saving tyres. These have a reduced rolling resistance and improved efficiency, which helps to keep fuel costs down.

8.69 Mark one answer — DBC s2, DGV s2

A tyre has been replaced on your vehicle. What precautions should be taken when tightening the wheel nuts?

- ☐ Tighten the nuts in a clockwise direction
- ☐ Fully tighten each nut before moving to the next
- ☐ Tighten the nuts evenly with a torque wrench
- ☐ Tighten the nuts with an air-operated power tool

When a wheel is refitted, the wheel nuts/studs should be tightened gradually and diagonally across the wheel. You should tighten the nuts to the specified torque setting, using a calibrated torque wrench.

8.70 Mark one answer — DBC s2, DGV s2

Why are energy-saving tyres effective?

- ☐ They have a reduced rolling resistance
- ☐ They cost less to produce
- ☐ They have no tread pattern
- ☐ They don't puncture at high speed

Energy-saving tyres keep fuel costs down because they have a lower rolling resistance and better grip than ordinary tyres. You should consider this when replacing your tyres.

8.71 Mark one answer — DBC s2, DGV s2

Where can you get advice to help carry out minor repairs?

- ☐ DVLA guidance notes
- ☐ EU and UK directives
- ☐ The Health and Safety Executive
- ☐ The vehicle handbook

The vehicle handbook will clearly explain what maintenance may be carried out by the driver. Always refer to it and comply with the safety guidance that it contains. However, you need to be aware of your employer's policy on making any minor repairs.

8.72 Mark one answer — DBC s2, DGV s2

Where can you get advice about carrying out repairs to your vehicle?

- ☐ Workshop manuals
- ☐ DVLA guidance notes
- ☐ EU and UK directives
- ☐ The Health and Safety Executive

Workshop manuals give detailed technical advice about the servicing, maintenance and repair of your vehicle. They're essential for anyone who wants to service or repair their own vehicle.

8.73 Mark one answer — DBC s2, DGV s2

What could happen if you overfill your engine with oil?

- ☐ Loss of power
- ☐ Lower emissions
- ☐ Better handling
- ☐ Longer service intervals

If the engine contains too much oil, the moving parts can hit the oil surface, causing a loss of power or even damage to the engine. Excess pressure may damage oil seals, causing oil to leak onto the road, creating a slippery and dangerous road surface.

8.74 Mark one answer — DBC s2, DGV s2

What's the purpose of the oil filter?

- ☐ To prevent the engine from overheating
- ☐ To give better fuel consumption
- ☐ To prevent the engine from over-revving
- ☐ To collect metal particles from the oil

As the lubricating oil is pumped around the engine, it collects tiny fragments of metal from the moving parts. The oil filter is designed to remove these fragments. Most filters are designed to be replaced at service intervals.

8.75 Mark one answer — DBC s2, DGV s2

Which tool is essential for fitting a road wheel?

- ☐ Torque wrench
- ☐ Ring spanner
- ☐ Open-end spanner
- ☐ Adjustable wrench

It's essential that all wheel nuts are tightened to the specified torque with a calibrated torque wrench. The wheel nuts, fixings and markers should be checked every day before you start your journey.

8.76 Mark one answer DBC s2, DGV s2

What's the most likely reason for a diesel-engined vehicle running erratically in very cold weather?

- ☐ The endurance brake engaging
- ☐ The air conditioning not working
- ☐ The speed limiter operating
- ☐ The fuel partly solidifying

If diesel fuel becomes very cold, it can partially solidify. Additives are used by the fuel companies to try to prevent this. Any solidifying (waxing) of the fuel can prevent it from flowing properly and cause the engine to run erratically or even stop.

8.77 Mark one answer DBC s2, DGV s2

What do heated fuel lines prevent?

- ☐ The cab temperature from dropping
- ☐ The radiator from freezing
- ☐ The windows from misting
- ☐ The diesel from solidifying

In cold weather, diesel fuel can solidify as it starts to freeze. This is known as waxing. Waxing prevents the fuel from flowing properly and this can stop the engine from running.

8.78 Mark one answer DBC s3, DGV s3

Why should you carry out a daily walk-round check before setting off?

- ☐ To check your route
- ☐ To check for any parking violations
- ☐ To check your schedule
- ☐ To check for any defects

You have a legal responsibility to make sure that your vehicle is fully roadworthy. Daily walk-round checks give you the chance to look for any defects with your vehicle.

8.79 Mark one answer DBC s2, DGV s2

What happens to diesel fuel when it gets hot?

- ☐ It expands
- ☐ It liquefies
- ☐ It shrinks
- ☐ It waxes

On a hot day, the fuel in your tank will expand. If this happens when the tank has been filled to the brim, fuel will spill onto the road. This can be very dangerous or even fatal for other road users – especially motorcyclists. Allow for this expansion when filling your tank.

8.80 Mark one answer DBC s3, DGV s3

What must vehicle operators provide for their drivers?

- ☐ A system for reporting vehicle defects
- ☐ A daily update on all motorway hold-ups
- ☐ A hands-free mobile communications system
- ☐ A fuel-saving incentive scheme

The operator must have a system in place so that drivers can report defects and have them repaired properly.

8.81 Mark one answer DBC s2

You're driving a three-axle double-deck bus. Why should you take extra care when using full steering lock?

- ☐ Passengers might alter the angle of tilt
- ☐ The power steering might fail
- ☐ You may damage the air suspension
- ☐ You may scrub the rear tyres

Very low speed is advisable to reduce damage to the rearmost tyres when the vehicle is on full lock. When you're driving around tight corners, the rear wheels will follow a different path from the front wheels. You should be alert to anything the rear wheels or bodywork might strike.

8.82 Mark one answer DBC s2

You're driving a three-axle double-deck bus and have the steering turned onto full lock. What should you do to avoid rear tyre scrub?

- ☐ Use the highest gear possible
- ☐ Drive at a very low speed
- ☐ Apply the exhaust brake (retarder)
- ☐ Reduce the pressure in the rear tyres

When turning tight corners, allow for the course the wheels will take. A very low speed is advisable when the steering is on full lock. This is to minimise any scrubbing effect on the rear tyres.

8.83 Mark one answer DBC s2

You're driving a double-deck bus. What's the minimum depth of tread required over three-quarters of each tyre's width?

- ☐ 0.8 mm
- ☐ 1 mm
- ☐ 1.6 mm
- ☐ 2 mm

It's essential that the tyres on your vehicle are in good condition. You must never forget that you have passengers on board. Their safety must be your priority. At no time should the depth of the tread be less than 1 mm over three-quarters of the width of the tyre.

8.84 Mark one answer DBC s3

What should you do if you notice that your coach's right rear indicator isn't working?

- ☐ Continue your journey using arm signals
- ☐ Get it repaired before continuing
- ☐ Get it repaired on your return to the depot
- ☐ Get your passengers to their destination and then repair it

All lights must be in working order, even in daylight. A faulty right rear indicator could cause a serious incident if another motorist attempts to overtake as you change lanes or turn right.

8.85 Mark one answer DBC s3

Before each journey, you should check all warning lights. What should you do if a warning light remains lit?

- ☐ Report the fault when you return
- ☐ Have the fault checked before setting off
- ☐ Have the fault checked at the next service
- ☐ Ignore it until the fault shows up

Many buses and coaches have a large panel of warning lights on the dashboard. A system check built into the ignition system will allow you to check that all the warning-light bulbs are working before you start your journey. If a warning light stays on, there's a fault. Have it checked before continuing. Remember, your passengers' lives are in your hands.

8.86 Mark one answer DBC s2

Why should you check your tyres more frequently on a coach with three axles?

- ☐ Punctures can be more difficult to detect
- ☐ Air pressure is more easily lost
- ☐ The wheels will need balancing more often
- ☐ You've no room for a spare wheel

Tyre checks should be made as a matter of routine. Inspect both the inside and outside walls, as well as the treads, for signs of wear, damage, bulges, separation and exposed cords.

8.87 Mark one answer DBC s2

Some buses have different-sized wheels on the front and rear. What do you need to be aware of when driving at high speeds on long journeys?

- ☐ The tyres on the larger wheels are more likely to overheat
- ☐ The tyres on the smaller wheels are more likely to overheat
- ☐ The tyres on the larger wheels are more likely to lose pressure
- ☐ The tyres on the smaller wheels are more likely to lose pressure

Because small wheels need to rotate faster, their tyres are more likely to overheat on long journeys at speed. Make sure you check them when you take rest stops.

8.88 Mark one answer DBC s3

How frequently should a driver carry out a walk-round check?

- ☐ Daily
- ☐ Weekly
- ☐ Every 100 miles
- ☐ Every 1000 miles

Drivers should carry out a walk-round check daily, as part of a routine. They have a legal responsibility to take all reasonable precautions to ensure that legal requirements are met before they drive any vehicle.

8.89 Mark one answer DBC s4

What should you do before stopping a turbocharged engine?

- ☐ Release the air-suspension valve
- ☐ Allow the engine to idle
- ☐ Select reverse gear
- ☐ Rev the engine sharply

You should always allow engines fitted with turbochargers to idle for about a minute before stopping the engine. This prevents the turbine bearings from being starved of oil.

8.90 Mark one answer DBC s3

A passenger tells you they've noticed that a wheel nut is missing from your bus. How often should you check your vehicle's wheel nuts?

- ☐ At the end of every week
- ☐ At the start of every week
- ☐ Every day before starting out
- ☐ Only at every service interval

It's essential to make sure that all wheel nuts are tightened with a correctly calibrated torque wrench. The wheel nuts should be checked every day before you start your journey. If a passenger tells you they've noticed that a wheel nut is missing, check it immediately. If there's a problem, don't move off until it's been rectified.

8.91 Mark one answer DBC s4

What should you do when you park a vehicle that's fitted with a turbocharger?

- ☐ Rev the engine up, then switch it off
- ☐ Switch the engine off immediately
- ☐ Allow the engine to idle, then switch it off
- ☐ Switch the engine off and on repeatedly

An engine fitted with a turbocharger should be allowed to idle for about a minute before it's switched off. This prevents the turbocharger bearings from being starved of oil.

8.92 Mark one answer DBC s4, HC r274–275

You're driving at night on a very busy road when your headlights fail. What should you do if the fuse box is on the outside of the bus, on the right-hand side?

- ☐ Ask a passenger to watch for traffic
- ☐ Drive on without lights
- ☐ Fix the problem yourself
- ☐ Wait for the breakdown services

Fast-moving traffic is passing close by, so wait for the breakdown services to arrive. Their vehicle will provide protection while your vehicle is repaired or recovered.

8.93 Mark one answer DBC s2,4

Your coach often tows a trailer. How often should you check the trailer tyres for pressure?

- ☐ At least once a week when they're cold
- ☐ At least once a month when they're hot
- ☐ At least once a week when they're hot
- ☐ At least once a month when the trailer is fully laden

Just like your bus tyres, your trailer tyres must be in good condition. You might not tow a trailer on every journey you make, but when you do, check the trailer tyres while they're cold and make sure they're suitable for the load they'll carry.

8.94 Mark one answer DBC s2

The bus you're driving is fitted with automatic transmission. When would you use kickdown?

- ☐ When stopping in an emergency
- ☐ When changing to a higher gear
- ☐ When driving at slow speed
- ☐ When needing brisk acceleration

The kickdown facility on automatic transmission allows a lower gear to be engaged to allow faster acceleration (for example, when overtaking). This is achieved by firmly pressing the accelerator to the floor.

8.95 Mark one answer DBC s2

Your coach is fully laden. What's the most likely reason for the steering feeling heavy?

- ☐ An icy road
- ☐ A burst rear tyre
- ☐ Faulty power steering
- ☐ Too many passengers

All modern vehicles have some form of power-assisted steering, which will be able to cope with everything the vehicle requires. If the steering feels heavy, the power-steering system could be faulty and needs to be checked before you continue your journey.

8.96 Mark one answer DBC s2

Why are frequent tyre checks advisable on a tri-axle double-deck vehicle?

- ☐ These tyres are more likely to deflate
- ☐ Punctures can be difficult to detect
- ☐ Blow-outs are more common on these vehicles
- ☐ Tyre pressures are difficult to maintain

On multi-axled vehicles, the inside wheels on the rear twin-axle are the most difficult to check for punctures and damage. A deflated tyre will transfer the weight to the second tyre of the twin-axle pair. This will overload it and could cause it to burst.

8.97 Mark one answer DBC s2

When should the driver of a coach wear gloves?

- ☐ To load and stow passengers' luggage
- ☐ To operate a lift for disabled passengers
- ☐ To handle the fares
- ☐ To top up the oil or water levels

Even the best-maintained vehicles accumulate a certain amount of grime around the engine. If this comes into contact with skin, it can cause irritation and may lead to conditions such as dermatitis. By wearing gloves, you can also keep your hands and cuffs clean to present a smart appearance to your customers.

8.98 Mark one answer DBC s2

You're driving a six-wheeled double-deck bus. What do you need to know about punctures on the rear tyres?

- ☐ They're much easier to detect
- ☐ They're more likely to happen
- ☐ They're more difficult to detect
- ☐ They're less likely to happen

The handling of a six-wheeled bus or coach is not much different from that of a two-axle vehicle, except that punctures can be more difficult to detect.

8.99 Mark one answer DBC s2

What should you regularly check on the tyres of your bus?

- ☐ The tyre manufacturer's details are displayed
- ☐ The tyre pressures are all correct
- ☐ The dust caps are all in place
- ☐ The tread depths are equal on all tyres

Tyre care is especially important. You should routinely check tyre pressures and general tyre condition. An unroadworthy vehicle will endanger the lives of your passengers.

8.100 Mark one answer DBC s2

What will result from having over-inflated tyres?

- ☐ You'll have improved acceleration
- ☐ The tyres will wear more quickly
- ☐ The tyres will give better grip
- ☐ The fuel consumption will be reduced

Your tyres must be in good condition and properly inflated to the manufacturer's recommended pressure. Research has shown that the tyres most likely to fail are those that have been over-inflated.

8.101 Mark one answer DBC s2

What should you do before uncoupling your trailer?

- ☐ Disconnect the battery
- ☐ Park on a firm, level surface
- ☐ Park in a well-lit location
- ☐ Unload some of the cargo

Before uncoupling a trailer, always make sure that you've parked on firm, level ground and applied the trailer parking brake.

Section nine

Leaving the vehicle

In this section, you'll learn about

- parking
- leaving the cab
- health and safety
- security.

Leaving the vehicle

Parking

When pulling up on the left-hand side of the road, be aware of pedestrians. They may be close to the edge of the road and your nearside mirror could hit them on the head.

DBC s4 **DGV** s1, 4

If your coach is fitted with air suspension, don't park too close to another vehicle or obstruction. The coach can move a considerable distance when you start the engine and the air suspension units inflate, and this movement could cause a collision.

DBC s4

If you're a bus or coach driver, there are four occasions when you can use your hazard warning lights

- when you're temporarily obstructing traffic
- to warn of an obstruction on a motorway or unrestricted dual carriageway
- when the vehicle has broken down
- if you're driving a school bus, when you're stationary and children are getting on or off.

DBC s1 **HC** r116

Leaving the cab

Before you leave the vehicle, make sure that

- you're parked in a safe place
- your parking brake is on
- the engine is switched off.

It's an offence to leave your vehicle unattended with the engine running and/or without applying the parking brake.

Before you open the right-hand (offside) door

- check your mirror and blind spot so that you don't put yourself, pedestrians or passing traffic in danger when you open the door
- climb down facing the vehicle, using the footholds and hand grips.

DBC s3 DGV s3 HC r123

Find information about high-visibility clothing and preventing falls from vehicles at this website.

hse.gov.uk/workplacetransport/factsheets.htm

hse.gov.uk/workplacetransport/publications.htm

Health and safety

When you're loading or unloading the vehicle, wear a high-visibility vest so that you're clearly visible.

DBC s3 DGV s1, 3

When lifting anything heavy, use **approved manual-handling techniques** to reduce your risk of injury.

DBC s1 DGV s1

Definition

approved manual-handling techniques
Approved methods for reducing the risks associated with moving a load manually.

Find out more information about health and safety in road haulage at this link.

hse.gov.uk/pubns/indg379.pdf

When returning to your vehicle, or taking over a different one, make sure that you carry out all routine safety checks.

DBC s2 DGV s2

When you unload from the platform of a flat-bed lorry

- use a suitable set of steps to reach the ground
- look out for danger from other vehicles, especially in a busy yard or loading bay.

DGV s3

When you open the tank hatches of a tanker, be careful of slippery walkways, overhead cables and overhead pipeways.

DGV s2

Security

When you arrive at your destination, passengers who will be returning to your vehicle later should take their personal property with them, unless it can be locked in a secure luggage compartment.

DBC s3

You're responsible for your vehicle and its load. If you're carrying a high-value load such as cigarettes, be aware of your personal safety.

TIP

Watch for anything unusual, such as people who seem to be taking a keen interest in you or your load.

Get security advice for commercial vehicles at this website.

rha.uk.net/training/compliance/vehicles/vehicle-security

Read more about bus and coach security best practice at this website.

www.gov.uk

For added security, think about having roof markings on your vehicle. These help police air-support units to identify stolen vehicles.

DGV s3

Meeting the standards

You must be able to

check for oncoming cyclists, pedestrians and other traffic before opening your door

minimise the risk of falling when you're exiting the vehicle

select a safe, legal and convenient place to stop and park. Once stationary, secure the vehicle on gradients, facing both up and down slope, as well as on the level

maximise the security of the vehicle and its load against theft or illegal access whenever you leave it. Check the security of the vehicle when you return.

You must know and understand

what factors to take into consideration when identifying a

- safe
- secure
- legal
- convenient

place to stop or park

that vehicles with air suspension may move a considerable amount when parked or when started

methods for securing different types of vehicle against theft or illegal access.

Notes

You can use this page to make your own notes or diagrams about the key points you need to remember.

Think about

- Am I aware of the obstruction caused by my left-hand mirror?
- How far does my vehicle move when the air suspension inflates?
- Do I know where my blind spots are?
- Have I received training in the approved manual-handling techniques for the loads I carry?
- If I carry a high-value load, do I know how to be safe and stay aware of potential risks to myself or my load?

Your notes

Things to discuss and practise with your instructor

These are just a few examples of what you could discuss and practise with your instructor. Some are for PCV, some are for LGV and some are for both. Read more about leaving the vehicle to come up with your own ideas.

Discuss with your instructor

- your safety when leaving the bus through an offside door
- responsibility for passengers' belongings when parked
- your safety when unloading luggage from the bus.

- your safety when dismounting from the cab at the roadside
- where you should park a semi-trailer.

- what you must do before leaving the vehicle
- when you should use hazard warning lights.

Practise with your instructor

- a safe technique for lifting heavy items
- the routine for leaving your vehicle.

9.1 Mark one answer DGV s4

What should you be careful of when stopping on the left in a busy place?

- ☐ There's good access to unload
- ☐ You've disconnected all the air lines
- ☐ Your nearside mirror doesn't strike the head of a pedestrian
- ☐ You change your tachograph mode

When stopping on the left, you should always be aware of pedestrians, particularly if they're close to the edge of the road. The height of your nearside mirror can vary depending on the size and type of vehicle you're driving. If in doubt, approach these situations with caution and stop if necessary.

9.2 Mark one answer DGV s3

You've just parked a lorry on the side of the road. What should you do just before dismounting from the cab?

- ☐ Make sure the radio is turned down
- ☐ Check the rear-view mirrors
- ☐ Make sure the hazard warning lights are on
- ☐ Check that all the windows are closed

Getting out of the cab on the offside, directly into the road, can be hazardous – especially if traffic is travelling at speed. Use your mirrors to check that it's safe to leave the vehicle. Be sure to use all proper footholds and hand grips when you leave the cab.

9.3 Mark one answer DGV s3

Where are you allowed to park a long, rigid vehicle?

- ☐ At a pedestrian crossing
- ☐ At a bus stop
- ☐ At a school entrance
- ☐ At a service area

It's important that you always select a place to park that's both safe and legal. Don't park on or too close to a pedestrian crossing or a school entrance; this would restrict the view for drivers and pedestrians. Don't park at a bus stop either, because buses would then have to stop in the road, creating a hazard for other traffic and for passengers getting on or off the bus.

9.4 Mark one answer DGV s3

Where should you park a semi-trailer if you plan to leave it unattended?

- ☐ On a residential street
- ☐ On a steep hill
- ☐ In a lorry park
- ☐ In a lay-by

If you need to park and leave a semi-trailer, find a safe place. Don't park it in a lay-by or anywhere it could be an obstruction or a danger. Leave lay-bys for drivers who wish to stop and rest. Find a place off the road – preferably a lorry park or somewhere safe that will also reduce the risk of theft.

9.5 Mark one answer DGV s3

Which of the following is important when you're getting out of a lorry cab?

- ☐ Checking your load is secure
- ☐ Using the mirrors
- ☐ Turning off the radio
- ☐ Disconnecting the air lines

When getting out of your lorry cab, it's essential that you check it's safe to dismount by checking your mirrors and looking out for traffic. When you're safely down, it's then a good idea to walk round and check your tyres, load, lights, brake lines, electrical connections, etc.

9.6 Mark one answer DGV s2

You're the driver of a fuel tanker. What danger should you be aware of when opening the tank hatches?

- ☐ Low air pressure
- ☐ Speed limiters
- ☐ Slippery walkways
- ☐ Emergency air lines

Take your time when you're using walkways at high levels. Spilt fuel can make the surface slippery and increase the risk of falling.

9.7 Mark one answer DGV s3

You're working on the platform of a flat-bed lorry. What's the safest way to get down to the ground?

- ☐ Use a suitable set of steps
- ☐ Jump down wearing non-slip shoes
- ☐ Use ropes to lower yourself down
- ☐ Climb down facing away from the vehicle

Every year, many people are seriously injured by falling from vehicles. Using suitable steps will lessen the chance of falling. Be aware of the dangers when loading or unloading vehicles.

9.8 Mark one answer DGV s3

What do you need to consider when transporting a high-value cargo of cigarettes?

- ☐ There'll be a risk of theft or hijack
- ☐ There'll be more Revenue and Customs checks
- ☐ You'll be allowed to take fewer rest breaks
- ☐ You'll need to maintain your regular route

If you're carrying a high-value load, you should consider your personal safety. There's always the possibility of an attempted theft. Watch out for and report anything unusual, such as people taking a special interest in you or your vehicle.

9.9 Mark one answer DGV s3

What should you consider having on your vehicle to combat theft?

- ☐ Roof markings
- ☐ Diff-lock
- ☐ Air horns
- ☐ Tinted windows

Roof markings on lorries can help police air-support units to identify stolen vehicles. Fleet operators, particularly those that regularly carry vulnerable or dangerous loads, are encouraged to use these markings.

9.10 Mark one answer DGV s3, 4

When should you use your mirrors?

- ☐ As you signal
- ☐ To check the blind spot
- ☐ Occasionally
- ☐ Before opening your door

You must be aware of the position of other road users, so that you know how they may be affected by your actions. Before opening your door, it's important to check the mirrors, as well as looking around for passing or approaching vehicles and pedestrians.

9.11 Mark one answer HC r239

What must you do before you leave your vehicle?

- ☐ Empty the air tanks
- ☐ Adjust your mirrors
- ☐ Apply the parking brake
- ☐ Check that the warning lights are working

The parking brake must always be set whenever you leave the vehicle. It's an offence not to leave your vehicle properly secured.

9.12 Mark one answer HC r239

What should you do before leaving your vehicle parked on a public road?

- ☐ Remove your tachograph chart
- ☐ Switch off the engine
- ☐ Adjust the mirrors
- ☐ Put the gear lever into reverse

It's an offence to leave a vehicle unattended on a public road with the engine running. Take care if you have to open the cab door directly into the road. Look for passing traffic – especially cyclists or motorcyclists – that may not be visible in your mirrors.

9.13 Mark one answer HC r239

What should you be aware of before opening your cab door?

- ☐ Vehicles passing near the door
- ☐ The height of your cab from the ground
- ☐ Loose grab rails near the door
- ☐ People crossing the road behind you

It can be dangerous for vehicles passing close by if you open the door carelessly. Always look properly to make sure that it's safe – checking the blind spots as well as using the mirrors – before you get out of the cab.

9.14 Mark one answer HC r116

When should you use hazard warning lights?

- ☐ To warn other drivers that you're towing
- ☐ When approaching queuing traffic on a motorway
- ☐ When parked illegally on a busy road
- ☐ To thank a driver for giving way to you

The only time you may use hazard warning lights while driving is when you're on an unrestricted dual carriageway or motorway and you want to warn other drivers of a hazard or obstruction ahead. Only use them for long enough to make sure your warning is seen.

9.15 Mark one answer HC r239

You need to stop and get out of your vehicle. When should you use the parking brake?

- ☐ After the service brake has been released
- ☐ Only if you park on a gradient
- ☐ Whenever you leave the vehicle
- ☐ After the air pressure has dropped

It's an offence to leave your vehicle unattended at any time without applying the parking brake.

9.16 Mark one answer HC r239

You've parked on the roadside. What must you do before leaving the vehicle?

- ☐ Reset the retarder
- ☐ Stop the engine
- ☐ Switch off the tachograph
- ☐ Remove your personal items

Before leaving your vehicle parked, you must stop the engine. The parking brake must be set and the vehicle should be safe and secure when you leave the cab.

9.17 Mark one answer — HC r116

When may hazard warning lights be used while driving?

- ☐ To thank a driver who has let you move back to the left after overtaking
- ☐ As a warning to drivers that you're towing another vehicle
- ☐ To show you want to go ahead when your position suggests otherwise
- ☐ When on motorways or dual carriageways, to warn drivers behind of a hazard ahead

Use your hazard warning lights on a motorway or unrestricted dual carriageway to warn following drivers of an obstruction or hazard ahead. This will alert drivers who may not be able to see the hazard due to the size of your vehicle.

9.18 Mark one answer — DBC s1, DGV s1

What should you try to do when you lift a heavy box or suitcase?

- ☐ Lift and twist together
- ☐ Look down all the time
- ☐ Lean sideways and lift
- ☐ Look ahead when the load is secure

Once you have a secure hold on the object, keep your head up and bend your knees.

9.19 Mark one answer — DBC s1, DGV s1

What's the best way to lift a heavy object?

- ☐ Twist your back while lifting
- ☐ Have a stable position
- ☐ Hold the load at arm's length
- ☐ Lift the load as quickly as possible

To lift a load manually, your feet should be apart, with one leg slightly forward to help you keep your balance. Place one foot alongside the load if it's on the ground. Be prepared to move your feet to keep a stable position.

9.20 Mark one answer — DBC s1, DGV s1

How should you stand when you're lifting a heavy object?

- ☐ Leaning sideways
- ☐ In a stable position
- ☐ With one leg straight
- ☐ Keeping your feet together

Always think of your own and others' physical safety before lifting any heavy object. Plan the lift and move any other items out of the way. Decide whether you need assistance.

9.21	Mark one answer	DBC s1

When can a bus driver use hazard warning lights?

- ☐ When children are getting off a school bus
- ☐ To thank a driver who has given way
- ☐ When stopping at a pedestrian crossing
- ☐ When parking a bus on yellow lines

All drivers may use hazard warning lights to warn other road users of danger on specific occasions. When you're driving a school bus, you may also use these lights when you've stopped and children are getting on or off the bus.

9.22	Mark one answer	DBC s4

When do you need to take account of your vehicle having air suspension?

- ☐ When you're passing animals or horse riders
- ☐ When you're travelling up a steep gradient
- ☐ When you're parked close to another vehicle
- ☐ When you're driving in falling snow

Vehicles fitted with air suspension can sometimes move a considerable amount when the airbags are inflated. If you're parked too close to another vehicle or obstruction, this could result in collision damage.

9.23	Mark one answer	DBC s3

What should your passengers do with their belongings when you arrive at your destination?

- ☐ Place them on luggage racks
- ☐ Take everything with them
- ☐ Place them on the seats
- ☐ Leave them with you

Passengers shouldn't leave any personal property on the vehicle, unless it can be locked in secure luggage compartments.

9.24 Mark one answer DBC s3

What should you do before getting out of your bus?

- ☐ Apply the parking brake
- ☐ Change the destination board
- ☐ Put the hazard warning lights on
- ☐ Check the air-pressure gauges read full

Don't park where you'll cause obstruction or inconvenience to other road users. It's an offence to leave your bus with the engine running and/or without applying the parking brake.

9.25 Mark one answer DBC s4

What must you do before leaving your bus?

- ☐ Apply the parking brake
- ☐ Count the fares
- ☐ Reset the odometer
- ☐ Operate the fuel cut-off switch

Before you leave your bus, you must always apply the parking brake.

9.26 Mark one answer DBC s3, 4

What should you wear while unloading luggage from your coach?

- ☐ High-visibility vest
- ☐ Heatproof gloves
- ☐ Safety goggles
- ☐ Ear protectors

Very often, you'll load or unload luggage by the roadside. For your own safety, make yourself visible to other traffic.

Section ten
Vehicle loading

In this section, you'll learn about

- security of loads
- weight distribution
- transporting loads.

Vehicle loading

Security of loads

It's your responsibility to make sure that your load is secure and will stay secure and stable when braking or cornering and in emergency situations. Your load shouldn't come loose even if there's a tyre failure.

If you realise that any part of your load has become insecure, you **MUST** stop as soon as it's safe to do so. Re-secure your load before continuing. If this isn't possible, you **MUST** seek assistance.

Make sure that you're familiar with all the different kinds of load restraint. Use the correct securing method for the load. For example, you shouldn't use ropes to secure metal loads as sharp edges may fray the ropes. Use chains on metal loads but don't use iron or split-link chains as they're less reliable than solid link or steel chains.

Tubular loads in particular can move forward if you brake hard. This can damage or even demolish the headboard, with serious consequences. Make sure that you properly secure loads like these and position them in contact with the headboard to prevent movement.

Secure a cover or net over loaded skips during transportation. For loads that need sheeting, such as loose, dry sand, start at the back and overlap each new sheet at the edges, with the forward sheet on top. This will minimise wind resistance and stop the wind lifting the sheeting off the load.

How to tie a dolly knot

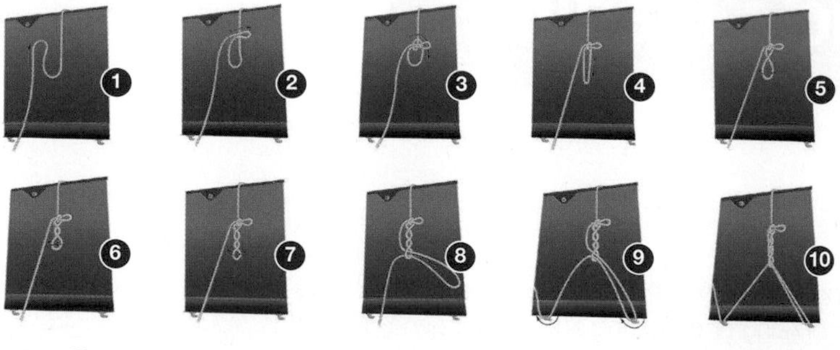

TIP

When roping down a sheeted load, it's best to use a dolly knot to secure the ropes.

When carrying spare sheets and ropes on an empty trailer, you **MUST** make sure they're tied down securely.

Ropes should

- be of three-strand construction
- be a minimum of 10 mm in diameter
- have ends that are spliced or otherwise treated to prevent fraying.

Wire ropes must be a minimum of 8 mm in diameter and free from damage and rust.

When using a ratchet strap to secure loads, make sure that you use the correct anchorage points built into the vehicle's main frame or chassis. **DO NOT** secure ratchet straps to rope hooks – they're not suitable and can distort, meaning the straps and the load could become unstable.

Only carry ISO containers on lorries or trailers with the appropriate securing points. These lock into the container body using the locking levers on the trailer.

Items on pallets, such as bricks, building supplies, etc, **MUST** be secured to the pallet, and the pallet **MUST** be secured safely to the vehicle. Make sure that your load distribution doesn't exceed your vehicle's axle weight limits.

DGV s2

For more advice on securing loads, visit this link.

www.gov.uk/government/publications/load-securing-vehicle-operator-guidance/load-securing-vehicle-operator-guidance

Weight distribution

When loading your vehicle, make sure that you spread the load evenly over the axles. You need to take into account

- the nature of the load
- the vehicle's suitability
- the stability of the load
- the type of restraint
- protection from weather
- prevention of theft
- prevention of damage to the load
- ease of delivery.

Find information about overloading at this link.

www.gov.uk/government/publications/hgv-overloading-the-basics

During multi-drop journeys, check the reducing payload after each drop. If necessary, redistribute and re-secure it on the vehicle. Make sure that the vehicle isn't top heavy, especially in windy weather.

When using tipper vehicles, find a firm, level area before tilting the vehicle body. The vehicle's **centre of gravity** will keep changing until the load is discharged.

DGV s2

Definition

centre of gravity

The point within an object at which gravity can be considered to act equally.

Transporting loads

Many load types have special training requirements, such as when transporting animals or hazardous goods.

DGV s1, 2

Vehicles that carry suspended meat carcasses can be affected by the **pendulum effect**. This can lead to wheel-lift or rollover when cornering or negotiating roundabouts.

Definition

pendulum effect

Where a hanging load swings in one direction, as a result of either braking or cornering, and then swings back in the opposite direction. This movement of the load has the potential to destabilise the vehicle.

Check the internal temperature of refrigerated vehicles from time to time to be sure it's correct so that your load isn't spoiled.

DGV s2

Some car transporters have a large overhang. When turning, there's a risk the vehicle may collide with

- lampposts and bus stops
- traffic signals and signs
- buildings and awnings
- other vehicles.

DGV s2

Tankers carrying liquid loads are fitted with baffle plates that stop the liquid from moving from side to side (known as the 'wave effect').

Adjust your speed well before cornering to avoid wheel-lift and rollover.

Tankers often carry hazardous goods. You **MUST** display diamond-shaped plates that detail any hazardous load you're transporting.

There are nine classes of hazardous goods. If you're carrying these goods, you must carry a valid **ADR vocational training certificate** with you at all times and be trained in how to deal with spillages.

Definition

ADR vocational training certificate
The certificate that must be held by drivers of vehicles carrying dangerous goods.

Drivers **MUST** also have with them all necessary protective clothing, safety equipment and any documentation relevant to the class of goods being carried. The appropriate firefighting equipment **MUST** be available and the driver **MUST** be trained in its use.

DGV s2

Find more information about the carriage of dangerous goods at this website.

hse.gov.uk/cdg/index.htm

Meeting the standards

You must be able to

make sure that, when lifting loads, you

- use safe manual-handling techniques
- make proper use of any equipment provided
- are suitably qualified

make sure that loads are secure and distributed correctly, depending on the vehicle

allow for the effect that loads have on the vehicle's stability

check the condition of the load at regular intervals.

You must know and understand

how to reduce the risk of injury when lifting loads

the risks associated with movement and even loss of a load

how to adjust the vehicle and your driving behaviour to allow for

- extra weight
- changed weight distribution

methods for protecting different types of loads.

Notes

You can use this page to make your own notes or diagrams about the key points you need to remember.

Think about

- Am I familiar with all the different types of load restraint?
- Can I choose the right restraint for different loads?
- Can I tie a dolly knot?
- Do I know the axle weight limits for my vehicle?
- Do I know how to evenly distribute a load?
- Do I display the correct hazard plate(s) for any hazardous loads I carry?
- Have I been trained in the carriage of hazardous loads and do I keep my certificate with me?
- Do I know what personal protective equipment (PPE) and firefighting equipment I should carry for any hazardous loads?

Your notes

Things to discuss and practise with your instructor

These are just a few examples of what you could discuss and practise with your LGV instructor. Read more about vehicle loading to come up with your own ideas.

Discuss with your instructor

- how to distribute a load over the axles of your lorry
- what an ISO container is and how it should be secured
- things to consider when driving a tanker, such as
  - the 'wave effect'
  - how your braking will affect the liquid that you're carrying
- symbols for various loads
  - compressed gases
  - corrosive materials
  - a load that's dangerous when wet
  - radioactive materials.

Practise with your instructor

- distributing loads evenly over the axles of your lorry
- how to sheet and secure a load
- coupling and uncoupling a trailer.

10.1 Mark one answer — DGV s2

How should you load a vehicle?

- ☐ Spread the load evenly
- ☐ Load it towards the rear
- ☐ Load it towards the front
- ☐ Arrange the load to make it easy for unloading

It's your responsibility to make sure that the vehicle's load is spread evenly, to avoid overloading individual axles. Overloading carries severe penalties for the driver and operator.

10.2 Mark one answer — DGV s2

You're going to drive an articulated lorry with an ISO container on the trailer. How should the container be secured to the trailer?

- ☐ The container must be secured by ropes
- ☐ The container must be secured using the locking levers
- ☐ The container must be held up against the headboard
- ☐ The container must be sealed

ISO (International Standards Organization) cargo containers should only be carried on lorries or trailers with the appropriate securing points. These are designed to lock into the container body.

10.3 Mark one answer — DGV s2

You're using three sheets to cover your load. Which is the correct overlap?

☐  ☐ 

☐  ☐ 

Keep a check in your mirrors as you're driving to make sure that the sheets are secure. Air can force itself under the sheets and work them loose.

10.4 Mark one answer DGV s2

You're covering a load using more than one sheet. Why should you start with the rearmost sheet, then work forwards?

- ☐ To stop you tripping when walking on the load
- ☐ To stop wind and rain from getting under the sheets
- ☐ To make it much easier to fold up the sheets
- ☐ To make it easier to carry longer loads

If the load is sheeted incorrectly, the wind may get underneath a sheet, making it flap about. This is dangerous, as it can catch unsuspecting pedestrians or cyclists, and it can also seriously reduce the driver's view of what's happening behind.

10.5 Mark one answer DGV s2, 7

What's the best knot to use when securing a load with ropes?

- ☐ A dolly knot
- ☐ A reef knot
- ☐ A slip knot
- ☐ A bowline knot

If a load is being secured with ropes, you must make sure it's tied securely to the body of the unit. The most effective method of tying is with 'dolly knots'. These are non-slip knots that hold firmly.

10.6 Mark one answer DGV s2

Why are ropes unsuitable to tie down a load of scrap metal?

- ☐ Ropes are hard to tie
- ☐ Ropes will loosen in rain
- ☐ Ropes are hard to untie
- ☐ Ropes can wear and snap

When securing a load, the driver must use the most suitable type of restraint. Scrap metal is likely to have sharp edges that could wear through straps or ropes. Security of the load is the driver's responsibility; a load that has been correctly secured shouldn't move if an emergency arises.

10.7 Mark one answer DGV s2

You're driving an open tipper lorry carrying loose dry sand. Why should you sheet this load?

- ☐ To stop the handling from being affected
- ☐ To stop the load from shifting
- ☐ To stop the load from blowing away
- ☐ To stop the tipper mechanism from accidentally operating

If you're carrying a load that consists of loose materials, it must be covered by suitable sheeting. You mustn't risk losing any part of your load. This could cause damage or injury to other road users, and you would be responsible.

10.8 Mark one answer DGV s2

What should you do if the load on your lorry becomes insecure?

- ☐ Continue at a slower speed to ensure the load doesn't fall off
- ☐ Attach 'hazard' boards to the load to warn other road users
- ☐ Park and resecure the load before continuing
- ☐ Inform base at the earliest opportunity

If you become aware that any part of your load is insecure, you must stop as soon as it's safe to do so. Resecure the load before continuing your journey. If this isn't possible, then you must seek assistance. Don't take risks.

10.9 Mark one answer DGV s2

Tankers carrying liquids can experience the 'wave effect'. What helps to reduce this problem?

- ☐ Spray guards
- ☐ Harsh braking
- ☐ Baffle plates
- ☐ Wind deflectors

Modern tankers are fitted with baffle plates inside the tank compartment. These help to minimise the movement of liquids and, therefore, reduce the 'wave effect'.

10.10 Mark one answer DGV s2

You're driving an articulated tanker on a straight road. How will the liquid load affect the vehicle as you brake to a stop?

- ☐ It will try to push the vehicle forward
- ☐ It will try to push the vehicle to the side
- ☐ It will make the trailer wheels bounce
- ☐ It will make the trailer wheels skid

When braking a vehicle carrying a liquid load, it's important to apply even pressure on the brake pedal. Don't relax the pedal pressure until the vehicle has stopped. Secure the vehicle with the parking brake before releasing the footbrake; this will minimise the risk of unintentional movement of the vehicle caused by 'surge' from the liquid load.

10.11 Mark one answer DGV s2

You're driving a tanker that's half full. The inside of the tank isn't divided into compartments. How should you use the footbrake when you're stopping?

- ☐ Keep pressure on the brake pedal until you've stopped
- ☐ Relax pressure on the brake pedal before you've stopped
- ☐ Pump the brake pedal rapidly
- ☐ Use the footbrake and parking brake together

When braking, always maintain steady pressure on the brake pedal until the vehicle has stopped. This helps to reduce the 'wave effect' that can build up as the liquid load moves when the vehicle changes speed.

10.12 Mark one answer DGV s2

Which type of load would benefit most from being carried on a lorry fitted with road-friendly suspension?

- ☐ Steel
- ☐ Timber
- ☐ Glass
- ☐ Cables

Fragile loads need extra care in loading and handling. The load must be secured using the most appropriate form of restraint. Air suspension reduces the vibration caused by the impact of the lorry wheels on road surfaces. This also reduces damage to the road, bridges and adjacent structures.

10.13 Mark one answer DGV s2

How wide does a load need to be before an attendant is required?

- ☐ 2.6 metres (8 feet 5 inches)
- ☐ 3.0 metres (9 feet 9 inches)
- ☐ 3.3 metres (10 feet 9 inches)
- ☐ 3.5 metres (11 feet 5 inches)

Wide loads are a hazard to other road users. In addition to having an attendant with you, you must notify the police if your load is wider than 3.5 metres (11 feet 5 inches). Side markers must also be used to show other road users the limits of your vehicle.

10.14 Mark one answer DGV s2

When is an articulated lorry most likely to jack-knife?

☐ When it's carrying a high load
☐ When its load is carried at the rear of the trailer
☐ When it's unloaded
☐ When it's fully loaded

When the trailer is loaded, the weight of the load presses the trailer tyres firmly onto the road surface, so improving the tyres' grip. Unladen, there is no load bearing down on the tyres, and under heavy braking the trailer wheels can lock or slide across the road surface, making the trailer more difficult to stop. Severe braking can result in jack-knifing: the trailer doesn't slow as quickly as the tractor unit and this causes it to pivot around the coupling.

10.15 Mark one answer DGV s6

You've finished recoupling. How should you make sure that the tractor and trailer are secure?

☐ Try to move forward with the trailer parking brake on
☐ Reverse with the trailer parking brake on
☐ Try to move forward with the trailer parking brake off
☐ Reverse with the trailer parking brake off

Make sure the locking mechanism is secure by selecting a low gear and attempting to move forward. Apply the parking brake again before leaving the cab. Connect the 'dog clip' to secure the kingpin release handle.

10.16 Mark one answer DGV s6

What must you check first when uncoupling or recoupling your trailer?

☐ The lights are working
☐ The tilt-cab mechanism is secure
☐ The trailer brake is applied
☐ The air lines are safely stowed

If the trailer begins moving while you're working on it, you could put yourself and others in great danger. You must make sure that it's properly secured by the brake before you start work.

10.17	Mark one answer	DGV s2

You're uncoupling a lorry and trailer. What should you do after disconnecting the electrical line?

- ☐ Stow it away safely
- ☐ Drive forward slowly
- ☐ Lower the landing gear
- ☐ Apply the trailer brake

It's important to stow away all your electrical cables and air lines safely to avoid causing injury to others. Connectors left lying around can be a trip hazard and can be damaged if they're run over by another vehicle.

10.18	Mark one answer	DGV s2

Your lorry has a demountable body. What should you check before demounting the body?

- ☐ The rear doors are open
- ☐ The legs are up
- ☐ The body is unloaded and empty
- ☐ The surface is firm and level

If you demount the body on a poor or soft surface, there's a danger of it sinking and becoming difficult to handle. Always think carefully before you demount the body and be sure that the site is suitable.

10.19	Mark one answer	DGV s2

How should you adjust your mirrors after recoupling your trailer?

- ☐ To give a full view of your load
- ☐ To give a clear view of both pairs of rear wheels
- ☐ To give a view down each side of the trailer
- ☐ To give a view of the road on the other side

The view you get in your mirrors may vary, depending on the size of the trailer and the load it carries. Always adjust your mirrors to ensure that you have the best possible view down each side of the trailer before you drive away.

10.20	Mark one answer	DGV s2

Which vehicle is most likely to be affected by 'vehicle bounce'?

- ☐ A long-wheelbase empty vehicle
- ☐ A short-wheelbase laden vehicle
- ☐ A short-wheelbase empty vehicle
- ☐ A long-wheelbase laden vehicle

A short-wheelbase empty vehicle will bounce more noticeably than some long-wheelbase vehicles. This can affect braking efficiency and all-round control. Don't be tempted to push this type of vehicle into bends or corners simply because the vehicle appears to be easier to drive.

10.21 Mark one answer DGV s2

Which type of load should be secured using chains?

- ☐ Timber planks
- ☐ Hay bales
- ☐ Steel plates
- ☐ Canvas sacks

The choice of restraint will depend on the load. For example, ropes are unsuitable for loads such as steel plates, as there's a danger of sharp edges shearing the ropes. For steel plates, chains must be used with a suitable tensioning device.

10.22 Mark one answer DGV s2

Which load is most likely to move forward with some force when you brake sharply?

- ☐ Heavy material in canvas sacks
- ☐ Loose sand
- ☐ Timber secured with dolly knots
- ☐ Tubular metal

When deciding which type of restraint to use, consider what may happen if you have to brake sharply. Tubular loads may move forward with some force if you have to brake sharply in an emergency. This could cause serious damage to the vehicle's headboard.

10.23 Mark one answer DGV s2

You're part-loading a lorry with an empty ISO container. Where should you position it on the trailer?

- ☐ Close to the fifth wheel
- ☐ Over the front axle
- ☐ Close to the trailer edge
- ☐ Over the rear axles

To increase stability and reduce the risk of the trailer wheels lifting when turning, it's preferable to locate part-loads over the rear axle(s).

10.24 Mark one answer DGV s2

How must spare sheets and ropes be carried on your trailer?

- ☐ Laid out flat
- ☐ Visible from the cab
- ☐ Tied down securely
- ☐ Stacked loosely

All spare sheets and ropes carried on the vehicle must be tied down securely to prevent them falling onto the road, where they may be a hazard for other vehicles.

10.25 Mark one answer DGV s2

Which articulated trailer is most at risk of jack-knifing?

- ☐ A loaded flat-bed trailer
- ☐ A laden fuel tanker
- ☐ A loaded high-sided trailer
- ☐ An unladen trailer

Jack-knifing occurs when the tractor unit is pushed by the semi-trailer pivoting around the coupling (fifth wheel). This is more likely to occur with an unladen vehicle.

10.26 Mark one answer DGV s2

In what circumstances will short-wheelbase vehicles bounce more noticeably than long-wheelbase vehicles?

- ☐ When laden
- ☐ When turning
- ☐ When empty
- ☐ When travelling slowly

Short-wheelbase vehicles will bounce more noticeably than some long-wheelbase vehicles when empty. This can affect braking efficiency and all-round control.

10.27 Mark one answer DGV s2

What's the main cause of a lorry shedding its load?

- ☐ Driving on motorways
- ☐ Driving too fast
- ☐ Driving over a level crossing
- ☐ Driving in wet weather

The main causes of lorries shedding their loads are driver errors, such as sudden changes of speed or direction, driving too fast for the conditions, and harsh braking.

10.28 Mark one answer DGV s2

What could happen if you drive a double-deck lorry with the top deck loaded and the lower deck empty?

- ☐ The lorry will become unstable under normal braking
- ☐ The brakes will be less effective
- ☐ The lorry may overturn when cornering
- ☐ You'll need to change gear more often

If you're only carrying half a load on a double-deck lorry, this should be carried on the lower deck whenever possible. With only the top deck loaded, your lorry is top-heavy and at risk of overturning as you drive around bends or corners.

10.29 Mark one answer DGV s2

Your lorry has a double-deck body. The top tier is loaded and the lower deck is empty. When will the vehicle be most at risk of overturning?

- ☐ In thick fog
- ☐ In high winds
- ☐ In hot weather
- ☐ In heavy rain

A lorry loaded like this will be top-heavy and more likely to overturn in high winds or when cornering. If you're only carrying half a load, you should try to carry it on the lower deck.

10.30 Mark one answer DGV s2

You're carrying another vehicle piggyback on your lorry. Chocks should be used to secure the wheels. What else should you use to secure the load?

- ☐ Baffle plates
- ☐ Axle stands
- ☐ Restraining straps
- ☐ A scaffolding bar

You should use a suitable restraint and chock the wheels when another vehicle is being carried piggyback. Don't just rely on the other vehicle's parking brake.

10.31 Mark one answer DGV s2

The lorry you're driving is heavily laden. How will the load affect the vehicle if you approach a left-hand bend too fast?

- ☐ The load will push the lorry to the left
- ☐ The load will pull the lorry to the right
- ☐ The load will push the lorry straight on
- ☐ The load will hold the lorry back

The forces acting on a lorry will try to move it in a straight line as it enters a bend. The greater the weight, the greater the force. If you're going too fast, there's a danger that you could lose your load or tip over.

10.32 Mark one answer DGV s2

You're driving a vehicle with an unladen trailer. What could happen if you change down to a low gear while travelling at speed?

- ☐ Your vehicle may suddenly accelerate
- ☐ The endurance brake will come on
- ☐ You won't be able to brake
- ☐ You could lose control of your vehicle

Changing down to an inappropriate gear at too high a speed will cause your vehicle to slow down suddenly. With no separate braking for the trailer, the tractor could be pushed to the point where it turns over.

10.33 Mark one answer DGV s2

Your lorry has a crane fitted. You're loading very heavy items. What should you do when you feel that the ropes or straps may break?

- ☐ Reposition the load
- ☐ Use chains and tensioners
- ☐ Tie two straps together
- ☐ Continue loading carefully

If there's any danger that the load you're lifting with a crane is too heavy for ropes or straps, or that sharp edges on the load may damage them, you should use chains with compatible tensioners.

10.34 Mark one answer DGV s2

When you're carrying other vehicles piggyback, you should use restraints and the parking brake on each vehicle. What else should you do?

- ☐ Make sure the vehicles are sheeted
- ☐ Place chocks under the wheels
- ☐ Put the heavy vehicles at the top
- ☐ Rope the vehicles together

Vehicles being carried piggyback must always have chocks applied to their wheels, in addition to a restraint. Never rely on just their parking brake.

10.35 Mark one answer DGV s2

When will an articulated car transporter be least stable?

- ☐ When only the lower deck is loaded
- ☐ When only the top deck is loaded
- ☐ When it's fully laden
- ☐ When it's unladen

Keeping the centre of gravity as low as possible will improve the handling of a car transporter. Top-heavy loads are more unstable and require more care when turning and cornering.

10.36 Mark one answer DGV s2

What do the legs on a demountable body enable you to do?

- ☐ Load and unload the body without stopping
- ☐ Stack one body on top of another
- ☐ Alter the overall height of the vehicle
- ☐ Demount the body without a crane or lift

Demountable bodies have legs that can be lowered so that the carrier vehicle can be driven out from underneath the body. This has an advantage over ISO containers, which require some form of crane or lift to remove the container from the trailer.

10.37 Mark one answer DGV s2

What shape are the labels that indicate a vehicle is carrying hazardous cargo?

☐ Diamond

☐ Triangle

☐ Circle

☐ Oval

Hazardous substances in transit are readily identified by a diamond-shaped warning label. This label carries other information to identify the type of hazard more specifically, such as

- flammable gas
- toxic gas
- corrosive agents.

These are just some of the types of hazard encountered. Strict regulations apply to the storage and carriage of these goods. All drivers of this type of load must receive specific certificated training.

10.38 Mark one answer DGV s2

Why are baffle plates fitted to tankers carrying liquids?

☐ To reduce wind resistance

☐ To reduce the wave effect

☐ To stop the brakes from locking

☐ To make the steering lighter

Fluids carried in tankers can move in waves as the vehicle's speed or direction changes. This happens particularly when you brake and then ease off the pedal; the moving wave of liquid causes the vehicle to surge forward. Baffle plates are designed to reduce this surging effect.

10.39 Mark one answer DGV s2

Which of these vehicles will be most at risk of 'roll-over' when laden?

☐ 

☐ 

☐ 

☐ 

'Roll-over' usually happens as a result of the inside rear wheels of an articulated vehicle starting to lift when the driver changes direction sharply. This tends to happen when a driver is changing direction to leave a roundabout. If the load moves during the change of direction, the vehicle is increasingly at risk of rolling over. The problem often involves vehicles carrying fluids in bulk.

10.40 Mark one answer — DGV s2

You're unloading an end-tipper lorry. What should you do before tipping the body?

- ☐ Make sure the vehicle is on a firm, level surface
- ☐ Park facing uphill to make unloading easy
- ☐ Make sure the vehicle is on a soft, sloping surface
- ☐ Park downhill for easier unloading

Unless you're on firm, level ground, there's a risk of your lorry overturning when the body is tipped. Before raising the body, check that you're well clear of overhead obstructions such as power lines.

10.41 Mark one answer — NI EXEMPT — DGV s2

Who must you inform if the load on your trailer hits a railway bridge?

- ☐ The local authority
- ☐ DVLA
- ☐ Highways England
- ☐ The railway authority

If your vehicle hits a bridge, you must report it to the police. If a railway bridge is involved, you must also report it to the railway authority. The phone number is usually shown on or near the bridge.

10.42 Mark one answer — DGV s2

Which of these best describes a vehicle's payload?

- ☐ The maximum load the vehicle can carry
- ☐ The maximum load over each axle
- ☐ The maximum load plus the weight of the vehicle
- ☐ The maximum load each tyre can take

The formula used to calculate the payload of a vehicle is the maximum authorised mass (MAM) minus the tare weight. The tare weight is the total weight of the vehicle, including the crew and extra equipment.

10.43 Mark one answer — DGV s1, 2

What additional training do you need before transporting frozen foods?

- ☐ Packaging procedures
- ☐ Weight-distribution procedures
- ☐ Waste-handling procedures
- ☐ Hygiene procedures

When transporting frozen food, you need to know how to operate your refrigeration unit correctly. It's also essential that you're aware of correct hygiene procedures when handling food.

10.44 Mark one answer DGV s2

You've parked your vehicle on level ground to deliver a load of building materials. What should you do before using the vehicle-mounted crane?

- ☐ Switch the engine off
- ☐ Turn the steering wheels to the left
- ☐ Lower the stabilising legs
- ☐ Set out warning cones

Before using the crane, it's vital to make sure the vehicle is parked on level ground and the stabilising legs are lowered. This will reduce the risk of the load becoming unstable during lifting. Also check for any overhead cables that may obstruct the crane.

10.45 Mark one answer DGV s2

You're loading goods of varying weights. How should they be distributed over the width of the vehicle?

- ☐ Heavy items at the front, light items at the rear
- ☐ Light items near the centre line, heavy items towards the sides
- ☐ Heavy items near the centre line, light items towards the sides
- ☐ Light items at the front, heavy items at the rear

To achieve maximum stability, the load should be placed to keep the centre of gravity as low as possible. To do this, heavy items should be placed close to the centre line and spread over the full length of the vehicle. Lighter items should be placed along the sides.

10.46 Mark one answer DGV s2

What should you do when you're working on a vehicle platform?

- ☐ Make sure you can jump down
- ☐ Walk forward near the edges
- ☐ Work in stockinged feet
- ☐ Walk backwards near the edges

When working on a flat-bed or open curtain-sided vehicle, you should always be aware of your proximity to the edge of the platform. It's too easy to concentrate on moving and positioning the load, and lose sight of your own position. Stepping backwards could be fatal, so always make sure you're facing the edge as you approach it.

10.47 Mark one answer DGV s1, 2

Which of these loads needs to be transported at a controlled temperature?

- ☐ Frozen foods
- ☐ Timber
- ☐ Bulk grain
- ☐ Cement

Some loads need to be transported in special conditions. For example, chilled or frozen foods need to be transported in a refrigeration unit. To drive one of these units, you'll need to be trained in how to use it. You'll also need relevant training in hygiene procedures.

10.48 Mark one answer DGV s1, 2

What will you need to transport livestock on journeys of less than 65 km (40 miles)?

- ☐ Practical experience in the care of animals
- ☐ Membership of the RSPCA
- ☐ No driving convictions
- ☐ The ability to drive abroad

The type of training required by drivers carrying livestock varies according to the distance, journey duration and species involved. For journeys of less than 65 km (40 miles), no formal qualification is required but you must have practical experience in the care of animals being transported.

10.49 Mark one answer DGV s1, 2

What should you do when transporting a small quantity of livestock that doesn't fill the vehicle?

- ☐ Allow no direct access to the animals
- ☐ Check them for disease before you load
- ☐ Give the animals plenty of space
- ☐ Create compartments using moveable panels

When transporting livestock, it's important to limit the amount of space they have to move around. If you allow them to move around freely, it could increase the risk of injury to the animals. It will also affect the stability of the vehicle when braking and cornering, and could even cause the vehicle to tip over.

10.50 | **Mark one answer** | **DGV s2**

Your vehicle has a maximum authorised mass of 40 tonnes. The kerbside weight is 15 tonnes. What's your maximum payload?

- ☐ 15 tonnes
- ☐ 25 tonnes
- ☐ 35 tonnes
- ☐ 45 tonnes

Overloading a vehicle is dangerous. Your stopping distance will increase, and your vehicle could become unstable and difficult to control. You'll need to work out the payload. Do this by taking away the kerbside weight from the maximum authorised mass (MAM). This will give you the maximum weight you can carry.

10.51 | **Mark one answer** | **DGV s2**

You're making several deliveries. What problems may the decreasing payload cause?

- ☐ You might overload an axle
- ☐ You'll always have heavy items remaining
- ☐ You might exceed your kerbside weight
- ☐ You'll damage the rest of the load

As items are unloaded from the rear of the vehicle, weight will be transferred to the front axle. Take care that this axle doesn't become overloaded. You may need to redistribute the load.

10.52 | **Mark one answer** | **DGV s2**

What can result from overloading an axle?

- ☐ Reduced braking efficiency
- ☐ Reduced braking distance
- ☐ Increased kerbside weight
- ☐ Increased fuel efficiency

Too much weight on an axle can reduce braking efficiency and cause brake fade. As well as being dangerous, exceeding the axle weight limit is an offence that can result in prosecution.

10.53 | **Mark one answer** | **DGV s2**

What could be the result of overloading an axle while loading a lorry?

- ☐ Reduced tyre temperature
- ☐ Damage to the road surface
- ☐ Damage to the tachograph
- ☐ Increased tyre life

Overloading an axle has an impact on the environment, causing damage to road surfaces. You also risk a fine and driving penalties.

10.54 Mark one answer DGV s2

You're transporting a skip carrying loose waste. How should the material be carried?

- ☐ Covered
- ☐ Shrink-wrapped
- ☐ While displaying a warning beacon
- ☐ Under a waterproof membrane

The skip should be covered to prevent any part of the load from blowing away. Loose material blown onto the road could be a danger to other road users and can cause environmental damage.

10.55 Mark one answer DGV s2

You're securing a very heavy load with a ratchet strap. Which anchorage point is unsafe to use with ratchet straps?

- ☐ Rope hook
- ☐ Eye bolt
- ☐ Shackle
- ☐ 'D' link

Rope hooks aren't designed to withstand high forces. They're usually just welded or bolted to the underside of the platform. Many are so weak that they can easily be distorted by a ratchet buckle when a webbing strap is tightened. Secure anchorage points should be built into the main frame or chassis of the vehicle.

10.56 Mark one answer DGV s2

You're securing a load using chains. Which type of chain is unsafe for securing a load?

- ☐ Short-link
- ☐ Round-link
- ☐ Oval-link
- ☐ Split-link

Split-link and iron chains shouldn't be used to secure loads, as they're less reliable than solid-link and steel chains. For more information on the suitability of various-sized steel chains, consult the relevant British Standard (BS) leaflets.

10.57 Mark one answer DGV s2

You're securing a load using ropes. What's the minimum diameter of rope that should be used?

- ☐ 5 mm
- ☐ 10 mm
- ☐ 15 mm
- ☐ 20 mm

When using ropes, the ends should be spliced or otherwise treated to prevent fraying. The rope should be of at least three-strand construction, with a normal diameter of at least 10 mm.

10.58 Mark one answer DGV s2

You're loading timber onto a flat-bed lorry. You want to cover it with sheets. Which sheet should be positioned first?

- ☐ Front
- ☐ Middle
- ☐ Rear
- ☐ Side

When more than one sheet is used to cover and protect a load, the rear sheet should be positioned first. This is so that the overlaps face backwards, rather than forwards, preventing wind and rain from getting between the sheets.

10.59 Mark one answer DGV s2

What's the main reason for covering a skip that you're transporting?

- ☐ To prevent any spillage
- ☐ To stop children climbing in
- ☐ To keep the contents dry
- ☐ To prevent theft

Debris falling from vehicles can be very dangerous to other road users, so skips need to be covered when they're transported. Falling debris may also cause environmental damage.

10.60 Mark one answer DGV s1, 2

Which cargo needs to be transported at a controlled temperature?

- ☐ Perishable foods
- ☐ Barrels of beer
- ☐ Cement
- ☐ Silage

Some foods need to be transported at controlled temperatures. Drivers will need to be trained in the use of refrigeration units and correct hygiene procedures.

10.61 Mark one answer DGV s2

Which load may need to be transported at a controlled temperature?

- ☐ Chemicals
- ☐ Bulk grain
- ☐ Sugar
- ☐ Beer barrels

Some highly dangerous chemicals have to be transported at prescribed temperatures. Drivers must be fully trained in the use of these specially designed, temperature-controlled vehicles.

10.62 Mark one answer DGV s7

Which symbol on a lorry means it's likely to be carrying compressed gases?

☐  ☐ 

☐  ☐ 

Vehicles carrying dangerous or hazardous goods need to have markings on them that clearly identify the load. This could, for example, help the emergency services to deal with any incident quickly and safely.

10.63 Mark one answer DGV s7

What's being carried on a lorry displaying this symbol?

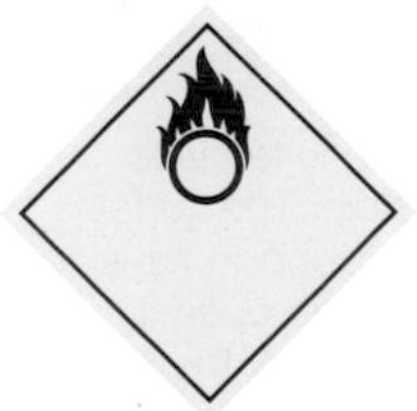

- ☐ Corrosive materials
- ☐ Compressed gases
- ☐ Oxidising agents
- ☐ Radioactive materials

The symbols on the back or side of a lorry should show the type of material that the vehicle will normally be carrying.

10.64 Mark one answer DGV s7

Which symbol on the back of a lorry means it's carrying a corrosive substance?

- [ ] 
- [ ] 
- [ ] 
- [ ] 

If you drive a vehicle carrying dangerous or hazardous material, you must make sure the correct symbol or mark is clearly visible.

10.65 Mark one answer DGV s7

Which symbol on a lorry shows its load is dangerous when wet?

- [ ] 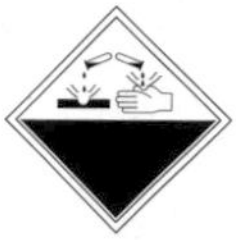
- [ ] 
- [ ] 
- [ ] 

If any vehicle displaying a warning symbol is involved in an incident, the consequences could be serious. The emergency services need to be aware of how a dangerous or hazardous material will behave in different situations; for example, when it's wet.

10.66 Mark one answer DGV s2

You're driving a loaded lorry with curtain sides. What should you do when you see the curtain bulging on one side?

- [ ] Check the load by carefully entering through the rear door or opposite curtain
- [ ] Check the load by carefully opening the curtain in the affected area
- [ ] Continue driving, as this is normal for these types of vehicles
- [ ] Continue driving at a much reduced speed

You must stop in a safe place as soon as possible. The curtain might be the only support for a slipped load, so don't open it before checking. Check the load by carefully opening the rear door or opposite curtain, if it shows no sign of bulging.

10.67 Mark one answer DGV s2

You're loading a curtain-sided vehicle. What are the curtains for?

- ☐ Restraint
- ☐ Protection from the weather
- ☐ Improved visibility
- ☐ To reduce noise

Unless the curtains are specifically designed, they must not be used to restrain the load. They should be used to protect it from the weather and provide a level of security. Most companies also use them as a mobile advertisement.

10.68 Mark one answer DGV s2

You're using a lorry-mounted crane to unload building materials. What safety features should you use?

- ☐ Wheel clamps
- ☐ Kingpin locks
- ☐ Jockey wheels
- ☐ Vehicle stabilisers

When using a lorry-mounted crane, stabilisers should always be used. All drivers should be fully trained and regularly tested on their competence in using stabilisers.

10.69 Mark one answer DGV s1

You intend to move a heavy object using a barrow or trolley. What's the best height for the handles?

- ☐ Between the shoulder and the waist
- ☐ As high as you can reach
- ☐ Between the knee and the waist
- ☐ As low as possible for comfort

When moving a load, make sure you don't have to bend down or stretch up to reach the handles. Also, try to push rather than pull, so you can see where you're going.

10.70 Mark one answer DGV s2

Your vehicle has a maximum authorised mass of 40 tonnes. The tare weight is 10 tonnes. What's your maximum payload?

- ☐ 20 tonnes
- ☐ 30 tonnes
- ☐ 40 tonnes
- ☐ 50 tonnes

To work out your vehicle's payload, use the following formula: maximum authorised mass (MAM), minus tare weight, equals payload. This is the maximum weight your vehicle can carry.

Section eleven

Restricted view

In this section, you'll learn about

- mirrors and blind spots
- awareness of your vehicle's size
- reversing large vehicles.

Restricted view

Mirrors and blind spots

Before you begin any journey, adjust your seat position and mirrors so that you can see clearly from your driving position.

Large vehicles will often have an additional nearside mirror fitted, especially for checking the area around the nearside wheel. Use this, particularly when

- pulling into the kerb
- reversing the vehicle
- you have to move closer to the left during normal driving.

DBC s4 DGV s4

Before moving off, stopping or turning, check all around you and especially down the sides of your vehicle, as cyclists or motorcyclists may be trying to pass you on either side.

Check out these links for more information on blind spots.

youtu.be/IV-rhiGRFTE

youtu.be/_Uf5WVfY_RY

Always use the Mirrors – Signal – Manoeuvre (MSM) routine before making any change of lane or direction, or when approaching any hazard.

Before turning right, look for motorcyclists who might be trying to overtake you.

Nearside mirror

Check before

- signalling
- turning left
- returning to the left after overtaking
- changing lanes
- moving to the left when leaving roundabouts
- pulling in to the kerb.

Offside mirror

Check before

- signalling
- turning right
- overtaking
- changing lanes
- moving to the right
- moving off.

DBC s4 **DGV** s4

If you're driving a passenger-carrying vehicle, you'll often have a good view from your raised driving position. However, you'll still have blind spots and this means you should always use your mirrors effectively.

DBC s4

TIP

Check carefully in front of your vehicle before moving off. Cyclists and pedestrians close to the front may be hidden from view.

A quick sideways glance is often helpful

- before moving off
- before changing lanes on a motorway
- where traffic joins from the right or left
- prior to merging from a motorway slip road.

DBC s4 DGV s4

Sleeper cabs can cause large blind spots. It's a good idea to take a quick sideways glance before any manoeuvre, as there may be something in one of your blind spots.

DGV s4

Awareness of your vehicle's size

When moving in near to the kerb or when turning tight corners, make sure any overhangs, including your mirrors, don't collide with

- pedestrians
- traffic signals
- lampposts
- street furniture
- walls and buildings.

DBC s4 DGV s4

Find guidance on driving with high loads at this website.

transport.gov.scot/our-approach/industry-guidance/freight-transport/#42449

Don't drive too closely behind another vehicle, as this will

- intimidate the other driver
- obstruct your view ahead
- make it impossible for you to stop safely in an emergency.

From your raised driving position, you'll be able to see over hedgerows or other obstructions and can scan ahead for potential hazards. However, you may not be able to see the area directly in front of your cab.

DBC s4 DGV s4

At junctions, be aware of your vehicle's limitations when waiting to join a main road. A large vehicle takes more time to emerge and build up speed, so it's essential that you make the correct decisions at junctions.

TIP

Approaching motorcyclists are more difficult to see than other vehicles and may be moving quickly.

In traffic queues, try not to restrict access from side roads into the main traffic flow.

DBC s4 DGV s4

When approaching a roundabout, some lanes may be wide enough for your vehicle, but you may need to straddle narrow lanes. Other road users may not realise what you plan to do or understand why you need to take up more than one lane.

- Indicate what you intend to do in good time.
- Check all mirrors before you manoeuvre.

DBC s4 DGV s4

Reversing large vehicles

Many large vehicles can sound an audible warning device when they're reversing. You **MUST NOT** use these between 11.30 pm and 7.00 am in a built-up area.

DBC s1, 4 DGV s1, 4 HC r112

Take care when reversing, for example into a confined space or loading bay. If you can't see adequately in your mirrors, ask someone to help guide you back.

DGV s4

Find more information about reversing safely at these links.

- **hse.gov.uk/workplacetransport/factsheets/reversing.htm**
- **hse.gov.uk/workplacetransport/information/reversing.htm**

Meeting the standards

You must be able to
position the vehicle correctly to carry out manoeuvres safely
make sure that if audible reversing warning systems are fitted they're used only when it's legal to do so
use reversing camera systems or proximity sensors effectively, where fitted.

You must know and understand
the blind spots for the vehicle and how to check them
how to allow for vulnerable road users when carrying out a manoeuvre
how the design of your vehicle may affect your field of vision, and how to overcome this
how to work with a signaller when reversing, where applicable.

Notes

You can use this page to make your own notes or diagrams about the key points you need to remember.

Think about

- Do I adjust my seat and mirrors before starting each journey?
- Do I know when to use my nearside and offside mirrors?
- Am I aware of any blind spots, including in front of the driver's cab?
- Am I aware of any overhangs, including mirrors, and do I watch out for these when manoeuvring?
- Do I know how long my vehicle takes to build up speed and allow for this at junctions?

Your notes

Things to discuss and practise with your instructor

These are just a few examples of what you could discuss and practise with your instructor. Some are for PCV, some are for LGV and some are for both. Read more about restricted view to come up with your own ideas.

Discuss with your instructor

- the checks you must make before pulling off in a high-sided coach
- when you would use a downward-angled mirror on the nearside of your bus.

- the safest way to reverse into a loading bay
- how a sleeper cab in your lorry may affect your view.

- what MSM stands for and how to use it
- when you may use an audible warning device.

Practise with your instructor

- the checks you need to make in your nearside mirror
- reversing into a loading bay
- identifying the blind spots in your vehicle and how to overcome them.

11.1 Mark one answer — DGV s4

Some lorries have an extra mirror angled down towards the nearside front wheel. What's this mirror especially useful for?

- ☐ Overtaking
- ☐ Parking
- ☐ Checking your trailer
- ☐ Turning right

Sitting in a high cab doesn't always give you all-round visibility. Other road users might assume that you can see them and position themselves out of sight close to your vehicle. Making good use of any extra mirrors can help prevent dangerous incidents, especially when parking.

11.2 Mark one answer — DGV s2

What should you look for when choosing a site to park your trailer?

- ☐ The site should be level
- ☐ The site should be made of concrete
- ☐ The site should be sheltered
- ☐ The site should have marked parking bays

Check that the ground is firm and level before you uncouple the trailer. If you need to, place a heavy plank under the legs to distribute the weight and stop the legs sinking into the ground. Don't park illegally. Make sure that your unattended trailer isn't blocking access for others.

11.3 Mark one answer — DGV s6

You're behind a parked car. What should you do when you're about to move off?

- ☐ Start to signal when you've started moving
- ☐ Move off when there's no oncoming traffic
- ☐ Check the blind spots before you start moving
- ☐ Only use your mirrors after you've started moving

Because the body of your vehicle is designed to take loads, your view around it will be restricted. Take extra care to look well out of the window to check the blind spots. Don't forget to check all the mirrors. Check ahead and signal, if necessary, before moving off.

11.4 Mark one answer DGV s4

You're driving a lorry. Why do you need to take extra care when emerging at this junction?

- ☐ There are bollards in the middle of the road
- ☐ It's a closed junction
- ☐ There's a motorcycle on the left
- ☐ There's a grass verge on the left

Make sure that you signal in good time, so that the motorcyclist behind you knows your intention. You may have to position yourself wide in order to negotiate the junction. Be aware that the motorcyclist may see this as an opportunity to filter through on your nearside. Check your left-hand mirror carefully before you start to turn.

11.5 Mark one answer DGV s4

You're driving this lorry (arrowed). What should you do about the emergency vehicle that's trying to emerge from the side road?

- ☐ Brake hard to a stop and wave it out
- ☐ Brake smoothly and allow it to emerge
- ☐ Drive on; you're on the major road
- ☐ Turn left to give it a clear view

Good forward planning will allow you to deal safely with this situation. You need to assess how your actions will affect other road users. Acting on impulse can have disastrous consequences.

11.6 Mark one answer DGV s4

What should you do when you're unable to see clearly while you're reversing into a loading bay?

- ☐ Get someone to guide you into the bay
- ☐ Use an audible warning signal while reversing
- ☐ Reverse into the bay until your bumper touches the end wall
- ☐ Open your door and lean well out while reversing

Don't take chances when reversing in a confined space. It's too easy to crush or kill someone without knowing they're in danger. If in doubt, get help from a reliable person.

11.7 Mark one answer DGV s4

This lorry is turning right from the side road. What should the driver be especially aware of?

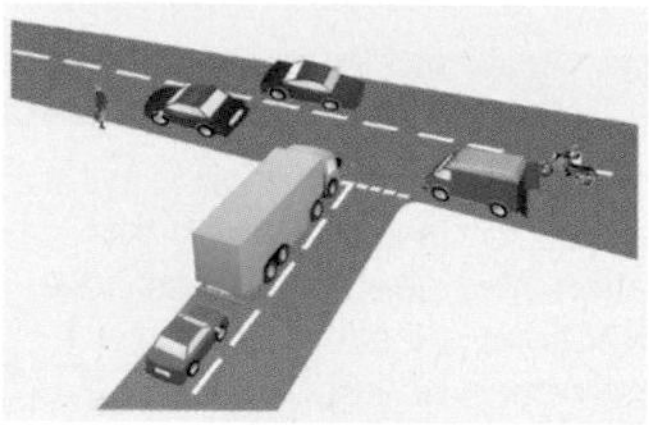

- ☐ The motorcyclist from the right passing the parked van
- ☐ The vehicles coming from the left along the main road
- ☐ The pedestrians on the footpath by the main road
- ☐ The vehicles behind on the minor road

Parked vehicles near junctions can hide smaller road users such as cyclists and motorcyclists. Double-check before emerging with a large vehicle. If in doubt, move forward slowly into a position where you can safely have another look.

11.8 Mark one answer DGV s4

You're parking your lorry at night. Where must you use parking lights?

- ☐ On the road
- ☐ In a motorway service area
- ☐ In a factory entrance
- ☐ In dock-authority areas

Goods vehicles exceeding 2500 kg laden weight must always have their parking lights switched on when they're parked on the road at night. This includes parking in lay-bys: other drivers entering the lay-by must be able to see the vehicle.

11.9	Mark one answer	DGV s4

You're driving a lorry with a sleeper cab. When would a quick sideways glance be helpful?

- ☐ After driving over a pedestrian crossing
- ☐ When traffic is merging from the right or left
- ☐ Before climbing a steep hill
- ☐ When driving round sharp bends

The size and design of some cabs can create blind spots. This is especially true of a sleeper cab. A quick sideways glance might show something you can't see in your mirrors, especially when traffic is merging.

11.10	Mark one answer	DGV s4

Your lorry has a sleeper cab. Why could this make your driving more difficult?

- ☐ It increases your blind spots
- ☐ It increases your view of the road ahead
- ☐ It increases your view in the right-hand mirror
- ☐ It increases your view in the left-hand mirror

A sleeper cab can create extra blind spots. Look out for vehicles just to the rear offside and nearside of your cab – they could soon become hidden in your blind spots.

11.11	Mark one answer	DGV s4

You're waiting to turn right in this lorry (arrowed). What danger should you be most aware of?

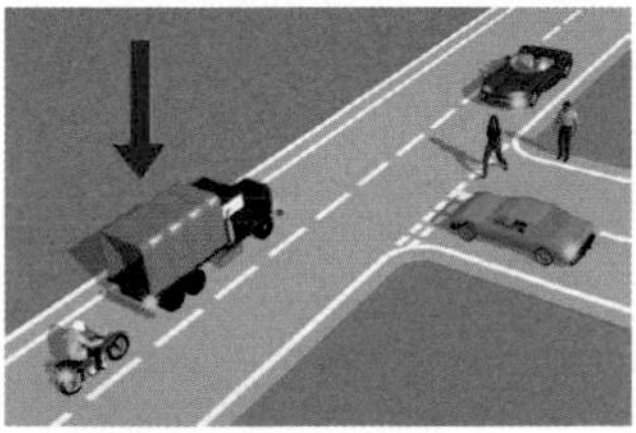

- ☐ The road surface
- ☐ The 'give way' lines
- ☐ The hazard line
- ☐ The motorcyclist

Indicate your intentions in good time, so that both the motorcyclist and pedestrians know you plan to turn right. You should note all potential hazards and keep checking as you turn.

11.12 Mark one answer — DBC s4, DGV s4

What's the first thing you should check before moving to the left?

- ☐ The nearside mirror
- ☐ The offside mirror
- ☐ Behind, over your right shoulder
- ☐ Behind, over your left shoulder

Before you make a turn or a change of direction, however slight, you should always check the mirrors. If you intend to turn left, check your left-hand (nearside) mirror first.

11.13 Mark one answer — DBC s4, DGV s4

What's the first thing you should check before moving to the right?

- ☐ The nearside mirror
- ☐ Behind, over your left shoulder
- ☐ Behind, over your right shoulder
- ☐ The offside mirror

If you intend to turn right, check your right-hand (offside) mirror first. All of your mirrors should be properly adjusted to give a clear view around and behind your vehicle. They should be kept clean at all times and shouldn't be cracked or broken.

11.14 Mark one answer — DBC s4, DGV s4

What should you do before moving off?

- ☐ Extend your right arm out of the window
- ☐ Just use the offside mirror and move away quickly
- ☐ Signal right with indicator and arm together
- ☐ Use your mirrors and look behind

To make sure it's safe, you must use the mirrors well before you make any manoeuvre. For example, you need to use them before moving off, turning, overtaking, slowing or stopping, opening a door, or changing direction, lanes or speed.

11.15 Mark one answer — DBC s4, DGV s4

When would you need to check your nearside mirror?

- ☐ As a final check before turning right
- ☐ After passing cars parked on the left
- ☐ Before moving into the lane on your right
- ☐ When getting out of your cab on the driver's side

As you pass pedestrians or vehicles on your left, you should use your nearside mirror and check that you've passed them safely. You need to allow a good safety margin before you move back to the left.

11.16 Mark one answer HC r180

The MSM routine should be used when you negotiate a hazard. What do the initials MSM stand for?

- ☐ Mirrors, signal, manoeuvre
- ☐ Manoeuvre, speed, mirrors
- ☐ Mirrors, speed, manoeuvre
- ☐ Manoeuvre, signal, mirrors

Always use the MSM routine when you're approaching a hazard.

M – Mirrors: check the position of the traffic behind you.

S – Signal: signal your intention to slow down or change course in good time.

M – Manoeuvre: a manoeuvre is any change in position, from slowing or stopping the vehicle to turning off a busy road.

11.17 Mark one answer KYTS p32

What does this sign mean?

- ☐ Contraflow bus and cycle lane
- ☐ With-flow bus and cycle lane
- ☐ No buses or cycles
- ☐ Priority to buses or cycles

Some lanes are set aside for certain types of vehicle. Signs and road markings show the authorised vehicles in words or pictures. There's usually also a sign showing the times that the lane is in operation. Some lanes might only be in operation for a short time. Check the sign and only use the lane if you're allowed to do so.

11.18 Mark one answer DBC s3, DGV s3

What must you do to ensure good all-round vision when driving?

- ☐ Keep your windows open
- ☐ Have a sun visor fitted
- ☐ Adjust your seat properly
- ☐ Keep your lights clean

You must make sure that you adjust the seat so that you're able to reach all the controls and see in all the mirrors before you start to drive any vehicle. You should also be able to lean out of the window and check your offside blind spots.

11.19 Mark one answer DBC s3, DGV s3

When may an audible warning device for reversing be used in areas with a 30 mph speed limit?

- ☐ Between 7.00 am and 11.30 pm
- ☐ Between 11.30 pm and 7.00 am
- ☐ During daylight hours only
- ☐ At any time of the day or night

Some vehicles are fitted with an audible warning device that sounds when the vehicle is being reversed. As these devices make a loud noise, they shouldn't be used between 11.30 pm and 7.00 am in a built-up area. Audible warning devices are an effective safety feature, but they don't take away the need to use effective observation around the vehicle before and while reversing.

11.20 Mark one answer DBC s1, DGV s1

What will happen if you drive too close to the vehicle in front?

- ☐ Your view ahead will be reduced
- ☐ Your view ahead will be increased
- ☐ Your view of following drivers will be increased
- ☐ Your view of following drivers will be reduced

The closer you are to the vehicle in front, the less you'll be able to see of the road ahead. You must be aware at all times that you won't be able to see all around your vehicle. This is why you must have an excellent mirror routine and constantly update yourself on what's going on around you.

11.21 Mark one answer DBC s4, DGV s4

Why might it be difficult to see motorcyclists at junctions?

- ☐ They can be hidden in blind spots
- ☐ They always ride in the gutter
- ☐ They wear bright clothing
- ☐ They ride with their headlights on

When driving large vehicles, windscreen pillars and large mirrors can easily create blind spots as you look left and right at junctions. Altering your body position slightly while looking into the new road can give you an improved view of the traffic. If in doubt, look again and reassess the situation. Don't take chances.

11.22 Mark one answer DBC s4, DGV s4

Why should you check your right-hand mirror just before turning right from a main road into a side road?

- ☐ There may be pedestrians stepping off the kerb
- ☐ You need to check your position
- ☐ Someone may be overtaking you
- ☐ Your rear view to the left is blocked

Driving a large vehicle will sometimes require you to position your vehicle well to the left before making a tight right turn. Inexperienced drivers or riders may not realise your intentions and may try to overtake you. Make sure you signal in good time and make a final mirror check before committing yourself to the turn.

11.23 Mark one answer DBC s4, DGV s4

What should you check just before you turn left?

- ☐ The left-hand mirror
- ☐ The interior mirror
- ☐ The right-hand mirror
- ☐ The overtaking mirror

When making a left turn with a long vehicle, you may have to adopt an unusual position, well over to the centre of the road. Always signal your intentions in good time, and make good use of the mirrors, especially the left-hand one. Be alert for less experienced road users putting themselves in danger by coming up on your nearside.

11.24 Mark one answer DBC s4, DGV s4

You're driving a long vehicle. What should you be especially careful of before turning left onto a main road?

- ☐ Pedestrians on the opposite kerb
- ☐ Cyclists on your left
- ☐ Solid white lines in the centre of the road
- ☐ Vehicles directly behind you

You should always check your nearside mirror before any change of direction. Be alert for cyclists and motorcyclists who don't realise the potential danger involved in trying to squeeze through on your nearside when you're waiting at a junction.

11.25 Mark one answer DBC s4, DGV s4

You want to turn right at a roundabout marked with two right-turn lanes. What should you do if there's ample room for your vehicle in either lane?

- ☐ Use the right-hand of the two lanes
- ☐ Use the left-hand of the two lanes
- ☐ Use the left-hand lane, then move to the right as you enter the roundabout
- ☐ Use the right-hand lane, then move to the left as you enter the roundabout

Using the left-hand lane will make it easier for you to leave the roundabout. If you use the right-hand lane, there could be traffic on your left and in your blind spot when you reach your exit and try to move back to the left.

11.26 Mark one answer DBC s3, DGV s3

Your vehicle is fitted with a reverse warning bleeper. How will this affect you when you're reversing?

- ☐ You'll be able to reverse more accurately
- ☐ You'll only need to use the offside mirror
- ☐ You'll need to take all-round observation
- ☐ You'll only need to use the nearside mirror

The reverse warning bleeper is fitted to your vehicle to warn others near the vehicle that it's reversing. However, it's no substitute for good all-round observation. Get someone to guide you if you can't see clearly where you're reversing.

11.27 Mark one answer DBC s4, DGV s4

What does 'blind spot' mean?

- ☐ An area of road covered by your right-hand mirror
- ☐ An area of road covered by your left-hand mirror
- ☐ An area of road that can't be seen, even with mirrors
- ☐ An area of road that isn't lit by your headlights

Blind spots can occur when bodywork – such as a door or window pillar – restricts your view. Different types of vehicle have different blind spots.

11.28 Mark one answer DBC s4, DGV s4

You're driving a large vehicle. How does its size and design affect the blind spots?

- ☐ It will have fewer blind spots than smaller vehicles
- ☐ It will have more blind spots than smaller vehicles
- ☐ It will have the same blind spots as other vehicles
- ☐ It will have no blind spots whatsoever

Take the time to check where the blind spots are when driving a vehicle that's unfamiliar to you. Being aware of the limitations of your mirrors will help you take steps to prevent collisions. Always check the blind spots before changing direction. Frequent and proper use of the mirrors will help you to be aware of a vehicle moving into your blind spot.

11.29 Mark one answer — DBC s4, DGV s4

You're turning right at a T-junction. What should you do when your view to the right and left is blocked by parked vehicles?

- ☐ Lean forward to get a better view without crossing the 'give way' lines
- ☐ Edge out until you're about 1 metre (3 feet 3 inches) over the 'give way' lines
- ☐ Ease forward until you can see clearly in all directions
- ☐ Ask a passenger to assist by waving you out when it's clear

When emerging, if you can't see, don't go. Be aware that there could be an approaching cyclist or motorcyclist hidden by parked vehicles. Edge out very slowly, checking in all directions, and be ready to stop.

11.30 Mark one answer — DBC s4, DGV s4

What must you do before you start reversing?

- ☐ Remove your seat belt
- ☐ Look all around
- ☐ Use an audible warning device
- ☐ Change the tachograph mode

Large or long vehicles have many blind spots. It's vital to check all these areas before starting to reverse and then make sure you keep checking all around while completing the manoeuvre.

11.31 Mark one answer — DBC s4, DGV s4

You're reversing your vehicle. What should you do while the audible warning device is operating?

- ☐ Rely on a clear path behind
- ☐ Expect others to be aware of your course
- ☐ Be cautious and keep looking all around
- ☐ Concentrate solely on your blind areas

Don't rely on an audible warning device to claim right of way. It's your responsibility to be cautious and take all-round observation while you're reversing.

11.32 Mark one answer — DBC s4

Some coaches have a mirror on their nearside, angled down to show the front nearside wheel. When should this extra mirror be used?

- ☐ When you're pulling in after overtaking
- ☐ When you're pulling in to park at the kerb
- ☐ Before you pass parked cars
- ☐ Before you change lanes on a motorway

This mirror offers the most benefit when you're manoeuvring in confined spaces. It shows you where your nearside front wheel is in relation to the kerb or a bay marking line.

11.33 Mark one answer DBC s4

What must you be aware of when you're driving a bus with a high driving position?

- ☐ The blind area close in front
- ☐ Passengers will have to look up to you
- ☐ Your judgement of speed will be impaired
- ☐ The lack of headroom in the cab

It's essential that you're constantly aware of other road users and pedestrians around you. A routine of effective mirror checking should be established. You must also know when it's essential to make checks in the blind spots; for example, just below the nearside front of the vehicle, where a pedestrian or a cyclist could be out of sight when you move off.

11.34 Mark one answer DBC s4

On a coach with high side windows, it can be difficult to see either side. What should you do before you move off in one of these vehicles?

- ☐ Get out of your vehicle and stop the traffic
- ☐ Ask a passenger to make sure it's safe to move off
- ☐ Indicate before checking all of your mirrors
- ☐ Open the window and look down and round to the right

High-sided coaches can create extra blind spots; you should be aware of these. You'll need to make additional checks on this type of vehicle, but don't forget your usual observations and mirror checks to make sure it's safe before moving off.

11.35 Mark one answer DBC s4

Your bus has a high seating position. What may be out of sight below the windscreen line?

- ☐ Following vehicles
- ☐ Overtaking vehicles
- ☐ Cyclists and pedestrians
- ☐ Other buses

If you've been stopped for some time at a bus stop or in a bus station, be aware that cyclists and pedestrians, especially children, can enter your blind spots. They may be out of sight below the windscreen line. Always take all-round observation before manoeuvring, whether you're moving away, parking or reversing.

11.36 Mark one answer DGV s4

Many modern vehicles are fitted with an additional nearside mirror. What does this help the driver to see?

- ☐ The front wheel in relation to the kerb
- ☐ The exhaust to check for emissions
- ☐ The rear of the vehicle when reversing
- ☐ The distance of the following vehicle

This mirror is specifically positioned so that the driver can see the position of the front nearside wheel in relation to the kerb. Use it when pulling in to park alongside the kerb or when you have to move close to the left in normal driving.

Section twelve
Essential documents

In this section, you'll learn about

- documentation
- regulations
- the driver's responsibility.

Essential documents

Documentation

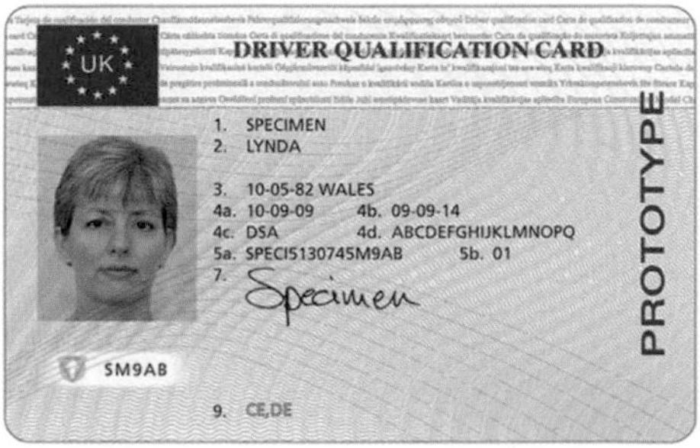

To obtain and keep your Driver Certificate of Professional Competence (Driver CPC), you'll need to complete a minimum of 35 hours' training every five years. When you complete this, you'll be issued with a Driver Qualification Card (DQC).

You'll need to hold a GB photocard licence to get your DQC. The DQC will be sent to the address on your driving licence, so it's important that you tell DVLA of any change of address.

DBC s1 **DGV** s1

If you're taking a coach abroad, you must have a passenger list with you when you cross any international borders. Some countries may also require other documents. Check each country's requirements before starting your journey.

DBC s3

When driving a goods vehicle in Europe, make sure that you have these documents in your possession

- driving licence
- insurance certificate
- vehicle registration document.

You'll also need to carry your passport with you at all times.

DGV s3

When carrying goods for hire or reward on international journeys, you **MUST** record these goods on four-part **CMR consignment notes**. Make sure that the right people keep or get their copy.

- **Red** The sender should keep this (the consignor).
- **Green** The sender should give you this to keep with the vehicle (the carrier).
- **Blue** The sender should give you this to hand to whoever receives the shipment (the consignee).
- **White with black border** This is a spare copy for other uses, and is kept by the consignor.

DGV s3

Definition

CMR consignment notes

A standard contract that's required to transport goods internationally by road. It confirms that the haulage company has received the goods and has a contract from the supplier to carry them.

Vehicle or fleet operators **MUST** have an effective system to prevent illegal immigration. Your operator should give you documents that detail

- how to secure the vehicle
- when and how to check for illegal immigrants
- what to do if illegal immigrants might be in the vehicle
- what to do if the vehicle's security is breached or compromised.

You should produce these documents immediately if an immigration officer asks to see them. If you don't, you'll incur a penalty.

DGV s3

Regulations

If you lose your licence because of drink-driving or certain other dangerous driving offences while driving your commercial vehicle, you'll automatically lose your PCV or LGV entitlement.

To get your licence back after a drink-driving offence, you'll need to convince the Drivers Medical Group at DVLA (Drivers Medical Section at DVA in Northern Ireland) that you don't have an alcohol problem. You'll also have to take an extended driving test after a ban for any offence. This may involve a separate extended test for every extra category of vehicle.

DBC s1, s3 DGV s1, s3 HC r95–96, p125–127

It's a serious offence to drive without insurance. If you're caught, you'll be fined an unlimited amount, you may face a driving ban and you'll get 6–8 penalty points on your licence.

DBC s3 DGV s3 HC p125–127

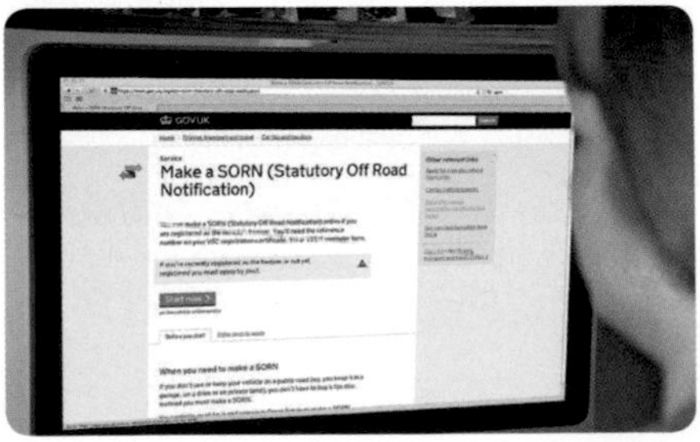

If you wish to take your vehicle off the road and not pay the vehicle tax, you **MUST** apply for a Statutory Off-Road Notification (SORN).

You can make a SORN online, by phone or by post. Your SORN is valid until the vehicle is taxed, sold or scrapped. If you want to use the vehicle on the road again you must pay the vehicle tax. You can only drive an untaxed vehicle on the public road if you're driving it to or from a

- licensed station for a pre-arranged MOT
- vehicle identity check
- weight or emissions test.

If you sell the vehicle, the SORN ends. The new owner **MUST** immediately either re-license it or apply for a SORN.

HC p122

Traffic commissioners enforce the regulations governing LGV and PCV operators' licences. They can take disciplinary action against any operator who doesn't obey the regulations and can impose **conditions** on its operating licence.

DBC s2, 7 **DGV** s2, 7

> Definition
>
> **conditions**
> Restrictions.

The driver's responsibility

If you have, or develop, any serious illness or disability that's likely to last more than three months and which could affect your driving, you need to immediately inform the Drivers Medical Group/Section at DVLA/DVA.

DBC s1, **DGV** s1

For the safety and security of your passengers, and to help prevent illegal immigration, load all luggage before you allow passengers to board for continental tours.

Only load bags that definitely belong to the people boarding the coach. Check that every bag has an 'owner' before you stow it.

Lock the coach if you leave it unattended. When you return, check it thoroughly, including the external luggage compartments.

DBC s3

Ensure that doors on LGVs have the appropriate locks. Cargo doors should use Transports Internationaux Routiers (TIR) security seals if travelling outside the EC. Check these seals after every stop. Customs officers will also check your TIR seals.

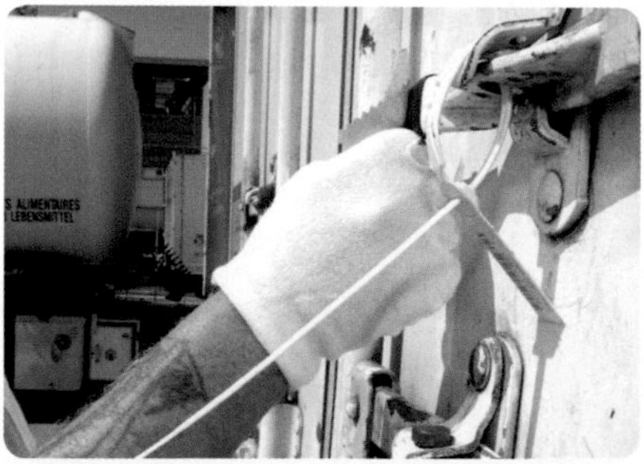

DGV s3

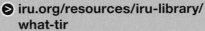

For more information about TIR, visit this link.

iru.org/resources/iru-library/what-tir

Meeting the standards

You must be able to

make sure that your driving licence is valid for the category of vehicle that you're driving
make sure that the vehicle is registered and the vehicle tax has been paid
make sure that you have valid insurance for your use of the vehicle
make sure that you have the required documentation if you're transporting dangerous goods.

You must know and understand

that you must have a valid driving licence for the vehicle you drive. You must also meet any restrictions on your licence

that the vehicle must be registered with the Driver and Vehicle Licensing Agency (DVLA)

that you must tell DVLA if you

- change your name
- change your address
- have or develop a medical condition that will affect your ability to drive

the MOT requirements for vehicles being used on the road.

Notes

You can use this page to make your own notes or diagrams about the key points you need to remember.

Think about

- Am I undertaking enough training to obtain or keep my Driver Qualification Card (DQC)?
- Does DVLA (DVA in NI) know my current address?
- Do I have all the documents that I need to keep with me for national or international travel?
- Has my operator provided me with documents on how to check for illegal immigrants and how to respond if I suspect anything, and do I know what to do?
- Am I adequately insured for the vehicle(s) I drive?
- Do my loading and unloading procedures and on-journey checks help me take responsibility for my passengers, luggage or cargo?

Your notes

Things to discuss and practise with your instructor

These are just a few examples of what you could discuss and practise with your instructor. Some are for PCV, some are for LGV and some are for both. Read more about essential documents to come up with your own ideas.

Discuss with your instructor

- what it means to be the holder of a full category D licence
- the responsibilities of a traffic commissioner.

- what it means to be the holder of a full category C licence
- what a CMR consignment note is and who gets a copy.

- the documents that you must have to drive in Europe
- what medical conditions may prevent you from driving a lorry or a bus
- what the letters 'SORN' stand for and what this means to the keeper of a vehicle
- what you'll need to do to keep your Driver Certificate of Professional Competence (Driver CPC).

Practise with your instructor

It's difficult to practise your knowledge and understanding of the documents relevant to driving a lorry or a bus. Just remember that if you, your documents and vehicle are in order and you drive safely and responsibly, you'll be much less likely to

- be involved in an accident
- be fined or accumulate penalty points
- damage your vehicle.

12.1 Mark one answer DGV s2, 3

You've been asked to drive a fully loaded petrol tanker back to the depot. It has a maximum authorised mass of 10 tonnes. What must you have before you can drive this vehicle?

- ☐ A request from the police to move a dangerous load
- ☐ A fully qualified tanker driver with you for the journey
- ☐ A valid ADR driver training certificate
- ☐ A copy of the vehicle's insurance certificate

Drivers of vehicles carrying dangerous goods are required to hold an ADR driver training certificate. This certificate is valid for five years and is issued by DVLA when you've attended an approved course and successfully completed the examination.

12.2 Mark one answer DGV s2, 3

Who's responsible for making sure that cargo arrives undamaged?

- ☐ The packer
- ☐ The driver
- ☐ The sender
- ☐ The loader

The driver mustn't allow any cargo to be damaged during transit, loading or unloading. This requires a responsible attitude to safeguard the load at all times.

12.3 Mark one answer DGV s3

How many copies of the CMR consignment note are required?

- ☐ Two
- ☐ Three
- ☐ Four
- ☐ Five

The full name of the CMR is the 'Convention on the Contract for the International Carriage of Goods by Road'. You'll need this document to transport goods internationally by road. Of the four copies required, the consignor, consignee and originator all have one copy, and the fourth travels with the vehicle.

12.4 Mark one answer DGV s1, 7

You hold a full category C licence. What does this entitle you to tow?

- ☐ A trailer of any weight
- ☐ A trailer with more than one axle
- ☐ A trailer up to 750 kg
- ☐ A curtain-sided trailer

When you pass a test to obtain a C or C1 licence, you can tow trailers up to 750 kg. You can do this without taking a further test.

12.5 Mark one answer — DGV s1

You're delivering boxes of chilled food to a supermarket. What specific training would you need?

- ☐ Hygiene procedures
- ☐ ADR procedures
- ☐ Waste-handling procedures
- ☐ Ecosafe driving procedures

Chilled foods are transported at temperatures above freezing point. Drivers need to be trained in hygiene procedures and how to operate refrigeration units.

12.6 Mark one answer — DBC s3, DGV s3

You'll be driving your vehicle in Europe. Which of these documents must you carry?

- ☐ The vehicle service record
- ☐ Your medical examination form
- ☐ Your national driving licence
- ☐ The vehicle workshop manual

As well as your driving licence, you'll need to carry the insurance certificate and vehicle registration document. Other documentation may also be needed for some countries. Check this before starting your journey.

12.7 Mark one answer — DBC s1, DGV s1

What condition could prevent you from holding a lorry or bus licence?

- ☐ Dyslexia
- ☐ Partial blindness
- ☐ Dermatitis
- ☐ Stomach upset

Any medical condition that could affect your ability to drive a lorry or a bus safely may prevent you from holding a licence. These conditions include serious visual impairment, heart disease and some mental disorders. For more information about medical standards, you should contact the Drivers Medical Branch, DVLA, Swansea.

12.8 Mark one answer — DBC s3, DGV s3

You've been convicted of a drink-drive offence and banned from driving. Which entitlement will this ban affect?

- ☐ Your entitlement to drive all motor vehicles
- ☐ Your car entitlement
- ☐ Your lorry entitlement
- ☐ Your bus entitlement

The dangers of drink-driving are well publicised. Anyone convicted of this offence will lose their entitlement to drive any motor vehicle on the road.

12.9 Mark one answer | NI EXEMPT | DBC s3, DGV s3

You're found to be two-and-a-half times over the legal drink-drive limit and are disqualified from driving. Before regaining your licence, who will you have to satisfy that you don't have an alcohol problem?

- ☐ The local hospital
- ☐ The Driver and Vehicle Licensing Agency
- ☐ Alcoholics Anonymous
- ☐ The Driver and Vehicle Standards Agency

If you're disqualified from driving because you were two-and-a-half times over the legal limit, you'll be classified as a high-risk offender. Before you can regain your licence, you'll have to satisfy the Driver and Vehicle Licensing Agency (DVLA) that you don't have an alcohol problem.

12.10 Mark one answer | HC p122

For how long is a Statutory Off-Road Notification (SORN) valid?

- ☐ Until the vehicle is taxed, sold or scrapped
- ☐ Until the vehicle is insured and MOT'd
- ☐ Until the vehicle is repaired or modified
- ☐ Until the vehicle is used on the road

A SORN allows you to keep a vehicle off the road and untaxed. The SORN remains valid until the vehicle is taxed, sold or scrapped. If you buy a vehicle that has a SORN, the SORN will finish as soon as you buy it. You'll need to make a new SORN if you want to keep the vehicle off the road and untaxed.

12.11 Mark one answer | HC p122

What's a Statutory Off-Road Notification (SORN)?

- ☐ A notification to tell DVSA that a vehicle doesn't have a current MOT
- ☐ Information kept by the police about the owner of the vehicle
- ☐ A notification to tell DVLA that a vehicle isn't being used on the road
- ☐ Information held by insurance companies to check the vehicle is insured

If you want to keep a vehicle untaxed and off the public road, you must make a SORN. It's an offence not to do so. When you've made a SORN, you won't have to pay vehicle tax, but you also won't be able to use the vehicle on the road. The SORN remains valid until your vehicle is taxed, sold, scrapped or exported.

12.12 Mark one answer | NI EXEMPT | HC p126

What's the maximum fine for driving without insurance?

- ☐ Unlimited
- ☐ £500
- ☐ £1000
- ☐ £5000

It's a serious offence to drive without insurance. As well as an unlimited fine, you may be disqualified or given penalty points.

12.13 Mark one answer — DBC s3, DGV s3

What's the current fine for each person illegally carried into the UK?

- ☐ Up to £1000
- ☐ Up to £2000
- ☐ Up to £3000
- ☐ Up to £4000

The importation of illegal immigrants into the UK carries heavy penalties. A fine of up to £2000 can be imposed for each illegal immigrant. This fine can be imposed on each 'responsible person', which includes the vehicle owner, hirer and driver.

12.14 Mark one answer — DBC s3, DGV s3

Drivers can be fined for bringing illegal immigrants into the UK. Who else can be fined?

- ☐ The vehicle repairer
- ☐ The vehicle insurer
- ☐ The vehicle manufacturer
- ☐ The vehicle hirer

There are heavy fines for those convicted of bringing illegal immigrants into the UK. This can mean a fine for each illegal person carried. The fine can be imposed on each 'responsible person', including the vehicle driver, owner and hirer.

12.15 Mark one answer — DBC s3, DGV s3

How can vehicle operators help to prevent the carriage of illegal immigrants?

- ☐ By using an effective vehicle security system
- ☐ By using vehicle tracking devices
- ☐ By taking out liability insurance
- ☐ By having effective mobile communications

Current legislation requires all operators to set up an 'effective system' to help prevent the carriage of illegal immigrants. This should include vehicle security and checking using a checklist. Failure to do this will incur penalties.

12.16 Mark one answer — DBC s3, DGV s3

You're returning to the UK and are about to board a ferry. An immigration officer asks to see your documentation. What must you show them immediately?

- ☐ Your vehicle registration document
- ☐ Your driver's hours record
- ☐ Your operator documentation
- ☐ Your driving licence

Operators must set up an 'effective system' to prevent the carriage of illegal immigrants. This should include documentation with advice on vehicle security and a list of checks to be made. These papers should be produced immediately if an immigration officer asks to see them.

12.17 Mark one answer — DBC s3, DGV s3

A driver is convicted of bringing illegal immigrants into the UK. How is the fine calculated?

- ☐ For each family group they bring in
- ☐ For each court attendance they make
- ☐ For each person they bring in
- ☐ For each journey they made in the last year

If convicted, the current fine for each illegal immigrant brought into the UK is £2000. This fine can be imposed on each responsible person; this includes the vehicle driver, owner and hirer.

12.18 Mark one answer — DBC s1, DGV s1

The Driver Certificate of Professional Competence (Driver CPC) requires you to take training every five years. How many hours of training must you take?

- ☐ 30 hours
- ☐ 35 hours
- ☐ 40 hours
- ☐ 45 hours

To keep your Driver CPC, you need to take 35 hours of training every five years. This can be split into sessions of at least seven hours each, so you can take one training day a year over the five years if that suits you. The penalties for non-compliance are the same as those for driving without a licence.

12.19 Mark one answer — DBC s1, DGV s1

You'll need to take 35 hours of training to maintain your Driver Certificate of Professional Competence (Driver CPC). Over what period must this be carried out?

- ☐ One year
- ☐ Three years
- ☐ Five years
- ☐ Ten years

You must take 35 hours of training every five years to maintain your Driver CPC. This can be done at any time during the five years: all at once or split into sessions of at least seven hours each. The penalties for non-compliance are the same as those for driving without a licence.

12.20 Mark one answer — DBC s1, DGV s1

Every five years, you must complete further training to keep your Driver Certificate of Professional Competence (Driver CPC). How many hours of training must you complete over those five years?

- ☐ 25 hours
- ☐ 30 hours
- ☐ 35 hours
- ☐ 40 hours

Drivers are required to take 35 hours of training every five years to maintain their Driver CPC. This can be split into shorter sessions of at least seven hours each. These can be spread over the five years, so effectively one day's training is taken each year. The penalties for non-compliance are the same as those for driving without a licence.

12.21 Mark one answer DBC s1, DGV s1

You're applying for an LGV or PCV licence for the first time. What will your medical examination include?

- ☐ An eyesight test
- ☐ A numeracy test
- ☐ A hygiene test
- ☐ A literacy test

If you apply for a PCV or LGV licence, you must demonstrate that your eyesight is good enough to drive safely. DVLA (or DVA in Northern Ireland) can give you specific information about this. It's your responsibility to tell DVLA or DVA if your eyesight changes or you can't meet the requirements.

12.22 Mark one answer DBC s1

Which category of licence do you need to drive a tri-axle double-deck coach?

- ☐ D1
- ☐ D
- ☐ D1+E
- ☐ C

A category D licence allows you to drive a coach or bus of any size. It also allows you to tow a trailer up to 750 kg. If you want to tow a trailer over 750 kg with a bus or coach, you'll need a category D+E licence.

12.23 Mark one answer DBC s3

Why may coaches be subject to a search by immigration authorities at international ports?

- ☐ For national security
- ☐ To check for red diesel
- ☐ For tachograph regulations
- ☐ To check vehicle condition

Because of the threat to national security, coaches may be searched at ports and the Channel Tunnel. If the authorities aren't satisfied, they may refuse to let you continue.

12.24 Mark one answer DBC s3

You're driving a coach, returning from a European trip. Why should you search the vehicle at the port?

- ☐ To look for missing property
- ☐ To ensure duty-free limits aren't exceeded
- ☐ To prevent the carriage of illegal immigrants
- ☐ To ensure the vehicle has the correct documentation

A coach driver can be fined up to £2000 for each illegal immigrant carried. The owner and hirer may also be fined the same amount.

12.25 Mark one answer DBC s3

You're making a journey that crosses international borders. Which document should you produce at immigration control?

- ☐ A list of passengers
- ☐ A breakdown insurance certificate
- ☐ A route plan
- ☐ A list of alcohol on board

When passing through immigration control, you should have a list of all your passengers. They must all be accounted for.

12.26 Mark one answer DBC s1

Which licence category do you need to drive an articulated bus ('bendy bus')?

- ☐ D1+E
- ☐ D
- ☐ D+E
- ☐ D1

An articulated bus, commonly known as a 'bendy bus', isn't considered to be a bus towing a trailer. Therefore, it can be driven on a category D licence.

12.27 Mark one answer DBC s7

What size of trailer can be towed by the holder of a full category D licence?

- ☐ 750 kg
- ☐ 1000 kg
- ☐ 1250 kg
- ☐ 1500 kg

A full category D licence entitles you to tow a trailer of up to 750 kg.

12.28 Mark one answer DBC s7

You hold a full category D licence. What entitlement does this give you for towing?

- ☐ You can't tow a trailer at all
- ☐ You can tow a trailer not exceeding 750 kg
- ☐ You can tow a trailer of any weight
- ☐ You can't tow a trailer with more than one axle

A full category D licence entitles you to tow a trailer of up to 750 kg. If you wish to tow a trailer over 750 kg, you must pass an additional test.

12.29 Mark one answer DBC s1, 7

You intend to drive a midibus for hire or reward. It has more than 16 passenger seats. What's the minimum licence entitlement you need?

- ☐ C
- ☐ C1
- ☐ D
- ☐ D1

Drivers of buses and coaches with more than 16 passenger seats require a category D licence. Make sure your licence has the correct entitlement for the vehicle you plan to drive.

12.30 Mark one answer DBC s1

Which licence category do you need to drive a bus with 25 passenger seats?

- ☐ D
- ☐ D1
- ☐ D+E
- ☐ D1+E

Drivers of coaches or buses with more than 16 passenger seats require a category D licence.

12.31 Mark one answer NI EXEMPT DBC s1

Which authority has the power to impose conditions on a passenger-carrying vehicle (PCV) operator's licence?

- ☐ The Traffic Commissioner
- ☐ DVLA
- ☐ DVSA
- ☐ Highways England

Traffic commissioners are responsible for administering the regulations governing a PCV operator's licence. They can take disciplinary action against an operator who falls foul of the legal requirements.

Section thirteen

Environmental issues

In this section, you'll learn about

- reducing fuel consumption and emissions
- road surfaces
- refuelling

Environmental issues

Reducing fuel consumption and emissions

Plan your route before you start your journey. Try to avoid busy times and areas of known congestion as this could

- shorten your journey time
- save fuel
- prevent your vehicle adding to the congestion.

TIP: It's a good idea to plan an alternative route, in case of delays or diversions.

DBC s4 DGV s4

To help reduce fuel consumption and pollution

- turn your engine off when the vehicle is going to be stationary for more than a few minutes
- don't brake in a sudden or harsh manner. Plan ahead so that you brake or accelerate smoothly and in good time
- maintain your vehicle in line with the manufacturer's maintenance schedules.

DBC s4 DGV s4

Officers from the Driver and Vehicle Standards Agency (DVSA) and local-authority environmental health department staff carry out roadside checks on vehicles and test their emission levels. Both authorities have the power to prosecute the driver and the operator if the vehicle emissions are found to be excessive.

DBC s3 **DGV** s3

Find out more about DVSA roadside checks at this link.

www.gov.uk/roadside-vehicle-checks-for-commercial-drivers

A bus can carry as many people as 20 cars, making buses environmentally friendly vehicles. As trams run on electricity, they emit no exhaust fumes and don't contribute to local roadside pollution.

DBC s3

The comfort of your passengers is important and in very hot weather using the vehicle's air conditioning can make their journey more comfortable. Running the air conditioning will increase the vehicle's fuel consumption, so try not to use it continuously.

DBC s3

If you drive a goods vehicle, keep the rev counter in the green band whenever possible.

Operators can save on fuel costs and reduce their vehicle's effect on the environment by

- using correctly fitted air deflectors to reduce wind resistance
- fitting tyres with reduced **rolling resistance**.

DGV s3, 4

Definition

rolling resistance
The force resisting the motion when a wheel rolls on a surface.

Road surfaces

Vehicles with faulty suspension can damage road surfaces. 'Road-friendly' air suspension is now fitted to many large vehicles. This helps to reduce damage to the road surface.

DBC s3 DGV s3

When carrying a load on a flat-bed lorry or trailer, make sure it's correctly covered or sheeted. Cover filled skips with a net to stop debris falling or being blown onto the road during your journey.

DGV s2

Refuelling

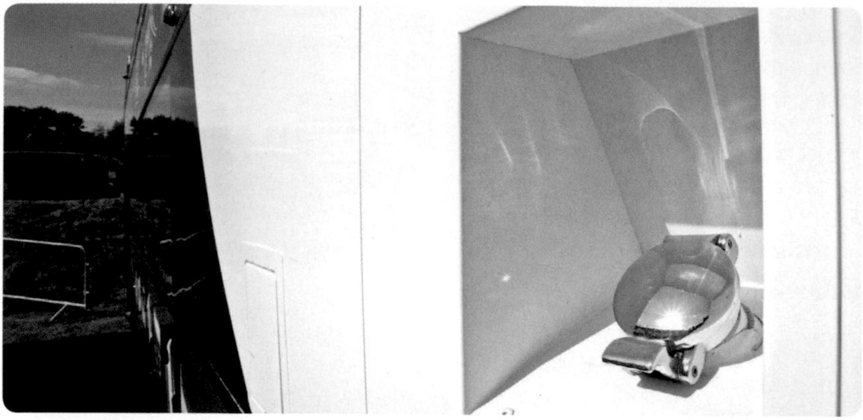

Check your filler cap is secure after refuelling. Spilt diesel is slippery and dangerous for other road users. If a filler cap is missing, you **MUST** replace it before continuing.

DBC s3 **DGV** s3

DVSA enforcement officers will check large vehicles to make sure that red diesel is being used for authorised purposes only. If you're found to be using this fuel illegally, you'll face severe penalties.

DBC s3 **DGV** s3, 4

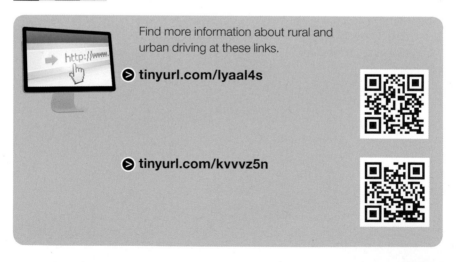

Find more information about rural and urban driving at these links.

tinyurl.com/lyaal4s

tinyurl.com/kvvvz5n

Meeting the standards

You must be able to

accelerate and decelerate smoothly and progressively, and effectively use aids such as cruise control

anticipate the need to stop, and use timely and smooth deceleration to reduce

- fuel consumption
- general vehicle wear and tear

drive in the highest responsive gear to keep full control and avoid labouring the engine.

You must know and understand

that correct use of the accelerator will have positive effects on the environment

the effect that unsuitable gear selection can have on the environment

that you should never reduce safety to improve economy.

Notes

You can use this page to make your own notes or diagrams about the key points you need to remember.

Think about

- Do I use road maps, sat-nav and traffic information to plan a main route and an alternative?
- Do I drive in ways that reduce my fuel consumption and pollution?
- Do I have the vehicle maintained for efficiency and to meet emissions standards?
- Do I check my filler cap after refuelling?

Your notes

Things to discuss and practise with your instructor

These are just a few examples of what you could discuss and practise with your instructor. Some are for PCV, some are for LGV and some are for both. Read more about environmental issues to come up with your own ideas.

Discuss with your instructor

- why it's important to switch off your engine while stationary for some time
- how a bus full of people helps to reduce pollution.

- how you should carry a loaded skip
- the purpose of wind deflectors.

- environmentally friendly means of transport
- how you can help to reduce the impact on the environment while driving your vehicle
- local-authority environmental health department roadside checks and what they'll be looking for.

Practise with your instructor

- switching off your engine when you stop for any length of time
- planning your journeys in a way that will reduce emissions
- refuelling carefully and replacing the filler cap securely.

13.1 | **Mark one answer** | **DGV s4**

Your lorry has been fitted with wind deflectors. How will they affect your vehicle?

- ☐ They'll increase the fuel consumption
- ☐ They'll reduce the stability
- ☐ They'll require increased tyre pressures
- ☐ They'll reduce the wind resistance

Wind deflectors direct the wind either around or over your vehicle, so reducing the wind resistance. The vehicle then requires less power to make progress against the wind, and so will use less fuel.

13.2 | **Mark one answer** | **DGV s4**

What can cab-mounted wind deflectors reduce?

- ☐ Journey times
- ☐ Load capacity
- ☐ Tyre wear
- ☐ Fuel consumption

A wind deflector redirects the air flow around the vehicle. This reduces the amount of energy required to drive the vehicle forward and so saves fuel.

13.3 | **Mark one answer** | **DGV s4**

How could you reduce your lorry's fuel consumption?

- ☐ Fit new brake linings
- ☐ Fit a high-level exhaust pipe
- ☐ Fit a larger fuel tank
- ☐ Fit wind deflectors

Wind deflectors reduce a vehicle's wind resistance and so help reduce fuel consumption.

13.4 | **Mark one answer** | **DGV s2**

You're a driver who's certified to carry dangerous goods. How long is your certificate valid?

- ☐ One year
- ☐ Two years
- ☐ Five years
- ☐ Ten years

The operator of a UK-registered vehicle engaged in the carriage of dangerous goods should make sure that the driver has a valid vocational training certificate. These certificates are valid for five years.

13.5 Mark one answer DGV s4

What would help to reduce the impact that your lorry has on the environment?

- ☐ Driving through town centres
- ☐ Braking in good time
- ☐ Planning routes to avoid motorways
- ☐ Racing to make up time

Noise, vibration, exhaust emissions and fuel consumption all contribute to the environmental impact made by a lorry.

Good anticipation and forward planning will help to reduce these effects. Braking in good time saves fuel and so helps reduce exhaust emissions. It also helps reduce wear on the vehicle, its tyres and the road surface.

13.6 Mark one answer DGV s2

What's reduced by fitting a fly sheet tightly over a tipper body?

- ☐ Drag effect
- ☐ Steering effort
- ☐ Legal load weight
- ☐ Load capacity

Drag is caused by wind entering the cavity of an uncovered tipper. If the cavity is covered, then the drag is reduced.

13.7 Mark one answer DGV s4

How could you reduce the fuel consumption of your lorry?

- ☐ Brake late as often as you can
- ☐ Fit a cab-mounted wind deflector
- ☐ Avoid sheeting any bulky loads
- ☐ Try to increase your overall speed

Wind deflectors lower a vehicle's wind resistance. The vehicle therefore requires less energy to drive it forward, and this reduces its fuel consumption. Ecosafe driving techniques, correct tyre pressures and proper maintenance also help to maximise fuel efficiency.

13.8 Mark one answer DGV s3

When may red diesel be used?

- ☐ For operators to reduce running costs
- ☐ For authorised purposes only
- ☐ For garages to test-drive repaired vehicles
- ☐ For environmental purposes

Red diesel is subject to less excise duty than standard diesel, so it's considerably cheaper. It may only be used for authorised purposes. Any driver whose vehicle is found to be using this fuel illegally faces penalties.

13.9 Mark one answer DGV s2

You're driving a lorry with a loaded skip. Why should the skip be covered with a net?

☐ To prevent rubbish from falling out of it
☐ To protect the contents from the weather
☐ To make it more visible to other traffic
☐ To stop others from adding to the load

If the load isn't covered, rubbish can be blown onto the carriageway. This can cause danger for other road users, as well as environmental damage. It may also lead to legal action against both the driver and the operator.

13.10 Mark one answer DGV s4, HC r246

What must you do when you park a lorry weighing more than 7.5 tonnes on a verge for essential loading?

☐ Sign the collection note
☐ Display an orange badge
☐ Obtain the owner's permission
☐ Make sure the lorry is always attended

Goods vehicles with a maximum authorised mass of more than 7.5 tonnes (including any trailer) mustn't be parked on a verge without police permission. The only exception is when this is essential for loading and unloading. In these cases, the vehicle mustn't be left unattended.

13.11 Mark one answer DGV s4, HC r246

What must you have before you can leave a lorry weighing more than 7.5 tonnes unattended on a verge?

☐ Police permission
☐ A warning triangle
☐ The owner's permission
☐ A loading permit

If your vehicle is to be left unattended on a verge, you must first get permission from the police.

13.12 Mark one answer DGV s4

What can help to reduce fuel consumption?

☐ Changing filters regularly
☐ Keeping tyres under-inflated
☐ Always using gears in sequence
☐ Keeping the rev counter in the amber band

Poorly maintained engines use more fuel and emit more exhaust fumes than those that are well maintained. Filters need to be changed on a regular basis. Remember that members of the public are encouraged to report vehicles emitting excessive exhaust fumes.

13.13 Mark one answer DGV s4

What will a correctly adjusted air deflector do?

- ☐ Save fuel
- ☐ Reduce tyre wear
- ☐ Reduce road-surface wear
- ☐ Cut loading time

Correctly adjusted air deflectors will save fuel. They should be adjusted to guide the airflow over the highest point at the front of the trailer or load and body.

13.14 Mark one answer DGV s3

What would staff from HM Revenue and Customs (HMRC) be looking for at a roadside check?

- ☐ Red diesel
- ☐ Vehicle defects
- ☐ Drivers' hours records
- ☐ Exhaust emissions

HMRC can check the type of fuel you're using, and the type and legality of your load. Red diesel is dyed gas oil with a lower tax than regular diesel. It can only be used in agricultural and construction vehicles (such as tractors). Red diesel mustn't be used in freight transport.

13.15 Mark one answer DGV s2

Why should the height of a load be kept to a minimum?

- ☐ To reduce aerodynamic drag
- ☐ To increase rolling resistance
- ☐ To increase momentum
- ☐ To reduce unloading time

Keeping the height of the load as low as possible will reduce the wind resistance of the vehicle. This will help to save fuel. It's particularly relevant when using a flat-bodied lorry or trailer.

13.16 Mark one answer DBC s3, DGV s3

What would be affected by a vehicle with faulty suspension?

- ☐ Overhead gantries
- ☐ Road surfaces
- ☐ Tyre pressures
- ☐ Engine performance

If your suspension is damaged, all the weight of the vehicle compresses the road and anything located below it. The vibrations travel through the ground and can also damage surrounding buildings.

13.17 Mark one answer — DBC s2, DGV s2

What makes the pictured vehicle environmentally friendly?

- ☐ It uses solar power
- ☐ It uses diesel fuel
- ☐ It uses electricity
- ☐ It uses unleaded fuel

This is the sign for a tram. These vehicles are powered by electricity and reduce the emissions released into the environment. As they can carry many people, they can help to reduce the number of other vehicles on the road.

13.18 Mark one answer — DBC s3, DGV s3

Which vehicles are most likely to cause severe damage to road surfaces?

- ☐ Lorries
- ☐ Cars
- ☐ Motorcycles
- ☐ Bicycles

Lorries and buses are much heavier than cars and will have a greater impact on the road surface.

13.19 Mark one answer — DBC s3, DGV s3

What does air suspension help to protect?

- ☐ The fuel system
- ☐ Tachograph accuracy
- ☐ Engine emissions
- ☐ Underground services

Air suspension reduces the pounding of the vehicle on an uneven road surface. This, in turn, reduces the vibrations transmitted to buildings and underground services along the route.

13.20 Mark one answer — DBC s4, DGV s4

How can you help to reduce the impact of road transport on the environment?

- ☐ By avoiding high gears
- ☐ By reducing rest periods
- ☐ By braking in good time
- ☐ By increasing your overall speed

Good forward planning will reduce fuel consumption and engine emissions. This will cause less wear on the vehicle and its tyres, as well as reducing environmental pollution.

13.21 Mark one answer DBC s4, DGV s4

How can you help to protect the environment?

- ☐ By driving faster to reduce travelling time
- ☐ By using bypasses and avoiding town centres
- ☐ By filling your fuel tank with red diesel fuel
- ☐ By leaving your engine running in traffic jams

By using a bypass, rather than joining the stop/start traffic in a town centre, you're less likely to be held up and can travel at a constant speed. Keeping to a constant speed will reduce the amount of pollution produced by your vehicle.

13.22 Mark one answer DBC s4, DGV s4

How could a professional driver improve their efficiency?

- ☐ Keep to maximum speeds for shorter journeys
- ☐ Plan routes to avoid busy times and congestion
- ☐ Avoid route-planning because of the time it takes
- ☐ Drive at a faster speed through hazardous areas

If you plan your journey to avoid busy times, you can avoid congestion. That way, your overall journey time may be reduced, so cutting costs and making you a more efficient driver.

13.23 Mark one answer DBC s2, DGV s2

You've just refilled your fuel tank. What must you check before driving away from the fuel pump?

- ☐ That the tank is completely full up to the filler neck
- ☐ That the filler cap is vented correctly by keeping it loose
- ☐ That the filler cap is properly closed and secure
- ☐ That the tank is nearly full and the filler cap is slightly loose

Leaving the filler cap off, or not properly securing it, could result in a spillage of diesel onto the road. Because diesel is slippery, this would be a major hazard, so make sure to correctly secure the filler cap after use.

13.24 Mark one answer DBC s2, DGV s2

What should you do to prevent fuel spillage?

- ☐ Stop refuelling when the tank is half full
- ☐ Use a filtered fuel system
- ☐ Close and secure the filler cap
- ☐ Place the drip tray correctly

Fuel spilled on the road is a serious danger to other road users, especially motorcyclists. To prevent this from happening, make sure the filler cap is secure after refuelling.

13.25 Mark one answer HC p130

You're following a lorry that's leaking fuel. What should you be especially aware of?

- ☐ The road may be very slippery
- ☐ The fuel may splash your vehicle
- ☐ Your brake linings will become slippery
- ☐ Your spray reducers won't be effective

Leaking fuel can make the road extremely slippery. Look out for this, particularly on roundabouts.

13.26 Mark one answer DBC s3, DGV s3

Which large vehicles are more likely to damage the road surface?

- ☐ Vehicles with a mixture of tyre makes
- ☐ Vehicles with a mixture of re-cut and new tyres
- ☐ Vehicles with faulty spray-suppression equipment
- ☐ Vehicles with faulty suspension

Suspension faults may result in road damage. Road-friendly suspension reduces the impact on the environment, the road surface and under-road services.

13.27 Mark one answer DBC s4, DGV s4

How could you save fuel?

- ☐ By extending the vehicle's service times
- ☐ By braking as late as you can
- ☐ By planning routes to avoid congestion
- ☐ By selecting lower gears

You should always plan your route. This will ensure you always drive on roads that are suitable for the size and weight of your vehicle. By planning to avoid peak travel times, you can also avoid congestion. This can save you fuel, as well as reducing your journey times.

13.28 Mark one answer DBC s4, DGV s4

Which fuel provides a reduction in exhaust emissions harmful to human health?

- ☐ High-sulphur diesel
- ☐ Red diesel
- ☐ Low-sulphur diesel
- ☐ Blue diesel

Low-sulphur diesel reduces the levels of sulphur-dioxide particles in exhaust emissions. It's widely available and makes a contribution to reducing emissions that are harmful to human health.

13.29 Mark one answer DBC s4, DGV s4

Which fuel is better for the environment?

- ☐ Anti-waxing diesel
- ☐ Low-sulphur diesel
- ☐ Red diesel
- ☐ Anti-foaming diesel

Sulphur produces acidic gases. Diesel fuels with lower sulphur content have been produced to reduce exhaust emissions that harm both people and the environment.

13.30 Mark one answer DBC s3, DGV s3

What must you do after filling your fuel tanks?

- ☐ Check your fuel gauge
- ☐ Check your tachograph
- ☐ Securely close the filler caps
- ☐ Complete the fuel log sheets

Make certain that your filler caps are properly closed, so that you don't spill fuel as you drive along the road.

13.31 Mark one answer DBC s3, DGV s3

You're driving between the hours of 11.30 pm and 7.00 am. When must you switch off your vehicle's reverse warning bleeper?

- ☐ Before reversing on a road that has a 30 mph speed limit
- ☐ Before reversing on a road that has a temporary speed limit
- ☐ Before reversing on a road that has the national speed limit
- ☐ Before reversing on a road that has a 40 mph speed limit

Your vehicle will make more noise than a car, so you should try to avoid making unnecessary noise at any time of the day or night. This is especially important when you're in a residential area at a time when people are likely to be sleeping.

13.32 Mark one answer DBC s4, DGV s4

What will cause your fuel consumption to increase?

- ☐ Continuous use of the air suspension
- ☐ Continuous use of the mirrors
- ☐ Continuous use of the air conditioning
- ☐ Continuous use of the electrical retarder

Use your air conditioning only when it's absolutely necessary. Running it continuously increases overall fuel consumption by up to 15%.

13.33 Mark one answer DBC s3, DGV s3

Why are vehicles fitted with air suspension?

- ☐ To reduce wear to roads
- ☐ To reduce tyre wear
- ☐ To improve fuel consumption
- ☐ To help the driver stay awake

Air suspension provides a smoother journey than traditional suspension and transmits less vibration to the road surface, under-road services and nearby structures, such as bridges. It's considered to be environmentally friendly because it reduces damage to the road network.

13.34 Mark one answer — DBC s3, DGV s3

You're driving in a built-up area. Your vehicle is fitted with an audible warning device that sounds when it's reversing. When must you turn off this alarm?

- ☐ Between midnight and 6.30 am
- ☐ Between 11.30 pm and 6.30 am
- ☐ Between 11.30 pm and 7.00 am
- ☐ Between 12.30 am and 7.00 am

Some vehicles are fitted with an audible warning that sounds when the vehicle is reversing. This is an effective way to warn pedestrians and other road users of a reversing vehicle, but it doesn't take away the need to use effective observation all around your vehicle before and while you're reversing. Don't use the device in built-up areas between 11.30 pm and 7.00 am. Have consideration for the residents and don't disturb them with excessive noise.

13.35 Mark one answer — DBC s3, DGV s3

What are the consequences of driving your vehicle when it's emitting excessive exhaust smoke?

- ☐ Its fuel consumption will improve
- ☐ Your vision ahead will be reduced
- ☐ The brakes will fade more easily
- ☐ You'll be breaking the law

It's illegal to drive a vehicle that's emitting excessive exhaust smoke, so you should get the problem fixed as soon as you're aware of it. Excessive exhaust smoke increases pollution and creates a hazard for other road users.

13.36 Mark one answer — DBC s3, DGV s3

What does 'road-friendly' suspension help to reduce?

- ☐ Damage to level crossings
- ☐ Damage to river banks
- ☐ Damage to the road surface
- ☐ Damage to overhead cables

Road-friendly suspension reduces the vibration caused by the impact of the wheels on the road surface.

13.37 Mark one answer — DBC s3, 4, DGV s3, 4

How can you help to ease traffic congestion?

- ☐ Plan journeys to avoid busy times
- ☐ Plan journeys to avoid quiet times
- ☐ Use motorways for all journeys
- ☐ Avoid motorways altogether

By planning your journey to avoid busy times, you'll avoid the build-up of traffic that often occurs.

13.38 Mark one answer DBC s3, DGV s3

You've parked for a short time in a town while you plan your route. What should you do while your vehicle is parked?

- ☐ Keep the engine running on tick-over
- ☐ Rev the engine occasionally to maintain air pressure
- ☐ Hold the engine at a fast tick-over to maintain air pressure
- ☐ Switch off the engine

Turning off the engine is safer and emits less fumes into the environment. Leaving the engine idling unnecessarily will also waste fuel.

13.39 Mark one answer DBC s3, 4, DGV s3, 4

What can cause unnecessary pollution to the environment?

- ☐ High-level exhaust systems
- ☐ Regular servicing
- ☐ Weak radiator coolant
- ☐ Poorly maintained vehicles

All drivers should play their part to protect the environment against traffic pollution. Make sure the vehicle you drive is properly maintained. Report any defects, such as excessive exhaust smoke, as soon as you can.

13.40 Mark one answer DBC s3, DGV s3

Diesel fuel has been spilled on the road. Which road users will this be a particular danger for?

- ☐ Lorry drivers
- ☐ Motorcyclists
- ☐ Horse riders
- ☐ Car drivers

Spilt diesel creates a serious risk to other road users, especially motorcyclists. Take care when refuelling and make sure that all filler caps and tank hatches are properly closed and secure.

13.41 Mark one answer DBC s4, DGV s4

You're driving in a built-up area between 11.30 pm and 7.00 am. For what reason can you sound the vehicle's horn?

- ☐ To greet a friend
- ☐ To get another driver to move out of your way
- ☐ To let another road user know you're giving way
- ☐ To warn another road user who poses a danger

The horn shouldn't be used between 11.30 pm and 7.00 am in a built-up area. The only exception is when another road user is unaware of you and poses a danger.

13.42 Mark one answer DBC s3, DGV s3

Spilled diesel makes the road very slippery. Which vehicles will be most at risk from this hazard?

- ☐ Motorcycles
- ☐ Tractors
- ☐ Buses
- ☐ Lorries

Take care to secure caps and tank hatches. Diesel is very slippery and, if spilled on the road, it's particularly dangerous to motorcyclists.

13.43 Mark one answer DBC s2, DGV s3

What should you do if you've lost the filler cap for your diesel tank?

- ☐ Get a replacement before driving
- ☐ Push a rag into the filler pipe
- ☐ Drive slowly back to your depot
- ☐ Only fill the tank half-full

Don't drive without a fuel filler cap in place, as fuel could spill onto the road, especially when you're cornering or turning at a roundabout. Diesel fuel will make the road surface extremely slippery. Any spillage should be reported, so that the emergency services can make the road safe.

13.44 Mark one answer DBC s4, DGV s4

Before starting a journey, it's wise to plan ahead. What would you do to help plan your route?

- ☐ Look at a map
- ☐ Contact your local garage
- ☐ Look in your vehicle service record
- ☐ Check your vehicle registration document

Planning your journey before you set out can make the drive much easier and reduce traffic congestion. Look at a map to help you do this. You may need maps of different scales, depending on where and how far you're going. Printing or writing down the route can also help. Don't look at the map while you're driving; stop somewhere safe before checking your route again.

13.45 Mark one answer DBC s4, DGV s4

Where can you get help to plan your route before starting a journey?

- ☐ Your local filling station
- ☐ A motoring organisation
- ☐ The Driver and Vehicle Licensing Agency
- ☐ Your vehicle manufacturer

Most motoring organisation websites allow you to create a detailed plan of your trip, showing directions and distances. Some also include advice on rest and fuel stops. The Traffic England website will give you information on roadworks and incidents, along with expected delay times.

13.46 Mark one answer — DBC s4, DGV s4

How can you plan your route before starting a long journey?

- ☐ Check your vehicle's workshop manual
- ☐ Ask your local garage
- ☐ Use a route planner on the internet
- ☐ Consult your travel agents

Various route planners are available on the internet. Most of them give you options, allowing you to choose the most direct route, the quickest route or a scenic route. The plans may also include rest and fuel stops. Print them off and take them with you, but don't look at them while you're actually driving.

13.47 Mark one answer — DBC s4, DGV s4

How can you plan your route before setting out?

- ☐ Look in a motoring magazine
- ☐ Only visit places you know
- ☐ Try to travel at busy times
- ☐ Print or write down the route

Print or write down your route before setting out. Some places aren't well signed, so using both place names and road numbers may help you avoid getting lost. Find out how far you'll be travelling and plan your rest stops into the route.

13.48 Mark one answer — DBC s4, DGV s4

Why is it a good idea to plan your journey to avoid travelling at busy times?

- ☐ Your vehicle will use more fuel
- ☐ There'll be fewer roadworks
- ☐ It will help to ease congestion
- ☐ You'll travel a much shorter distance

Avoiding busy times means that you're not adding needlessly to traffic congestion. Other advantages are that you'll use less fuel and you may have an easier journey, making you feel less stressed.

13.49 Mark one answer — DBC s4, DGV s4

What's likely to happen if you avoid travelling during peak times?

- ☐ You're more likely to be held up
- ☐ Your journey time is likely to be longer
- ☐ You'll travel a much shorter distance
- ☐ You're less likely to be delayed

If possible, avoid the early morning, late afternoon and early evening peak times. Doing this should allow you to miss the worst daily traffic jams and to travel in a more relaxed frame of mind, arriving at your destination feeling less stressed.

13.50 Mark one answer DBC s4, DGV s4

Planning your route before starting a journey is a good idea. Why should you also plan an alternative route?

☐ Your original route may be blocked

☐ Your maps may have different scales

☐ It may be more scenic

☐ You may get held up by a tractor

It can be frustrating and worrying to find your planned route is blocked by roadworks or diversions. If you've planned an alternative, you'll feel less stressed and able to concentrate fully on your driving. If your original route is mostly on motorways, it's a good idea to plan an alternative using non-motorway roads. Always carry a map with you just in case you need to refer to it – but stop driving before you do so.

13.51 Mark one answer DBC s3, 4, DGV s4

What will happen if you drive smoothly?

☐ You'll reduce journey times by about 15%

☐ You'll increase fuel consumption by about 15%

☐ You'll reduce fuel consumption by about 15%

☐ You'll increase journey times by about 15%

By driving smoothly, you'll reduce fuel consumption by about 15%, and you'll reduce the amount of wear and tear on your vehicle. You'll also feel more relaxed and have a more pleasant journey.

13.52 Mark one answer DBC s4, DGV s4

Which driving technique can save you fuel?

☐ Using lower gears as often as possible

☐ Accelerating sharply in each gear

☐ Using each gear in turn

☐ Missing out some gears

Missing out intermediate gears, when appropriate, reduces the amount of time you spend accelerating. Your vehicle uses the most fuel when you're accelerating, so reducing this will reduce your vehicle's fuel consumption.

13.53 Mark one answer DBC s2, 3, DGV s2

Usually, a rev counter is divided into coloured bands. Which band should you stay in for maximum fuel economy?

☐ Blue

☐ Green

☐ Amber

☐ Red

Most large vehicles have a rev counter (tachometer) that's divided into coloured bands. This helps you to drive in the most fuel-efficient way and avoid damage to the engine. The green band is the one that gives adequate torque and power with optimum fuel economy.

13.54 Mark one answer DBC s2, DGV s2

What would be the most likely effect of driving with the rev counter in the red band?

- ☐ You would have optimum fuel economy
- ☐ You would improve engine efficiency
- ☐ You would damage the engine
- ☐ You would create a false tachograph reading

Driving with the rev counter in the red band will cause the engine to wear prematurely and can lead to expensive repairs. This could result in the vehicle being off the road for some time.

13.55 Mark one answer DBC s1, DGV s1

What can be expected after drivers complete a Driver Certificate of Professional Competence (Driver CPC)?

- ☐ Increased vehicle breakdowns
- ☐ Increased fuel consumption
- ☐ Reduced traffic volume
- ☐ Reduced engine emissions

Better knowledge and use of ecosafe driving techniques can result in lower fuel consumption and emissions. Other benefits include improved road safety and greater professionalism.

13.56 Mark one answer DBC s2, DGV s2

Why do energy-saving tyres contribute to better fuel economy?

- ☐ They have a reduced rolling resistance
- ☐ They're much easier to manufacture
- ☐ They allow you to travel at higher speeds
- ☐ They allow heat to disperse more quickly

Less fuel will be used to move your vehicle at the same speed if you use a tyre with reduced rolling resistance, rather than one of normal construction. Low-rolling-resistance tyres aren't all the same; they have ratings for rolling resistance, wet-weather grip and noise. Specialists can advise on the best tyres for specific requirements.

13.57 Mark one answer DBC s2, DGV s2

Your vehicle is fitted with an engine management system. What will happen if you press the accelerator when you're starting the engine?

- ☐ An excessive amount of fuel will be used
- ☐ There'll be an excessive build-up of air pressure
- ☐ Exhaust emissions will be reduced
- ☐ Oil consumption will be reduced

When starting an engine fitted with an engine management system, you don't need to press the accelerator, because the engine management system automatically regulates the fuel/air supply. Using the accelerator at start-up wastes fuel, causes increased engine noise and is harmful to the environment.

13.58 Mark one answer DBC s1, DGV s1

What benefit is expected to come from having a Driver Certificate of Professional Competence (Driver CPC) qualification?

- ☐ A reduction in fuel consumption
- ☐ An exemption from tachograph regulations
- ☐ A reduction in vehicle tax
- ☐ An exemption from all congestion charges

The Driver CPC qualification and training syllabus is designed to lead to enhanced professionalism and improvements in road safety. Other expected benefits are a general reduction in fuel consumption and emissions, resulting from a better knowledge of ecosafe driving techniques.

13.59 Mark one answer DBC s3, DGV s3

You've been stopped at a roadside check. What would staff from the environmental health department be checking?

- ☐ Licence entitlement
- ☐ Exhaust emissions
- ☐ Tachograph changes
- ☐ Illegal immigrants

Local-authority environmental health departments check vehicles' exhaust emissions. They have the power to prosecute the driver and the operator, as do enforcement bodies such as DVSA.

13.60 Mark one answer DBC s4, DGV s4

What helps you to become an ecosafe driver?

- ☐ Being aware of hazards
- ☐ Avoiding block changing
- ☐ Accelerating rapidly
- ☐ Using air conditioning

Driving in an ecosafe manner means that you plan well ahead for hazards and react in good time. This can save fuel and is therefore good for the environment.

13.61 Mark one answer DBC s4, DGV s4

What can help to maximise fuel economy?

- ☐ Cruise control
- ☐ Air suspension
- ☐ Re-grooved tyres
- ☐ Diff-lock

Using cruise control can help save fuel. The sensitive electronics make continuous micro-adjustments that improve efficiency. It should only be used when you can see well ahead in good road conditions. Be prepared to switch it off immediately if the traffic situation changes.

13.62 Mark one answer DBC s4, DGV s4

How would you benefit from forward planning and early recognition of potential hazards?

- ☐ You'll use less fuel
- ☐ You'll use more fuel
- ☐ You'll be able to stop more quickly
- ☐ You'll reduce journey times

Poor planning leads to late braking and harsh acceleration. This causes higher fuel consumption and increased emissions, which is harmful to the environment.

13.63 Mark one answer DBC s4, DGV s4

What will cause a noticeable increase in your vehicle's fuel consumption?

- ☐ Using electric windows
- ☐ Using a manual sunroof
- ☐ Using air conditioning
- ☐ Using the endurance brake

Using air conditioning continuously can increase fuel consumption by about 15%. Think of the environment and only use it when it's necessary.

13.64 Mark one answer DBC s4, DGV s4

Which ancillary equipment will cause the greatest increase in fuel consumption?

- ☐ Air suspension
- ☐ Air brakes
- ☐ Air bags
- ☐ Air conditioning

A number of ancillary systems use power from the engine to operate. The downside to this is that fuel consumption can increase. Air conditioning can increase fuel consumption by up to 15%.

13.65 Mark one answer DBC s4

You'll be waiting at a terminus for some time. How can you reduce pollution?

- ☐ By revving your engine
- ☐ By switching off your engine
- ☐ By leaving your engine on tick-over
- ☐ By keeping your engine at high revs

If you have to wait for long periods of time, you should turn your engine off. Noise and exhaust fumes can cause annoyance as well as pollution.

13.66 Mark one answer DBC s3

You're waiting for some time in a stationary traffic queue. Why should you switch off your engine?

- ☐ To keep your passengers cool
- ☐ To save on vehicle air pressure
- ☐ To reduce exhaust emissions
- ☐ To reduce radio interference

When you're stationary for any length of time (for example, in a traffic jam), switch off your engine. This will cut exhaust emissions and so help to protect the environment.

13.67 Mark one answer DBC s3

You've been waiting in a traffic queue for several minutes. What should you do if it's clear that you won't be moving soon?

- ☐ Keep your engine at tick-over speed
- ☐ Rev your engine occasionally
- ☐ Switch off your engine
- ☐ Run the engine at a constant higher speed

Switch your engine off when you're stationary for some time, especially where noise and exhaust fumes may cause annoyance. This can help to protect the environment and save fuel.

13.68 Mark one answer DBC s3

What effect will continuous use of the air-conditioning system have on fuel consumption?

- ☐ Increase it by about 15%
- ☐ Increase it by about 30%
- ☐ Increase it by about 50%
- ☐ Increase it by about 75%

Avoid using air-conditioning systems for long periods, as they can increase fuel consumption by about 15%. Try to drive with fuel economy and the environment in mind.

13.69 Mark one answer DBC s3

When starting your bus, how should air pressure be built up?

- ☐ With the doors open
- ☐ With the service brake on
- ☐ With the engine ticking over
- ☐ With the engine revving up

Vehicle designers and bus drivers have a duty to help reduce the effects of pollution on the environment. The most environmentally friendly way to allow air pressure to build up is with the engine running at tick-over.

13.70 Mark one answer DBC s2

What must you check after refuelling your bus?

- ☐ Your filler caps are securely closed
- ☐ Your tank is full to the top
- ☐ The position of the emergency fuel cut-off switch
- ☐ The low-fuel warning light is working

It's a legal requirement that you check all filler caps are properly closed and secure after refuelling your vehicle.

Section fourteen
Other road users

In this section, you'll learn about

- being aware of other road users
- showing patience and care.

Other road users

Being aware of other road users

When turning into side roads, pedestrians who might be crossing have priority.

Before turning left, check your nearside mirror to make sure that cyclists aren't passing on your left, between your vehicle and the kerb. If they are, allow them to get clear before you turn.

DBC s4 **DGV** s4

If you're on a road where there are parked vehicles, stay aware of what's going on around you.

- Look for pedestrians, who might step out between the vehicles without first checking it's safe.
- If you see a bicycle wheel sticking out from between two vehicles, there may be a cyclist about to cross the road.

DBC s4 **DGV** s4

At night, in low light or in bad weather, **vulnerable road users** can be especially difficult to see. Remain alert and look for these road users.

DBC s4 **DGV** s4

Definition

vulnerable road users

Road users who have a high casualty rate, either because of a lack of protection (such as pedestrians, cyclists and motorcyclists) or because of a lack of task capability (such as novice drivers).

Near schools there will be yellow zigzag markings on the road. These prohibit parking or stopping. You may also see amber flashing lights at certain times.

- Keep your speed low.
- Look for any children who may be crossing in the area.

DBC s1, 4 DGV s4 HC r208–209

There are several kinds of **light-controlled pedestrian crossings**. At **toucan crossings**, cyclists are allowed to ride across the road. Be alert for cyclists approaching very quickly; they may be unable to stop if the lights change.

DBC s6 DGV s6 HC r191–199

light-controlled pedestrian crossing
A place for pedestrians to cross the road, controlled by traffic lights.

toucan crossing
A light-controlled pedestrian crossing that also allows bicycles to be ridden across.

You may need to straddle lanes when turning either left or right.

- Check carefully in your mirrors before making the manoeuvre.
- Look for cyclists or motorcyclists who may try to pass you without realising the danger they've created for themselves.

DBC s4 DGV s4

Cyclists or horse riders turning right at a roundabout may signal right but use the left lane for safety. Be aware of how these road users may behave. Give them the time and room they need to do this in safety.

DBC s4 DGV s4 HC r187

Motorcyclists often improve their visibility by using dipped headlights in daylight and wearing bright or fluorescent clothing.

Always look out for motorcyclists before you turn or emerge. They can be difficult to see and can approach far more quickly than you think.

DBC s4 **DGV** s4

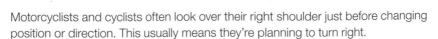

Motorcyclists and cyclists often look over their right shoulder just before changing position or direction. This usually means they're planning to turn right.

- Watch out for motorcyclists or cyclists who do this.
- Give them the time and room they need to signal and manoeuvre in safety.

DBC s4 **DGV** s4

When driving through bus/coach stations, look for passengers who have left other vehicles. They may step out from behind a parked vehicle without realising the danger from moving vehicles.

DBC s1

Showing patience and care

Take care when manoeuvring a large vehicle.

- Be aware of the vehicle's blind spots.
- If you're not sure it's safe to manoeuvre, ask for help.

DBC s4 **DGV** s4

Learner drivers may drive slowly and cautiously, or make mistakes. Treat them with patience and consideration, and give them plenty of room.

DBC s4 **DGV** s4

Be patient when stopping at zebra crossings and give people plenty of time to cross in safety. Some people, particularly older people or those with a disability, may need more time to finish crossing.

Remember that people who have sight or hearing problems (or both) may not be able to see or hear you.

DBC s6 **DGV** s6 **HC** r191–199, 207

Take care near horse riders, or in rural places where you could meet livestock in the road.

- Pass them very slowly.
- Stop if the person herding the animals signals for you to do so.

Don't rev your engine, cause the air brakes to hiss or use the horn. This could frighten the animals and make them behave unpredictably.

DBC s4 **DGV** s4 **HC** r214–215

Slow-moving vehicles, abnormal loads and recovery vehicles will often display an amber flashing light. Take care when passing them and allow plenty of room. Only pass when the road is clear ahead and it's safe and legal to do so.

DBC s4 **DGV** s4 **HC** r220, 225

See this link for more information on rural driving.

tinyurl.com/lyaal4s

Meeting the standards

You must be able to

actively scan for more vulnerable road users – for example, cyclists and motorcyclists – at junctions, roundabouts and crossings

give other road users enough time and space to perform manoeuvres

manage the risk that other road users may not give you enough space to manoeuvre

monitor and manage your own reaction to other road users.

You must know and understand

the rules that apply to other road users, particularly

- drivers of other large vehicles
- vulnerable road users such as cyclists and motorcyclists

and the position that they may select on the road as a result

the importance of predicting the likely actions of other road users, particularly vulnerable road users such as

- cyclists
- motorcyclists
- children
- older people

that other road users may not understand that your vehicle needs more space to manoeuvre, particularly

- when cornering
- at junctions
- on roundabouts

that driving without due care and attention and reasonable consideration for other road users is an offence.

Notes

You can use this page to make your own notes or diagrams about the key points you need to remember.

Think about

- Do I stay aware of vulnerable road users?
- Do I check for other road users before turning or straddling lanes?
- Do I look out for pedestrians who may have priority?
- Do I look out for motorcyclists and anticipate how they might manoeuvre?
- Do I show patience and care towards pedestrians, cyclists, learner drivers, horse riders or livestock?
- Do I respond appropriately when I see warning lights on other vehicles or at the side of the road?

Your notes

Things to discuss and practise with your instructor

These are just a few examples of what you could discuss and practise with your instructor. Read more about other road users to come up with your own ideas.

Discuss with your instructor

- how to behave when following a learner driver
- which sticks are used by people with different disabilities, eg a white stick with a red band
- the times when motorcyclists and cyclists may be particularly vulnerable and how you should behave towards them.

Practise with your instructor

- driving down country lanes where you may encounter horse riders or pedestrians walking along the road
- driving through a town centre and responding to the different types of pedestrian crossing
- identifying the signs warning you of vulnerable road users, eg a red triangle with a picture of a bicycle.

14.1 Mark one answer | HC p130

Your vehicle leaks diesel fuel on a roundabout. Who will this affect most?

☐ Drivers of three-wheeled vehicles
☐ Motorcyclists
☐ Towed vehicles
☐ Car drivers

When diesel fuel spills onto the road, it makes the road surface extremely slippery. This can cause danger for all types of vehicle, but particularly for those with two wheels, which rely on their tyres' grip to remain upright.

14.2 Mark one answer | DBC s4, DGV s4

Which vehicles are most likely to be affected by side wind?

☐ Flat-bed lorries
☐ Tractors
☐ Motorcycles
☐ Estate cars

You should be aware that motorcyclists are likely to be affected by side wind and can be blown into your path. Adjust your speed so that you can stop or take avoiding action safely if necessary.

14.3 Mark one answer | DBC s4, DGV s4

What should you do if the vehicle ahead is being driven by a learner?

☐ Keep calm and be patient
☐ Drive up close behind
☐ Put your headlights on full beam
☐ Sound your horn and overtake

Learners might take longer to react to traffic situations. Don't unnerve and intimidate them by driving up close behind.

14.4 Mark one answer | DBC s4, DGV s4, HC r215

You're about to overtake horse riders. What could scare the horses?

☐ Sounding your horn
☐ Giving arm signals
☐ Driving slowly
☐ Keeping well back

When passing horses, allow plenty of space, slow down and stop if you need to. Animals are easily frightened by sudden or loud noises and can quickly become uncontrollable, so don't sound your horn or rev the engine.

14.5 Mark one answer HC r198

You've stopped at a pelican crossing. What should you do if the lights have changed to green but a disabled person is still crossing?

- ☐ Give way to them
- ☐ Drive in front of them
- ☐ Drive behind them
- ☐ Sound your horn at them

At a pelican crossing, the green light means you may go, but you must give way to anyone still crossing. If someone hasn't finished crossing, be patient. People with mobility difficulties can struggle to cross in the time allowed by the lights.

14.6 Mark one answer HC p109, KYTS p13

Which sign means that there may be people walking along the road?

- ☐ 
- ☐ 
- ☐ 
- ☐ 

Always check the road signs. They'll inform you of hazards ahead so you know what problems to expect. A number of different signs show pedestrians in different situations. Learn them all so you'll understand the hazards you're likely to meet.

14.7 Mark one answer HC r170, 206

You're turning left from a main road into a side road. What should you do if people are already crossing the road into which you're turning?

- ☐ Continue, as you have right of way
- ☐ Signal to them to continue crossing
- ☐ Wait and allow them to cross
- ☐ Sound your horn to warn them of your presence

Approaching a junction at the correct speed will allow you enough time to observe and react. Pedestrians crossing here have priority and you must give way to them.

14.8 Mark one answer — HC r170, 206

You're at a road junction, turning into a minor road. What should you do if there are pedestrians crossing the minor road?

- ☐ Stop and wave the pedestrians across
- ☐ Sound your horn to let the pedestrians know that you're there
- ☐ Give way to the pedestrians who are already crossing
- ☐ Carry on; the pedestrians should give way to you

Always look into the road you're entering. If pedestrians are already crossing, be considerate and give way to them. Don't wave or signal them to hurry; they have priority here.

14.9 Mark one answer — DBC s4, DGV s4, HC r180

Why should you check for motorcyclists just before turning right into a side road?

- ☐ They might be overtaking on your left
- ☐ They might be following you closely
- ☐ They might be emerging from the side road
- ☐ They might be overtaking on your right

Never attempt to change direction to the right without checking your right-hand mirror. A motorcyclist might not have seen your signal and could have decided to overtake. This action should be a matter of routine.

14.10 Mark one answer — HC r25

Why is a toucan crossing different from other crossings?

- ☐ Moped riders can use it
- ☐ It's controlled by a traffic warden
- ☐ It's controlled by two flashing lights
- ☐ Cyclists can use it

Pedestrians and cyclists are shown the green light together. The signals are push-button-operated and there's no flashing amber phase.

14.11 Mark one answer — HC r25, KYTS p124

What happens at toucan crossings?

- ☐ You only stop when a pedestrian is waiting
- ☐ Cyclists aren't permitted
- ☐ There's a continuously flashing amber beacon
- ☐ Pedestrians and cyclists may cross

Toucan crossings can be used by cyclists and pedestrians. Some cycle routes lead cyclists to these crossings. Always look out for cyclists, as they're likely to be approaching much faster than pedestrians.

14.12 Mark one answer — HC p117

Where would you see this sign?

- ☐ In the window of a car taking children to school
- ☐ At the side of the road
- ☐ At playground areas
- ☐ On the rear of a school bus or coach

Vehicles that are used to carry children to and from school will be travelling at busy times of the day. Be prepared for a vehicle with this sign to make frequent stops. It might pick up or set down passengers in places other than normal bus stops.

14.13 Mark one answer — KYTS p36

What does this sign mean?

- ☐ No route for pedestrians and cyclists
- ☐ A route for pedestrians only
- ☐ A route for cyclists only
- ☐ A route for pedestrians and cyclists

This shared route is for pedestrians and cyclists only. Be aware that, when it ends, the cyclists will be rejoining the main road.

14.14 Mark one answer — DBC s4, DGV s4

What action would you take when elderly people are crossing the road?

- ☐ Wave them across so they know that you've seen them
- ☐ Be patient and allow them to cross in their own time
- ☐ Rev the engine to let them know that you're waiting
- ☐ Tap the horn in case they're hard of hearing

Don't hurry elderly people across the road by getting too close to them or revving the engine. Be aware that they might take longer to cross. They might also have hearing difficulties and not hear you approaching.

14.15 Mark one answer DBC s4, DGV s4

What should you do if you see two elderly pedestrians about to cross the road ahead?

- ☐ Expect them to wait for you to pass
- ☐ Speed up to get past them quickly
- ☐ Stop and wave them across the road
- ☐ Be careful; they might misjudge your speed

Elderly pedestrians may have impaired judgement, hearing and/or vision. Be aware that, if they do start to cross, they could take more time to do so.

14.16 Mark one answer DBC s4, DGV s4, HC r187

You're approaching a roundabout. What should you do if you see a cyclist signalling to turn right?

- ☐ Overtake on the right
- ☐ Give a warning with your horn
- ☐ Signal the cyclist to move across
- ☐ Give the cyclist plenty of room

Leave plenty of room if you're following a cyclist who's signalling to turn right at a roundabout. Give them space and time to get into the correct lane.

14.17 Mark one answer DBC s4, DGV s4, HC r187

You're approaching this roundabout and see the cyclist signal right. Why is the cyclist keeping to the left?

- ☐ It's a quicker route
- ☐ They're going to turn left instead
- ☐ The Highway Code doesn't apply to cyclists
- ☐ They may use the left-hand lane to turn right

Cycling in today's heavy traffic can be hazardous. Some cyclists might not feel happy about crossing the path of traffic to take up a position in an outside lane. Be aware of this and understand that, although they're in the left-hand lane, the cyclist might be turning right.

14.18 Mark one answer — DBC s4, DGV s4, HC r211

You're waiting to come out of a side road. Why should you look carefully for motorcycles?

- ☐ Motorcycles are usually faster than cars
- ☐ Police patrols often use motorcycles
- ☐ Motorcycles are small and hard to see
- ☐ Motorcycles have right of way

If you're waiting to emerge from a side road, remember to look for motorcycles. Their small size makes them easy to overlook. Be especially careful if parked vehicles are restricting your view. If you can't see, don't go.

14.19 Mark one answer — HC r86

Why do motorcyclists use dipped headlights in the daytime?

- ☐ To make themselves more easily seen
- ☐ To stop the battery overcharging
- ☐ To improve their vision
- ☐ To scare wildlife off the road

A motorcycle can be lost out of sight behind another vehicle. The use of the headlights helps to make the motorcycle more easily seen.

14.20 Mark one answer — HC r86

Why are motorcyclists advised to wear bright clothing?

- ☐ They must do so by law
- ☐ It helps to keep them cool in summer
- ☐ The colours are popular
- ☐ To help other road users to see them

Although they're advised to wear clothing that's bright or reflective, motorcyclists often wear clothing that's mainly black. This can make them difficult to see, so you must look carefully for them.

14.21 Mark one answer — DBC s4, DGV s4, HC r88

Why do motorcyclists often look over their right shoulder just before turning right?

- ☐ They need to listen for following traffic
- ☐ Motorcycles don't have mirrors
- ☐ Looking around helps them balance as they turn
- ☐ They need to check for traffic in their blind area

Expect a motorcyclist who makes a quick glance over their shoulder to be changing direction. They should do this before turning. By observing this, you'll get an early clue of their intention.

14.22 Mark one answer HC r214–215

You're approaching a roundabout. What should you do if there are horses being ridden ahead of you?

- ☐ Give them plenty of room
- ☐ Stop and wait for them to get off the road
- ☐ Sound your horn as a warning
- ☐ Accelerate past as quickly as possible

Horse riders often keep to the outside of the roundabout even if they're turning right. Give them room in case they have to cross lanes of traffic.

14.23 Mark one answer HC r214

What should you do when passing sheep on a road?

- ☐ Keep close to avoid oncoming traffic
- ☐ Drive very slowly
- ☐ Pass quickly but quietly
- ☐ Briefly sound your horn

Slow down and be ready to stop if you see animals in the road ahead. Animals are easily frightened by noise and by vehicles passing too close to them. Stop if signalled to do so by the person in charge of the animals.

14.24 Mark one answer HC r5

You're driving at night. What does it mean if you see a pedestrian wearing reflective clothing and carrying a red light?

- ☐ You're approaching roadworks
- ☐ You're approaching an organised walk
- ☐ You're approaching a slow-moving vehicle
- ☐ You're approaching an accident black spot

At night, a pedestrian carrying a red light should be at the back of an organised walk. The walkers should keep to the left, but this won't always be the case. Pass them very slowly. Be aware that the pedestrians may have their backs to you and might not know you're approaching.

14.25 Mark one answer HC r208

What action should you take if there are flashing amber lights under a school warning sign?

- ☐ Reduce speed until you're clear of the area
- ☐ Keep up your speed and sound the horn
- ☐ Increase your speed to clear the area quickly
- ☐ Wait at the lights until they change to green

The flashing amber lights are switched on to warn you that children may be crossing near a school. Drive slowly and be prepared for children running into the road.

14.26 Mark one answer HC r195

What should you do when you're approaching this zebra crossing?

- ☐ Prepare to slow down and stop
- ☐ Stop and wave the pedestrians across
- ☐ Speed up and pass by quickly
- ☐ Drive on unless the pedestrians step out

Be courteous and prepare to stop. Don't wave people across, as this could be dangerous if another vehicle is approaching the crossing.

14.27 Mark one answer HC p116, r243, KYTS p56

Why must these road markings be kept clear?

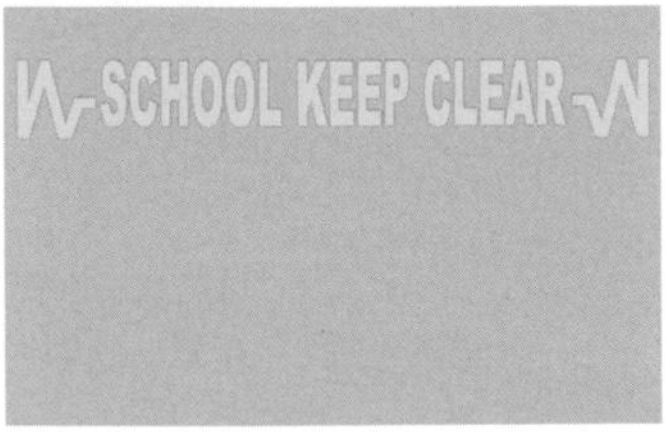

- ☐ To allow schoolchildren to be dropped off
- ☐ To allow teachers to park
- ☐ To allow schoolchildren to be picked up
- ☐ To allow an unobstructed view of the area

Keeping the markings clear ensures that drivers and riders passing and children crossing have a clear, unrestricted view of each other.

14.28 Mark one answer HC p116, r243, KYTS p56

Why mustn't you stop on these road markings?

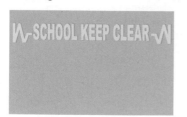

- ☐ It would obstruct everyone's view of the area
- ☐ It would obstruct teachers' access to the school
- ☐ It would obstruct delivery vehicles' access to the school
- ☐ It would obstruct emergency vehicles' access to the school

These markings are found on the road outside schools. Don't stop or park on them, even to set down or pick up children. The markings are to make sure that drivers, riders and children have a clear view.

14.29 Mark one answer HC r187

You're following two cyclists as they approach a roundabout in the left-hand lane. In which direction should you expect them to go?

- ☐ Left
- ☐ Right
- ☐ Any direction
- ☐ Straight ahead

If you're following cyclists onto a roundabout, be aware that they might not take the exit you expect. Cyclists approaching in the left-hand lane could be turning right but may not have been able to get into the right-hand lane. Give them plenty of room.

14.30 Mark one answer HC r182

You're travelling behind a moped. What should you do if you want to turn left a short distance ahead?

- ☐ Overtake the moped before the junction
- ☐ Pull alongside the moped and stay level until just before the junction
- ☐ Sound your horn as a warning and turn in front of the moped
- ☐ Stay behind until the moped has passed the junction

Passing the moped and then turning into the junction could mean that you cut across in front of the rider. This might cause them to slow down, stop or even lose control.

14.31 Mark one answer DBC s4, DGV s4, HC r187

You see a horse rider as you approach a roundabout. What should you do if they're signalling right but keeping well to the left?

- ☐ Proceed as normal
- ☐ Keep close to them
- ☐ Cut in front of them
- ☐ Stay well back

Allow horse riders to enter and exit the roundabout in their own time. Don't drive up close behind or alongside them; this could disturb the horse.

14.32 Mark one answer DBC s4, DGV s4

How would you react to drivers who appear to be inexperienced?

- ☐ Sound your horn to warn them of your presence
- ☐ Be patient and prepare for them to react more slowly
- ☐ Flash your headlights to tell them it's safe for them to proceed
- ☐ Overtake them as soon as possible

Many learners aren't confident when they first start to drive. Allow them plenty of room and don't react impatiently when they take their time. Remember that new drivers are inexperienced in dealing with traffic situations, so be patient.

14.33 Mark one answer DBC s4, DGV s4

What should you do if you're following a learner driver who stalls at a junction?

- ☐ Be patient, as you expect them to make mistakes
- ☐ Stay very close behind and flash your headlights
- ☐ Start to rev your engine if they take too long to restart
- ☐ Immediately steer around them and drive on

Learning is a process of practice and experience. Try to understand this and tolerate others who are at the beginning of this process.

14.34 Mark one answer HC r154

You're on a country road. What should you expect to see coming towards you on your side of the road?

- ☐ Motorcycles
- ☐ Bicycles
- ☐ Pedestrians
- ☐ Horse riders

On a quiet country road, always be aware that there may be a hazard just around the next bend, such as a slow-moving vehicle or pedestrians. There might not be a pavement and people may be walking on your side of the road.

14.35 Mark one answer HC r216

What should you do when you're following a car driven by an elderly driver?

- ☐ Expect the driver to drive badly
- ☐ Flash your lights and overtake
- ☐ Be aware that the driver's reactions may not be as fast as yours
- ☐ Stay very close behind but be careful

You must show consideration to other road users. Elderly drivers may have slower reactions and they might need more time to deal with a situation. Be tolerant and don't lose patience or show your annoyance.

14.36 Mark one answer DBC s4, DGV s4, HC r182

You're following a cyclist. What should you do if you wish to turn left a short distance ahead?

- ☐ Overtake the cyclist before the junction
- ☐ Pull alongside the cyclist and stay level until after the junction
- ☐ Hold back until the cyclist has passed the junction
- ☐ Go around the cyclist on the junction

Make allowances for cyclists. Allow them plenty of room, and be aware that they also have to deal with hazards. They might swerve or change direction suddenly to avoid uneven road surfaces.

14.37 Mark one answer DBC s4, DGV s4, HC r187

A horse rider is in the left-hand lane approaching a roundabout. What should you expect the rider to do?

- ☐ Go in any direction
- ☐ Turn right
- ☐ Turn left
- ☐ Go ahead

Horses and their riders will move more slowly than other road users. They might not have time to cut across heavy traffic to take up a position in the offside lane when they're turning right; it could also be hazardous for them to do so. Therefore, a horse and rider may approach a roundabout in the left-hand lane, even though they're turning right.

14.38 Mark one answer DBC s4, DGV s4

You're at the front of a queue of traffic waiting to turn right into a side road. Why should you check your right-hand mirror just before turning?

- ☐ To look for pedestrians about to cross
- ☐ To check for overtaking vehicles
- ☐ To make sure the side road is clear
- ☐ To check for emerging traffic

Always check your right-hand mirror and blind spot just before you make a right turn, as another vehicle could be overtaking you. Motorcyclists often filter along the outside of traffic queues. Keep checking all your mirrors while you're waiting to turn, because situations behind you can change quickly.

14.39 Mark one answer DBC s4, DGV s4

You're driving past a line of parked cars. What should you do if you notice a ball bouncing out into the road ahead?

- ☐ Continue driving at the same speed and sound your horn
- ☐ Continue driving at the same speed and flash your headlights
- ☐ Slow down and be prepared to stop for children
- ☐ Stop and wave the children across to fetch their ball

Beware of children playing in the street and running out into the road. If a ball bounces out from the pavement, slow down and stop if necessary. Don't encourage anyone to fetch the ball from the road. Other road users might not see your signal and you might lead a child into danger.

14.40 Mark one answer — DBC s4, DGV s4, HC r180

What should you do just before turning right from a main road into a side road?

- ☐ Cancel your right-turn signal
- ☐ Select first gear
- ☐ Check for traffic passing on your right
- ☐ Stop and set the handbrake

Motorcyclists can often pass slow or stationary traffic because of the relatively small size of their vehicle. Take a last look in your right-hand mirror to check for motorcyclists and so avoid turning across their path.

14.41 Mark one answer — HC r223

You're driving past a bus that's waiting at a bus stop on the other side of the road. Why should you be careful?

- ☐ The bus may have broken down
- ☐ Pedestrians may come from behind the bus
- ☐ The bus may move off suddenly
- ☐ The bus may remain stationary

Watch out for pedestrians if you see a bus ahead. They may not be able to see you if they're crossing behind the bus.

14.42 Mark one answer — DBC s4, DGV s4, HC r215

How should you overtake horse riders?

- ☐ Drive up close and overtake as soon as possible
- ☐ Speed isn't important but allow plenty of room
- ☐ Use your horn just once to warn them
- ☐ Drive slowly and leave plenty of room

Horses are easily frightened, so pass them slowly and carefully, giving them as much room as possible. On country roads, expect farm animals, horses, pedestrians and farm vehicles. Always be prepared to slow down or stop.

14.43 Mark one answer — DBC s4, DGV s4

Where in particular should you look out for motorcyclists?

- ☐ At a filling station
- ☐ At a road junction
- ☐ Near a service area
- ☐ When entering a car park

Motorcyclists and cyclists are smaller and usually more difficult to see than other vehicles. Look out for them at junctions; failing to see them can lead to a collision.

14.44 — Mark one answer — HC r5

While driving at night, you see a pedestrian ahead. What does it mean if they're wearing reflective clothing and carrying a red light?

- ☐ You're approaching men at work
- ☐ You're approaching an organised walk
- ☐ You're approaching an incident blackspot
- ☐ You're approaching slow-moving vehicles

Pedestrians who are part of an organised walk using the road at night should wear bright or reflective clothing. The walker in front should display a white light, while the one at the back should display a red light. Be particularly careful, slow down and give the walkers plenty of room.

14.45 — Mark one answer — DBC s4, DGV s4

Which road users are more vulnerable at night in built-up areas?

- ☐ Drivers of black taxi cabs
- ☐ Drivers of double-deck vehicles
- ☐ Cyclists
- ☐ Ambulance drivers

Look out for cyclists who don't have lights on. Also be aware that pedestrians – especially those in dark clothing – may be difficult to see.

14.46 — Mark one answer — DBC s4, DGV s4, HC r88

You're about to overtake a motorcyclist when they look around over their right shoulder. What's the most likely reason for them doing this?

- ☐ The rider intends to move to the right
- ☐ Something has fallen from the machine
- ☐ The drive chain is slack
- ☐ The rear tyre is flat

Understanding other road users' behaviour can help you anticipate their next action. On this occasion it could prevent a serious incident. This type of assessment is part of the forward planning of a professional driver.

14.47 — Mark one answer — DBC s4, DGV s4

Why should you allow extra room for motorcyclists when they're riding through roadworks?

- ☐ There may be a reduced speed limit
- ☐ There may be temporary traffic lights
- ☐ They may swerve to avoid potholes
- ☐ The traffic may be in single file

Motorcyclists can ride through narrow gaps in traffic and generally don't take up as much room as a car. However, other factors, including the road surface, can affect their stability. Give them an extra safety margin whenever possible.

14.48 Mark one answer DBC s4, DGV s4

You're following a motorcyclist along a potholed road. How should you adjust your driving to take account of this situation?

- ☐ Give them extra room
- ☐ Drive alongside them
- ☐ Try to pass them
- ☐ Stay close behind them

Good forward planning can help you keep other road users out of trouble. Information to help you do this is available if you look for it. Watch for clues: for example, a motorcyclist taking a 'lifesaver' look over their shoulder could be about to change direction.

14.49 Mark one answer DBC s4, DGV s4

Which vehicles are most at risk from spilt diesel fuel?

- ☐ Motorcycles
- ☐ Empty tankers
- ☐ Towed vehicles
- ☐ Fire engines

Where the road surface has been made slippery (for example, by spilt diesel fuel), motorcyclists are particularly vulnerable to skidding.

14.50 Mark one answer HC p117, r209

You're driving in a town. What should you do when you see this sign on a stationary bus ahead?

- ☐ Accelerate quickly
- ☐ Wait until it moves off
- ☐ Drive past slowly
- ☐ Expect the driver will be looking after the children

Children getting off the bus will be hidden from your view, and they may not be able to see you. Even if they can see you, they may have poor judgement of danger. You should drive slowly as you pass the bus and be ready to stop.

14.51 Mark one answer DBC s4, DGV s4

What should you do when you're following a scooter on a poor road surface?

- ☐ Overtake without any delay
- ☐ Stay close behind until you can pass
- ☐ Stay well back and allow them room
- ☐ Sound your horn as you get close

On a poor road surface, the rider may need to move out to avoid potholes. You may not get much warning. A look to the right or a 'lifesaver' check may warn you that they're about to move out. Stay well back and allow the rider plenty of room until you can pass safely.

14.52 Mark one answer HC r199

At toucan crossings, pedestrians share the crossing with which other road users?

- ☐ Horse riders
- ☐ Cyclists
- ☐ Motorcyclists
- ☐ Trams

A toucan crossing is signal-controlled, but unlike a pelican crossing it doesn't have a flashing amber light in the sequence. Cyclists and pedestrians cross together, and cyclists are allowed to ride their bicycles over the crossing.

14.53 Mark one answer HC r220

You're driving on a dual carriageway. What type of vehicle must use a flashing amber beacon on this road?

- ☐ An ambulance
- ☐ A fire engine
- ☐ A doctor's car on call
- ☐ A mobility scooter

Small battery-powered vehicles are often used by disabled people. On unrestricted dual carriageways, these vehicles must show a flashing amber light. Be very careful, as they have a maximum speed of 8 mph. You may see other slow-moving vehicles with amber flashing lights, such as agricultural vehicles.

14.54 Mark one answer DBC s4, DGV s4

You're driving near a school in heavy traffic. What should you do when there are children walking close to the kerb on your side of the road?

- ☐ Move to the other side of the road
- ☐ Wave at them to move back from the kerb
- ☐ Stop for a moment to see what they do
- ☐ Drive slowly until you're clear of the area

Children can be unpredictable, and when they're in groups they may fool around and push each other into the road. You should drive with extreme caution when you're near them. You need to be ready to stop suddenly.

14.55 Mark one answer DBC s4, DGV s4

You're driving on a road with an uneven surface. What should you do if you're following a motorcyclist along this road?

- ☐ Allow extra room in case the rider swerves to avoid potholes
- ☐ Leave less room so the rider can see you in their mirrors
- ☐ Drive close behind and get ready to overtake
- ☐ Drive closely to shield the rider

Never follow any road user too closely. This is particularly important when you're following scooter or motorcycle riders, or cyclists. They might brake or swerve suddenly; for example, to avoid a pothole on a road with a poor surface.

14.56 Mark one answer HC r182

You wish to turn left into a side road. What should you do if there's a cyclist in front as you approach the junction?

- ☐ Overtake the cyclist before the junction
- ☐ Wait until the cyclist has passed the junction
- ☐ Sound your horn to warn the cyclist to wait
- ☐ Drive alongside and watch the cyclist in your mirrors

Stay back and allow the cyclist to proceed. You mustn't cut across in front of them. Also be aware of any cyclists who may try to pass on your left as you slow for the turn.

14.57 Mark one answer | DBC s4, DGV s4

You're waiting to turn left at a junction. In your mirror you can see a cyclist moving up between the kerb and the nearside of your vehicle. How should you deal with the cyclist?

- ☐ Allow them to move in front of you
- ☐ Move off and make them wait for you
- ☐ Steer to the left to make them dismount
- ☐ Tell them to move out of your way

Always be aware of your limited vision around your vehicle, due to its size and shape. Never move off without checking along the nearside of your vehicle. You should be aware of the possibility of cyclists and motorcyclists getting trapped between your vehicle and the kerb.

14.58 Mark one answer | DBC s4, DGV s4, HC r187

You're approaching a roundabout. You see a cyclist signal right. Why do they keep to the left?

- ☐ It's a quicker route for a cyclist
- ☐ They're going to turn left
- ☐ It's a safer route for a cyclist
- ☐ The Highway Code doesn't apply to cyclists

The cyclist may not be able to get into a right-hand lane due to heavy traffic. Give them room, as they may not be taking the exit you expect.

14.59 Mark one answer | DBC s4, DGV s4, HC r187

You're entering a roundabout. What should you do if a cyclist in front of you is signalling to turn right?

- ☐ Overtake on the right
- ☐ Sound the horn
- ☐ Overtake on the left
- ☐ Allow plenty of room

Allow the cyclist plenty of room. Give them space and be prepared for them to change direction. To stay safe, they may travel around the roundabout in the left-hand lane.

14.60 Mark one answer | DBC s4, DGV s4, HC r215

A group of horse riders comes towards you. What should you do if the leading rider's horse becomes nervous of your presence?

- ☐ Brake gently to a stop until they've passed
- ☐ Brake quickly to a stop, applying the parking brake
- ☐ Continue driving, keeping well to the left
- ☐ Increase speed to pass the riders quickly

If any animal you pass on the road becomes unsettled, you should brake gently and come to a stop. A nervous animal is unpredictable, so you should wait until it has settled or passed by. Other road users behind you may have a limited view of the hazard, so use your mirrors and give clear signals if necessary.

14.61 Mark one answer DBC s4, DGV s4

You're emerging from a side road into a queue of traffic. Which of these vehicles is especially hard to see?

- ☐ Bicycle
- ☐ Tractor
- ☐ Milk float
- ☐ Car

Always check carefully for cyclists when emerging from a junction. Bicycles, being much narrower than any other vehicle, can be quite difficult to see.

14.62 Mark one answer DBC s4, DGV s4

Why are motorcyclists more at risk from other road users?

- ☐ They're easy to see
- ☐ They often break down
- ☐ They can't give arm signals
- ☐ They're easily overlooked

Always look for the more vulnerable road users, such as motorcyclists, cyclists, pedestrians (especially children), animals and horse riders. All need your consideration. Motorcycles are relatively small, but they can move just as fast as other vehicles, making them easy to overlook.

14.63 Mark one answer DBC s4, DGV s4

Which of these is the main cause of collisions involving motorcycles?

- ☐ Other drivers
- ☐ Other motorcyclists
- ☐ Wet roads
- ☐ Icy roads

The main reason why there are so many collisions at junctions is that drivers fail to see motorcyclists. A bright jacket or single headlight may help you identify an approaching motorcyclist, but you should be aware that not all motorcyclists wear bright clothing or use their headlights during the day.

14.64 Mark one answer DBC s4, DGV s4

What should you be on the lookout for before changing lanes in slow-moving queuing traffic?

- ☐ Private-hire taxis
- ☐ Breakdown vehicles
- ☐ Filtering motorcyclists
- ☐ Traffic cones

In slow-moving traffic, particularly on dual carriageways or motorways, always be on the lookout for motorcyclists who approach from behind and filter through narrow gaps between vehicles.

14.65 Mark one answer DBC s4, DGV s4

Which road users are especially at risk at a road junction?

- ☐ Motorcyclists
- ☐ Lorry drivers
- ☐ Car drivers
- ☐ Bus drivers

Junctions are especially risky for some road users – particularly motorcyclists, cyclists, pedestrians and mobility-scooter users. Their small size means they're easily hidden from view. At busy junctions, scan the near, middle and far distance to identify all hazards before emerging with a long vehicle.

14.66 Mark one answer DBC s4, DGV s4

When are motorcyclists particularly vulnerable?

- ☐ Near zebra crossings
- ☐ In gusting winds
- ☐ When leaving motorways
- ☐ At traffic lights

Always be concerned for the safety of other road users. Be aware of the various factors and situations where you may have to take action to keep more vulnerable road users safe.

14.67 Mark one answer HC r195

You're driving towards a zebra crossing. What should you do if you see that a person in a wheelchair is waiting to cross?

- ☐ Continue on your way
- ☐ Wave to the person to cross
- ☐ Wave to the person to wait
- ☐ Be prepared to stop

Prepare to slow down and stop, as you would for a person without a disability. Don't wave them across, as other traffic may not stop.

14.68 Mark one answer HC p105, r105

How will a school crossing patrol signal you to stop?

- ☐ By pointing to children on the opposite pavement
- ☐ By displaying a red light
- ☐ By displaying a 'stop' sign
- ☐ By giving you an arm signal

If a school crossing patrol steps out into the road with a 'stop' sign, you must stop. Don't wave anyone across the road or get impatient and rev your engine.

14.69 | Mark one answer | DBC s4, DGV s4

There's a slow-moving motorcyclist ahead of you. What should you do if you're unsure what the rider is going to do?

- ☐ Pass on the left
- ☐ Pass on the right
- ☐ Stay behind
- ☐ Move closer

The motorcyclist might be turning right, changing direction or unsure of their destination. Stay behind until the rider has made a move or you're sure you can pass safely. Don't intimidate them by moving too close.

14.70 | Mark one answer | DBC s4, DGV s4

Which road users are more vulnerable at road junctions?

- ☐ Cyclists
- ☐ Car drivers
- ☐ Lorry drivers
- ☐ Taxi drivers

Road junctions can be especially dangerous for some road users, especially those that are small and more easily overlooked. Effective observation is necessary to ensure you don't overlook anything.

14.71 | Mark one answer | DBC s4, DGV s4, HC r215

You're driving on a single carriageway road. What should you do when you see horse riders ahead?

- ☐ Move into the middle of the road
- ☐ Check your mirrors and prepare to slow down
- ☐ Switch on your vehicle's hazard warning lights
- ☐ Signal right to show you're going to pass them

Look well ahead and be ready to deal with any hazards you see coming up, such as animals in the road. Horses are easily scared by the noise of an engine and the size of your vehicle may also be intimidating. Approach them as slowly and quietly as you can.

14.72 Mark one answer DBC s4, DGV s4

As you approach a pelican crossing, the lights change to green. What should you do if an elderly couple are halfway across?

- ☐ Wave at them to cross as quickly as they can
- ☐ Rev your engine to make them hurry
- ☐ Flash your lights in case they haven't heard you
- ☐ Wait patiently and give them time to finish crossing

When the lights turn to green, you may drive on, but only if the crossing is clear. You should wait for any pedestrians to finish crossing, regardless of their age. Allow them to cross the road at their own pace; don't try to hurry them in any way.

14.73 Mark one answer DBC s4, DGV s4

What should you do when you're following a motorcyclist on an uneven road?

- ☐ Allow less room, so you can be seen in their mirrors
- ☐ Overtake them immediately
- ☐ Allow extra room, in case they swerve to avoid potholes
- ☐ Allow the same room as you would any other road user

Potholes and raised drain covers in the road can unsteady a motorcyclist. They may suddenly swerve to avoid these uneven surfaces. Watch out at places where the road is in poor condition and this is likely to occur.

14.74 Mark one answer DBC s4, DGV s4

Why do you have to be alert for motorcyclists in queues of traffic?

- ☐ They may have panniers fitted to their motorcycle
- ☐ They may ride in single file
- ☐ They may be filtering between the lanes
- ☐ They may ride with their headlights on dipped beam

In queuing traffic, motorcyclists may take the opportunity to filter between the lanes. Be aware that they may pass on either side and very close. Stay alert and always check your mirrors and blind spots before moving off, changing lanes or turning.

14.75 — Mark one answer — DBC s4, DGV s4

You're driving past parked cars. What should you do if you notice a bicycle wheel sticking out between them?

- ☐ Accelerate past quickly and sound your horn
- ☐ Slow down and wave the cyclist across
- ☐ Brake sharply and flash your headlights
- ☐ Slow down and be prepared to stop for a cyclist

Scan the road as you drive. Try to anticipate hazards by being aware of the places where they're likely to occur; for example, pedestrians and cyclists emerging between parked cars on a busy street. If you're prepared, you'll be able to slow down or stop in good time.

14.76 — Mark one answer — HC p116, r243, KYTS p56

What do yellow zigzag lines on the road outside a school mean?

- ☐ Sound your horn to alert other road users
- ☐ Stop to allow children to cross
- ☐ You mustn't wait or park on these lines
- ☐ You mustn't drive over these lines

Parking on the yellow zigzag lines will block the view of the school entrance. This will endanger the lives of children wanting to cross the road on their way to and from school.

14.77 — Mark one answer — DBC s4, DGV s4, HC r180

You're driving on a main road. What should you do just before turning right into a side road?

- ☐ Adjust your interior mirror
- ☐ Flash your headlights
- ☐ Steer over to the left
- ☐ Check for traffic overtaking on your right

Just before turning, a check in the offside mirror will help you to see any cyclist or motorcyclist passing on your offside. This check may help to prevent a serious incident from developing.

14.78 Mark one answer DBC s4, DGV s4, HC r170

Where should you take particular care to look for motorcyclists and cyclists?

- ☐ On dual carriageways
- ☐ At junctions
- ☐ At zebra crossings
- ☐ In one-way streets

Motorcyclists and cyclists may be more difficult to see, especially at junctions. They can be easily hidden by obstructions such as other vehicles, and the small size of an approaching motorcycle may make it difficult to judge its speed and distance. It's vital that you don't emerge until you're sure it's safe.

14.79 Mark one answer DBC s1

What should you do when you're driving through a bus station?

- ☐ Change your destination boards
- ☐ Drive through quickly if you're not stopping
- ☐ Look for people leaving buses
- ☐ Sound your horn as you enter

Bus stations are busy places. If people are in a rush, they may not always look properly before getting off a bus. Keep your speed down and look for manoeuvring vehicles or pedestrians stepping into your path.

Section fifteen

Road and traffic signs

In this section, you'll learn about

- signs
- road markings
- lane markings
- traffic lights and warning lights
- signals given by drivers and the police
- road lanes.

Road and traffic signs

Road and traffic signs give important information to keep you safe on the road. It's important that you know what they mean and what you need to do when you see them.

Signs

The shape and colour of a road sign tell you about its meaning.

Circular signs give orders.

Blue circles give an instruction or show what sort of road user can use a route; for example, 'buses and cycles only'.

Red rings or circles tell you what you mustn't do; for example, 'no goods vehicles over weight shown'.

Triangular signs give warnings.

Rectangular signs give information.

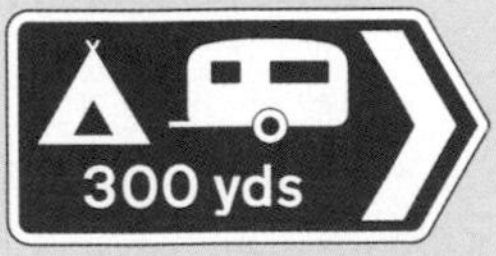

Signs with a brown background give tourist information.

KYTS p9, 84, 100–104

'Stop' signs don't follow the shape rule. 'Stop' signs are octagonal so that they stand out and can be understood even if partly covered, eg by snow.

HC r109 KYTS p9

Maximum speed limits are shown inside red circles. You **MUST NOT** go faster than the speed shown. Where no speed limit is shown, the national speed limits apply. Speed-limit signs may be combined with other signs, such as those that show a traffic-calmed area.

HC p106 KYTS p20

It's impossible to mention all the signs here. *Know Your Traffic Signs* shows all the signs you're likely to see and *The Official Highway Code* contains important advice, information on current laws in Great Britain, and best practice in road safety.

It's important to get to know these signs, to make sure you don't break the law or find that your large vehicle is too long, wide, tall or heavy for the road ahead.

HC p106–116 KYTS p9–61, 77–89

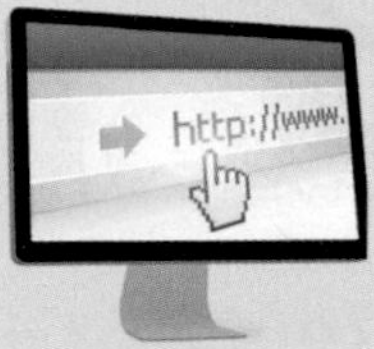

Test your knowledge of signs using the activity on the Safe Driving for Life website.

safedrivingforlife.info/road-signs-quiz

Road markings

Markings on the road give information, orders or warnings. As a general rule, the more paint there is, the more important the message.

DBC s4 DGV s4 HC r127–131 KYTS p62–64

There are three types of road markings: those along the middle of the road, those along the side of the road, and lines on or across the road.

Along the middle of the road

Short broken white lines mark the centre of the road.

Longer broken white lines show that a hazard is ahead: only overtake if the road ahead is clear.

You **MUST NOT** cross or straddle double white lines with a solid white line on your side of the road unless

- you're turning into a junction or an entrance
- you need to pass a stationary vehicle
- you need to overtake a cyclist, horse or road maintenance vehicle if they're moving at 10 mph or less.

White diagonal stripes or chevrons separate lanes of traffic or protect traffic turning right.

Sometimes red tarmac is used within a block of white lines or diagonals. This highlights the area that separates traffic flowing in opposite directions.

HC p114 KYTS p62–64

Along the side of the road

A white line shows the edge of the carriageway.

Yellow lines show waiting and stopping restrictions.

Zigzag lines (white at pedestrian crossings, yellow outside schools) mean no stopping or parking at any time.

HC p115, 116 **KYTS** p39–44, 56, 65, 122

Lines on or across the road

Broken lines across the road mean 'give way'. At a roundabout, give way to traffic from the right.

A single solid line means 'stop'.

Various markings on the road, for example 'give way' triangles, road-hump markings and rumble strips, warn of a hazard.

HC p114–116 **KYTS** p62–75

Look at *Know Your Traffic Signs* and *The Official Highway Code* to learn more about road markings.

You may see reflective studs on motorways and other roads. These are especially useful at night and when visibility is poor, as they help to make the lanes and edges of the road easier to see. Different coloured studs are used on motorways to help drivers identify which lane they're using.

Lane markings

Reflective studs help you to see where you are on the carriageway, especially at night or in fog. Different colours are used in different places.

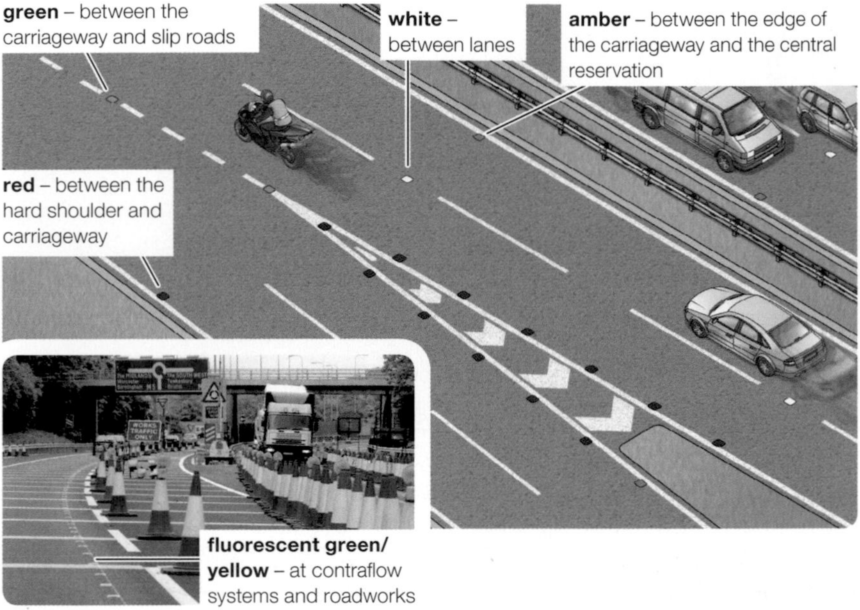

HC r132

Traffic lights and warning lights

Traffic lights work in a sequence.

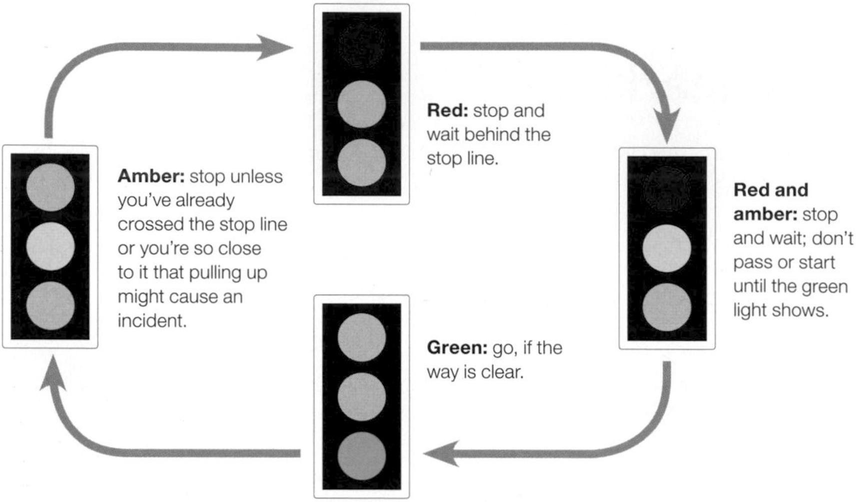
Red: stop and wait behind the stop line.
Red and amber: stop and wait; don't pass or start until the green light shows.
Green: go, if the way is clear.
Amber: stop unless you've already crossed the stop line or you're so close to it that pulling up might cause an incident.

HC p102 **KYTS** p119–120

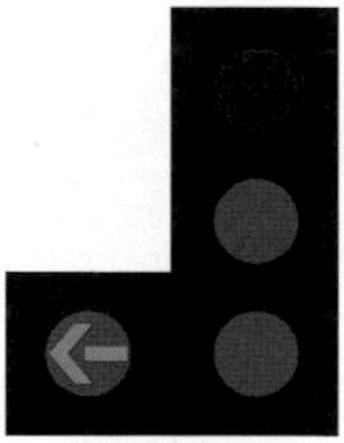

On some traffic lights there's a green filter arrow. This means you can go in the direction of the arrow, even if the main light isn't showing green.

If a set of traffic lights is out of order, drive very carefully: nobody has priority. There may be a sign telling you that the lights aren't working.

DBC s4 **DGV** s4

Red flashing lights are used at level crossings and other locations, such as lifting bridges and outside some fire stations. You **MUST** stop when these show.

HC r293, p102 **KYTS** p13, 26–29, 120

At roadworks, traffic can be controlled by

- a police officer
- traffic lights
- a 'stop/go' board.

HC r288 **KYTS** p136

On motorways, signals on the overhead gantries or at the roadside may also have flashing lights.

Amber warns you of a hazard (eg lane closures, to leave at the next exit, fog) or a temporary maximum speed advised for the conditions.

Red above your lane tells you that the lane is closed beyond this point, and you should move into another lane.

Red above all the lanes, on the central reservation or on the roadside tells you to stop. You **MUST NOT** go beyond that point in any lane.

HC r255–258, p102 **KYTS** p89–91

Signals given by drivers and the police

Road users normally use their indicators to signal where they're intending to turn. Make sure that your indicators are cancelled after you've turned, to avoid confusing other road users. Be aware that another driver may have left their indicators on by mistake.

If you're emerging from a junction and a driver coming along the main road from the right is close to you and indicating left, wait until the vehicle starts to turn before you emerge.

HC r103–104

If you're slowing down and stopping just after a junction, wait to signal until you're passing the junction, or just after it.

You can use the horn to warn others that you're there. You **MUST NOT** use it between 11.30 pm and 7.00 am when driving in a built-up area, except when another road user puts you in danger. You **MUST NOT** use your horn when stationary unless another vehicle is likely to cause a danger.

The only reason you should flash your headlights is to warn other road users that you're there.

DBC s4 DGV s4 HC r110–112

> **TIP** If you're driving on a motorway or unrestricted dual carriageway, you can briefly use your hazard warning lights to warn drivers behind you that there's an obstruction ahead.
>
> **HC r116**

Police or traffic officers may signal to you if they're carrying out roadside checks or directing traffic. Make sure you know all the official arm signals in case you need to react to them.

HC p103–105

A police or traffic officer following you in a patrol vehicle may flash their headlights, indicate left and point to the left to direct you to stop. Pull up on the left as soon as it's safe to do so.

HC r106

Remember, you **MUST** obey any signals given by police or traffic officers, traffic wardens and signs used by school crossing patrols.

HC r105–108, p104–105

Road lanes

Contraflow lanes are lanes that flow in the opposite direction to most of the traffic. Bus and cycle contraflow lanes may be found in one-way streets. They'll be signed and marked on the road.

HC r140–141, 143

You may also see contraflow lanes at roadworks. When you see the signs

- reduce your speed and comply with any temporary speed limits
- look for width or weight restrictions and choose an appropriate lane in good time
- keep a safe distance from the vehicle in front.

KYTS p128–133

Large vehicles fitted with a speed limiter aren't allowed to use the right-hand lane of a motorway with more than two lanes. Always move back to a lane on your left after overtaking, to allow other vehicles to overtake. On a free-flowing motorway or dual carriageway, you mustn't overtake other vehicles on their left.

DBC s4 DGV s4 HC r265

Meeting the standards

You must be able to

respond correctly to all

- permanent traffic signals, signs and road markings
- temporary traffic signals, signs and road markings.

You must know and understand

the meaning of all mandatory traffic signs and how to respond to them

the meaning of all warning signs and how to respond to them

the meaning of all road markings and how to respond to them

the meaning of the particular signs that apply to the vehicle that you're driving (such as height, width and weight restrictions).

Notes

You can use this page to make your own notes or diagrams about the key points you need to remember.

Think about

- Do I know what each road-sign shape means, so that I can respond accordingly?
- Have I learned what each road sign means?
- Do I understand what the road markings in the middle, at the side, and across the road all mean?
- Can I recognise different indicator lights and driver or police signals?
- Do I know how to safely use the lanes on the road?

Your notes

Things to discuss and practise with your instructor

These are just a few examples of what you could discuss and practise with your instructor. Read more about road and traffic signs to come up with your own ideas.

Discuss with your instructor

- the signs that apply only to buses and lorries, respectively
- which is the only octagonal road sign and why it's unique
- what different patterns of road markings mean. For example, double white lines, yellow boxes, etc.

Practise with your instructor

- identifying signs from *The Official Highway Code* and *Know Your Traffic Signs*
- driving through a busy town centre and identifying all the warning signs that you see
- motorway driving and identifying the signs that you see. For example
  - active traffic management (also known as smart motorways)
  - emergency refuge areas
  - reflective studs and what different colours mean.

15.1 | Mark one answer | NI EXEMPT | HC p106, KYTS p20

You're driving a lorry with a maximum authorised mass of 38 tonnes on a single carriageway road in England. What's the speed limit for your vehicle after passing this sign?

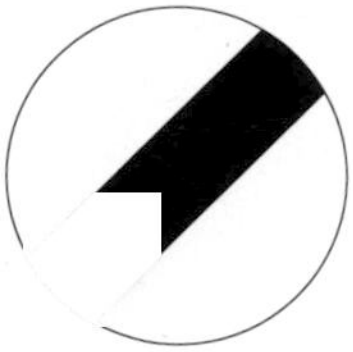

- ☐ 40 mph
- ☐ 50 mph
- ☐ 60 mph
- ☐ 70 mph

The national speed limit for a goods vehicle exceeding 7.5 tonnes on a single carriageway road in England and Wales is 50 mph. In Scotland, the limit is 40 mph. The national speed limit varies for different vehicle categories. You must know the limit that applies to the vehicle you're driving.

15.2 | Mark one answer | NI EXEMPT | HC p106, KYTS p20

You're driving a laden articulated lorry with a maximum authorised mass of 38 tonnes on a dual carriageway in England. What speed limit applies to your vehicle after passing this sign?

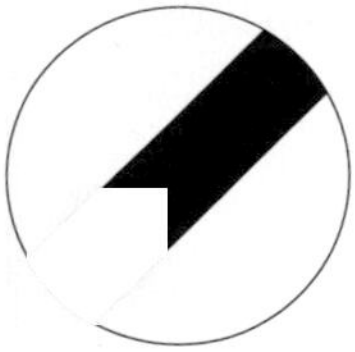

- ☐ 40 mph
- ☐ 50 mph
- ☐ 60 mph
- ☐ 70 mph

60 mph may be the legal speed limit for a goods vehicle on a dual carriageway in England and Wales, but it doesn't mean that it's safe to drive at that speed. You should always take into account the road, weather and traffic conditions, and maintain safety margins.

15.3 — Mark one answer — KYTS p32

You're driving an articulated lorry. What should you do when you see this sign?

- ☐ Turn round and find an alternative route
- ☐ Park safely and arrange alternative transport for the goods
- ☐ Inform your vehicle operator and await further instructions
- ☐ Drive past the sign because it doesn't apply to you

The sign prohibits buses and coaches with more than eight passenger seats and so doesn't apply to you.

15.4 — Mark one answer — KYTS p17

Which of these signs means you can't drive your lorry past this point?

☐  ☐ 

☐  ☐ 

No motor vehicles are allowed past this sign. Although a motorcycle and car are shown, the restriction applies to all motor vehicles.

15.5 Mark one answer KYTS p17

You're driving a lorry 30 feet long and towing a trailer 15 feet long. What should you do when you see this sign?

- ☐ Find an alternative route to your destination
- ☐ Stop and wait for a police escort
- ☐ Continue past the sign but reduce your speed
- ☐ Carry on; the sign applies to the towing vehicle only

The maximum permitted length of vehicle includes any trailer being towed. Make sure you know the total length of your vehicle.

15.6 Mark one answer KYTS p12

You're driving a lorry that's carrying a heavy load. What should you be prepared to do when you see this sign ahead?

- ☐ Brake to a lower speed
- ☐ Change to a higher gear
- ☐ Stop to check your load
- ☐ Change to a lower gear

A heavy vehicle will lose speed when climbing a steep hill. Changing to a lower gear will help you to maintain an appropriate speed.

15.7 Mark one answer KYTS p129–131

You're on a motorway. Your lorry has a maximum authorised mass of more than 7.5 tonnes. What does this sign mean to you?

- ☐ You mustn't use the right-hand lane
- ☐ You can only use the right-hand lane
- ☐ You can't leave the motorway at this junction
- ☐ You can use the middle or right-hand lane

At motorway roadworks, some lanes may be narrower than normal and large vehicles may not be allowed to use those lanes. Look for weight-limit signs to check for any lanes that you can't use. Move to the appropriate lane in good time.

15.8 Mark one answer HC p40

What's the national speed limit on a motorway for a lorry weighing over 7.5 tonnes?

- ☐ 50 mph (80 km/h)
- ☐ 55 mph (88 km/h)
- ☐ 60 mph (96 km/h)
- ☐ 70 mph (112 km/h)

Be aware of, and obey, all speed limits. On a motorway, any lorry that's articulated, towing a trailer, or over 7.5 tonnes mustn't exceed 60 mph (96 km/h).

15.9 Mark one answer HC p117

When must these two signs be used?

- ☐ When you're carrying a load that overhangs the front or rear of the vehicle by more than 1 metre (3 feet 3 inches)
- ☐ Whenever your vehicle is being towed
- ☐ Whenever a police escort is required
- ☐ When you're carrying a load that overhangs the front or rear of the vehicle by more than 2 metres (6 feet 6 inches)

The law requires you to use projection markers for long or wide loads. It's the driver's responsibility to keep the markers clean and securely fixed. They must also be independently lit at night.

15.10 Mark one answer HC p40, 106

You're driving a lorry weighing more than 7.5 tonnes on a motorway. What does this sign mean?

- ☐ Maximum speed 40 mph
- ☐ Maximum speed 50 mph
- ☐ Maximum speed 60 mph
- ☐ Maximum speed 70 mph

The maximum speed for lorries over 7.5 tonnes maximum authorised mass on a motorway is 60 mph. However, it may not always be appropriate or possible to drive at this speed, because of the weather, volume of traffic and other factors.

15.11 Mark one answer | NI EXEMPT | HC p40, 106

You're driving a lorry weighing more than 7.5 tonnes on a dual carriageway in England. What does this sign mean?

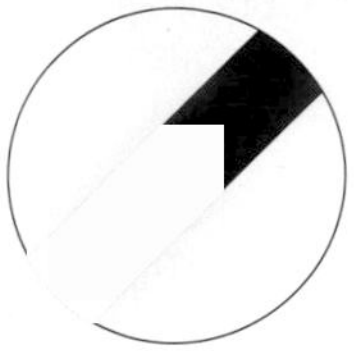

- ☐ Maximum speed 40 mph
- ☐ Maximum speed 50 mph
- ☐ Maximum speed 60 mph
- ☐ Maximum speed 70 mph

The maximum speed for lorries over 7.5 tonnes maximum authorised mass on a dual carriageway in England and Wales is 60 mph. In Scotland, the limit is 50 mph. Be aware of the speed limits applying to different types of vehicle on particular roads. This is the maximum speed allowed, but it may not always be safe to drive at this speed, because of the weather, volume of traffic and other factors.

15.12 Mark one answer | NI EXEMPT | HC p40, 106

You're driving a lorry weighing more than 7.5 tonnes on a single carriageway road in England. What does this sign mean?

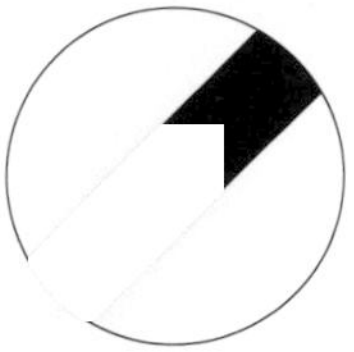

- ☐ Maximum speed 30 mph
- ☐ Maximum speed 40 mph
- ☐ Maximum speed 50 mph
- ☐ Maximum speed 60 mph

In England and Wales, the national speed limit for lorries over 7.5 tonnes maximum authorised mass on a single carriageway road is 50 mph. In Scotland, the limit is 40 mph.

15.13 Mark one answer

DGV s3, HC p115, KYTS p56–57

What's prohibited when a Red Route is in operation?

- ☐ Stopping or parking
- ☐ Overtaking
- ☐ Changing lanes
- ☐ Straddling the lines

The hours of operation of Red Routes vary from one area to another. As a rule, you mustn't stop on a Red Route, but there may be special marked boxes where loading and unloading can be carried out at certain times. Look out for signs giving information about the restrictions in place.

15.14 Mark one answer

KYTS p38

What does this sign mean?

- ☐ End of restricted speed area
- ☐ End of restricted parking area
- ☐ End of clearway
- ☐ End of cycle route

Even though there are no restrictions, make sure that you park where you won't cause an obstruction or endanger other road users.

15.15 Mark one answer

KYTS p52–55

Which sign means 'No stopping'?

☐  ☐ 

☐  ☐ 

Stopping where you see this sign is likely to cause an obstruction, which could lead to traffic delays. Obeying the sign will help keep the traffic flowing.

15.16 Mark one answer KYTS p73

What does this sign mean?

- ☐ Give way to traffic from the opposite direction
- ☐ You're entering a one-way street
- ☐ Two-way traffic ahead
- ☐ You have priority over vehicles from the opposite direction

You have priority over vehicles from the opposite direction. However, if a vehicle approaching from the opposite direction isn't going to stop, you should slow down and give way to avoid confrontation or an incident.

15.17 Mark one answer KYTS p9, 16, 29

At a junction, you see this sign partly covered by snow. What does it mean?

- ☐ Crossroads
- ☐ Give way
- ☐ Stop
- ☐ Turn right

The 'Stop' sign is the only sign that's octagonal. This helps you to recognise it even when the wording can't be seen.

15.18 Mark one answer KYTS p21

What does this sign mean?

- ☐ Service area 30 miles ahead
- ☐ Maximum speed 30 mph
- ☐ Minimum speed 30 mph
- ☐ Lay-by 30 miles ahead

This sign is shown where slow-moving vehicles would block the flow of traffic. However, if you need to slow down to avoid an incident, you should do so.

15.19 Mark one answer KYTS p19

Which of these signs means turn left ahead?

☐ 
☐ 
☐ 
☐ 

This sign gives a clear instruction. You should prepare to negotiate a left-hand turn.

15.20 Mark one answer KYTS p30

What does this sign mean?

- ☐ Route for trams
- ☐ Give way to trams
- ☐ Route for buses
- ☐ Give way to buses

Take extra care when you're driving on roads where trams operate. Look out for road markings and signs that tell you to expect them. Modern trams are very quiet and you might not hear them approaching.

15.21 Mark one answer KYTS p18–19, 73

Which of these signs means that you're entering a one-way street?

☐ 
☐ 
☐ 
☐ 

If a one-way street has two lanes, you can use either lane and overtake on either side. Use the lane that's more convenient for your destination unless directed by road signs or markings.

15.22 — Mark one answer — KYTS p33

What does this sign mean?

- ☐ Bus station on the right
- ☐ Contraflow bus lane
- ☐ With-flow bus lane
- ☐ Give way to buses

If you see a contraflow sign, there'll also be markings on the road to indicate the bus lane. Don't use this lane for parking or overtaking.

15.23 — Mark one answer — KYTS p10–15

Which hazard would be shown in a triangular road sign?

- ☐ Road narrows
- ☐ One-way street
- ☐ Bus lane
- ☐ Minimum speed

A triangular sign is a warning sign. Knowing this, you can prepare for the hazard ahead.

15.24 — Mark one answer — KYTS p13–14

Which sign means that pedestrians may be walking along the road?

☐ 

☐ 

☐ 

☐ 

Be cautious, especially if there's a bend in the road and you're unable to see well ahead. If you have to pass pedestrians, leave them plenty of room. You might have to use the right-hand side of the road, so look well ahead and in your mirrors before you pull out.

15.25 Mark one answer KYTS p12

What does this sign mean?

- ☐ Side wind
- ☐ Road noise
- ☐ Airport
- ☐ Adverse camber

Side-wind warning signs are often found on roads that are frequently affected by strong wind. High-sided vehicles, caravans and motorcycles are more susceptible to windy weather, and you should be aware of the difficulties they may be having.

15.26 Mark one answer KYTS p13

What does this traffic sign mean?

- ☐ Slippery road ahead
- ☐ Tyres liable to puncture ahead
- ☐ Danger ahead
- ☐ Service area ahead

A sign showing an exclamation mark (!) tells you that there may be danger ahead. Be ready to slow down for whatever that may be.

15.27 Mark one answer KYTS p13

What should you do if you see this sign when you're about to overtake?

- ☐ Overtake the other driver as quickly as possible
- ☐ Move to the right to get a better view
- ☐ Switch your headlights on before overtaking
- ☐ Hold back until you can see clearly ahead

You won't be able to see any hazards that might be out of sight in the dip. As well as oncoming traffic, there may be cyclists, horse riders, parked vehicles or pedestrians hidden from view. Only start to overtake when you can see the road ahead is clear, giving you enough time to overtake safely.

15.28 Mark one answer KYTS p12

What does this sign mean?

- ☐ Uneven road surface
- ☐ Bridge over the road
- ☐ Road ahead ends
- ☐ Water across the road

This sign is found where a shallow stream crosses the road. Heavy rainfall could increase the flow of water. If the water looks too deep or the stream has swelled over a large distance, stop and find another route.

15.29 Mark one answer KYTS p102

You see this traffic light ahead. Which light or lights will come on next?

- ☐ Red alone
- ☐ Red and amber together
- ☐ Green and amber together
- ☐ Green alone

The amber light means stop. You may only go on if the light changes to amber after you've crossed the stop line or when you're so close to the line that stopping might cause a collision. The amber light is followed by the red light.

You must wait behind the stop line until the green light shows.

15.30 Mark one answer KYTS p102

Where would you find these flashing red lights?

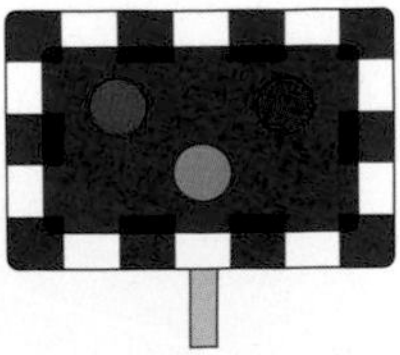

- ☐ At a pelican crossing
- ☐ At a level crossing
- ☐ At a zebra crossing
- ☐ At a school crossing

These signals can be found at level crossings, some airfields, bridges and emergency access sites. You must stop if the red lights are flashing. At level crossings, if the lights continue to flash after a train has passed, continue to wait because another train is coming.

15.31 Mark one answer KYTS p131

You're driving through roadworks. What information does this sign give you?

- ☐ Large vehicles must go straight ahead
- ☐ Traffic is joining from the left
- ☐ All traffic must leave at the next exit
- ☐ The distance to the next exit

This sign gives you advance warning of a junction where the permanent sign is obscured by roadworks.

15.32 Mark one answer KYTS p52–58

What does this sign mean?

- ☐ Stop only to pick up passengers
- ☐ No stopping at any time
- ☐ Stop only to set down passengers
- ☐ No stopping at peak times

This traffic sign means no stopping on the main carriageway at any time, not even to set down passengers. However, you may stop in a lay-by.

15.33 Mark one answer HC r264, KYTS p90

You're driving on a motorway and there's no traffic ahead. You see this sign. Which lane should you use?

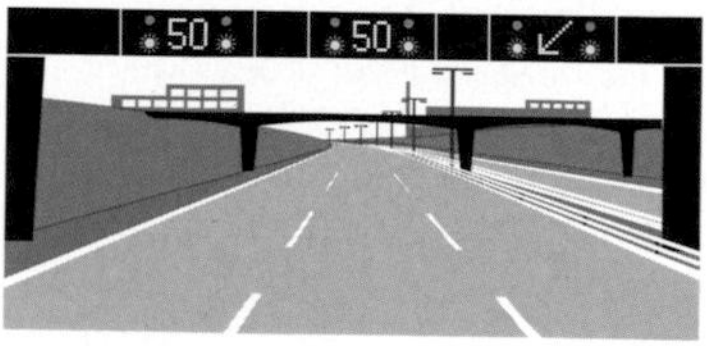

- ☐ The right-hand lane
- ☐ The hard shoulder
- ☐ The left-hand lane
- ☐ The middle lane

When there's no traffic ahead of you, you should always drive in the left-hand lane.

15.34 Mark one answer KYTS p12

Which of these signs shows an uphill gradient?

- [ ] 

- [ ] 
- [ ] 

- [ ] 

You'll need to identify the sign in time, so that you can select an appropriate gear.

15.35 Mark one answer KYTS p12

Which of these signs means 'uneven road'?

- [ ] 
- [ ] 
- [ ] 
- [ ] 

Some signs can look similar to others but each one has a different meaning. For example, this one looks a little like the sign for a hump bridge. Learn the meaning of every sign to prepare yourself for how to react to them.

15.36 Mark one answer HC r178, KYTS p36

What are these advanced stop lines for?

- [ ] To allow room for pedestrians to cross the road
- [ ] To allow space for large vehicles to turn
- [ ] To allow cyclists to position in front of other traffic
- [ ] To allow you to select where to stop

Advanced stop lines allow cyclists to take a position ahead of other traffic. When the green signal shows, they then have the time and space to move off in front of the following traffic.

15.37 Mark one answer HC p103

What does this arm signal mean?

- ☐ The driver intends to turn left
- ☐ The driver intends to slow down
- ☐ The driver wants you to keep back
- ☐ The driver wants you to go past

Sometimes, drivers or riders use arm signals as well as direction indicators and brake lights; for example, in bright sunshine or when approaching a zebra crossing. Using the arm signals shown in The Highway Code can help other road users to understand your intentions.

15.38 Mark one answer KYTS p91

What does this motorway sign mean?

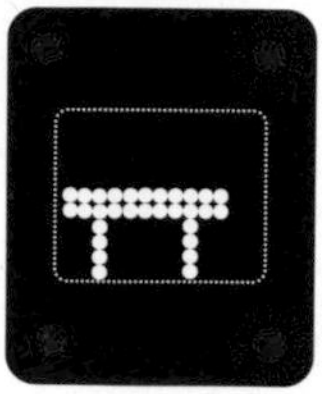

- ☐ Use the hard shoulder
- ☐ Contraflow system ahead
- ☐ Overhead bridge repairs
- ☐ All lanes closed ahead

When the red lights are flashing, don't go beyond this signal. This sign shows that all lanes ahead are closed and you mustn't enter the motorway. Motorway signals are located on the central reservation, to the left of the hard shoulder, on overhead gantries and at the entrance to the motorway.

15.39 Mark one answer DBC s3, DGV s2, KYTS p26

What does this sign warn of?

- ☐ Slippery road
- ☐ Double bend
- ☐ Overhead electrified cable
- ☐ Series of bends

This sign is used to warn of overhead electrified cables and is usually accompanied by a plate indicating the safe height limit.

15.40 Mark one answer KYTS p73

Who has priority when you're approaching this sign?

- ☐ Larger vehicles
- ☐ Oncoming traffic
- ☐ Smaller vehicles
- ☐ You have priority

When you see this sign, you must give way to traffic from the opposite direction, regardless of the size of your vehicle.

15.41 Mark one answer KYTS p13

What does this sign mean?

- ☐ Car lane only
- ☐ Single file only
- ☐ Queues likely
- ☐ Keep your distance

When you see this sign, beware of traffic queues ahead. Check your mirrors and reduce your speed. Research shows that you make poor decisions when you're angry, so you're less likely to spot and respond safely to hazards. Be patient when you're delayed and reduce the possibility of being involved in an incident.

15.42 Mark one answer KYTS p137

What does this sign mean?

- ☐ Road flooded
- ☐ Risk of punctures
- ☐ Loose chippings
- ☐ Uneven surface

This is a warning sign to indicate loose chippings ahead. The sign is usually accompanied by an advisory speed limit. Vehicles travelling too fast for the conditions can throw up chippings or stones. Avoid causing damage to paintwork and windscreens by keeping your speed down.

15.43 Mark one answer KYTS p18

What does this sign mean?

- ☐ You're allowed to carry on, but only with a police escort
- ☐ You should continue slowly if your weight is above the limit
- ☐ Don't cross unless the bridge is clear of other vehicles
- ☐ Don't cross the bridge if your weight exceeds the limit

Vehicles over the weight shown on the sign may not use the bridge ahead. You must know the weight of your vehicle, including any load. With this knowledge you should plan your route in advance, so that you don't come to a bridge you can't cross. That way, you'll avoid delays and save time and fuel.

15.44 Mark one answer KYTS p27

What does this marker count down to?

- ☐ A motorway slip road
- ☐ A primary road junction
- ☐ A concealed level crossing
- ☐ A roadside rest area

When you see this sign, reduce your speed. You may well have to stop at the level crossing, or there may already be traffic queuing ahead.

15.45 Mark one answer DBC s3, DGV s3, HC p115, KYTS p56–57

What does it mean when there are double red lines running along the edge of a road?

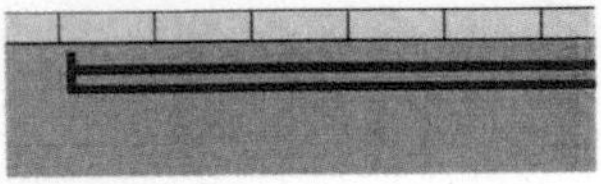

- ☐ Limited loading
- ☐ No stopping
- ☐ Bus route
- ☐ Short-term parking

Double red lines indicate that you're on a Red Route and in a no-stopping area. Red Routes also have single red lines, with signs showing the times that restrictions are in force. There are also parking and loading boxes, which have signs explaining the restrictions that apply.

15.46 Mark one answer KYTS p24

Where would you expect to see these road markings?

- ☐ At the entrance to a car park
- ☐ On the approach to an arched bridge
- ☐ At the start of a cycle lane
- ☐ On the approach to a lifting barrier

High vehicles are often directed to the centre of the road to go under an arched bridge. Check that your vehicle doesn't exceed the height restriction indicated. Plan your route in advance to avoid any bridges that your vehicle won't be able to clear.

15.47 Mark one answer HC p102

You're approaching traffic lights. Only the red light is showing. Which series of lights will show next?

- ☐ Red and amber, then green
- ☐ Green, then amber
- ☐ Amber, then green
- ☐ Green and amber, then green

If you know which light is going to show next, you can plan your approach. This will help prevent excessive braking or hesitation at the junction.

15.48 Mark one answer HC r129, KYTS p64

The double white line along the centre of the road is unbroken on your side. When may you cross the line?

- ☐ To overtake any maintenance vehicle in front
- ☐ To overtake a pedal cycle travelling at 10 mph or less
- ☐ To check it's safe to overtake the vehicle in front
- ☐ To overtake a driver travelling at 20 mph or less

Solid white lines are used where it would be dangerous to overtake. If the solid line is on your side of the road, you may cross it to pass a stationary vehicle, or to overtake a pedal cycle, horse or road-maintenance vehicle that's travelling at 10 mph or less. You may also cross the solid line to enter premises or a side road.

15.49 Mark one answer HC p113, KYTS p73

What should you do when you see this sign ahead?

- ☐ Accelerate because you have priority
- ☐ Slow down; the road may be narrow ahead
- ☐ Stop and give way to oncoming traffic
- ☐ Maintain your speed; it's just an information sign

When the road narrows, there'll be less room than normal and others may not obey the rules, so slow down and approach with caution. Even when you have priority, you should still be ready to stop.

15.50 Mark one answer HC p109, KYTS p14

Why should you slow down when you see this sign?

- ☐ Because pedestrians have right of way
- ☐ There's a 'road race' in progress
- ☐ There's a pedestrian crossing ahead
- ☐ Because children may be crossing

Whenever you see this sign, take extra care with these very vulnerable road users.

15.51 Mark one answer HC p111, KYTS p107

What does this sign mean?

- ☐ Vehicle carrying dangerous goods in packages
- ☐ Vehicle broken down ahead
- ☐ Holiday route
- ☐ Emergency diversion route for motorway traffic

Major roadworks often cause complicated diversions for large vehicles. Drivers may be advised to follow a special symbol until the original road can be rejoined. Look for the yellow-and-black symbols in these shapes:

- square
- triangle
- diamond
- circle.

They'll often be shown with additional information and will help guide you around the diversion.

15.52 Mark one answer — HC r132

You're driving in fog on a three-lane motorway. Which lane are you in when you can see red reflective studs on your left and white reflective studs on your right?

- ☐ The hard shoulder
- ☐ The middle lane
- ☐ The right-hand lane
- ☐ The left-hand lane

White reflective studs separate each of the lanes. Red studs mark the left-hand edge of the carriageway, where the hard shoulder begins.

15.53 Mark one answer — KYTS p31

Which drivers are given instructions by diamond-shaped signs?

- ☐ Drivers of lorries
- ☐ Drivers of trams
- ☐ Drivers of buses
- ☐ Drivers of tractors

You need to show caution when driving in areas where trams operate. You might not hear their approach and they can't change direction to avoid you. There may also be crossing points where you'll need to give way to them, or areas specifically reserved for trams, which you aren't allowed to enter.

15.54 Mark one answer — DBC s3, DGV s3

What's a 'Red Route'?

- ☐ An area where special waiting restrictions apply
- ☐ An area where part-time traffic lights operate
- ☐ An area where drivers have to pay a toll
- ☐ An area where night-time and weekend weight limits apply

Red Routes operate in many busy cities and are designed to ensure the route has no obstructions to cause delays during peak periods. Road signs and red road markings identify roads where Red Route waiting and loading restrictions apply.

15.55 Mark one answer HC r132

You're driving on a motorway. What colour are the reflective studs on the right-hand edge of the carriageway?

☐ Amber
☐ Green
☐ Red
☐ White

Reflective studs on motorways and dual carriageways are provided to help drivers in bad visibility. Reflective studs are coloured

- red on the left-hand edge of the carriageway
- white to indicate lane markings
- green at slip roads and lay-bys
- amber on the right-hand edge of the carriageway, marking the central reservation.

15.56 Mark one answer HC r132

What's the purpose of green-and-yellow fluorescent studs on a motorway?

☐ They mark the lanes in a contraflow system
☐ They separate the slip road from the motorway
☐ They mark access points for emergency services
☐ They separate the edge of the hard shoulder from the grass verge

Roadworks can involve complex lane layouts. Reflective green-and-yellow fluorescent studs are used to separate the lanes in a contraflow traffic system. As you approach roadworks, look for traffic signs giving you advance information, which may relate to the size or type of vehicle you're driving.

15.57 Mark one answer KYTS p31

What does this sign mean?

☐ Tramway speed limit
☐ Distance to level crossing
☐ Maximum passenger capacity
☐ Goods-vehicle weight limit

Trams are becoming increasingly common in large towns and cities, as the move to more environmentally friendly transport continues. They may either cross the road you're driving on or share it with you. Always be aware of their virtually silent approach, and look out for places where you may be required to give way to them.

15.58 Mark one answer HC r23

Which light won't show to a driver at a puffin crossing?

- ☐ Flashing amber
- ☐ Red
- ☐ Steady amber
- ☐ Green

A flashing amber light is shown at pelican crossings, but not at puffin crossings. Puffin crossings have sensors that detect when pedestrians are using the crossing.

15.59 Mark one answer KYTS p123

You're approaching a red light at a puffin crossing. Pedestrians are on the crossing. How long will the red light stay on?

- ☐ Until you start to edge forward onto the crossing
- ☐ Until the pedestrians have reached a safe position
- ☐ Until the pedestrians are clear of the front of your vehicle
- ☐ Until a driver from the opposite direction reaches the crossing

The traffic light will remain on red to give pedestrians on the crossing time to reach safety. Don't proceed until the green light shows and it's safe to do so.

15.60 Mark one answer HC r141

A bus-lane sign shows no hours of operation. When is the bus lane in operation?

- ☐ The bus lane isn't in operation at all
- ☐ The bus lane is only in operation at peak times
- ☐ The bus lane is in operation 24 hours a day
- ☐ The bus lane is only in operation in daylight hours

When a bus-lane sign shows no hours of operation, the lane operates all the time.

15.61 Mark one answer KYTS p123

What colour follows the green signal at a puffin crossing?

- ☐ Steady red
- ☐ Flashing amber
- ☐ Steady amber
- ☐ Flashing green

Puffin crossings have sensors that detect when pedestrians are using the crossing. The sensors control the timing of the traffic lights, and this means that there's no flashing amber phase as there is with a pelican crossing.

15.62 Mark one answer KYTS p126

What must you do when you're approaching a pelican crossing in its flashing amber phase?

- ☐ Give way to pedestrians on the crossing
- ☐ Encourage pedestrians to cross
- ☐ Wait until the green light appears
- ☐ Stop even if the crossing is clear

When the amber light is flashing, you must give way to any pedestrians on the crossing. Don't be impatient or encourage pedestrians to cross by waving or flashing your headlights; they or other road users may misunderstand your signal.

15.63 Mark one answer HC r264

What should the hard shoulder on a motorway be used for?

- ☐ As a place to stop to answer a mobile phone
- ☐ As a place to stop when an emergency arises
- ☐ As a place to stop for a short rest when you're tired
- ☐ As a place to stop to check a road atlas

When an emergency arises, pull onto the hard shoulder and use the emergency telephone to report your problem. You'll be connected to an operator, who will put you through to a breakdown service. Never cross the carriageway or a slip road to use a telephone.

15.64 Mark one answer HC r258

When must you stop on a motorway?

- ☐ When you're tired and need to walk in the fresh air
- ☐ When you're picking up hitchhikers
- ☐ When you're signalled to do so by flashing red lights
- ☐ When you need to answer your mobile phone

If red lights flash on the overhead signals above your lane, you mustn't go past the signal in that lane. If other lanes aren't displaying flashing red lights, you may go ahead in those lanes if it's clear and you can move into them safely. If red lights flash on a signal in the central reservation or at the side of the road, you mustn't go past the signal in any lane.

15.65 Mark one answer KYTS p18

Which sign means 'no overtaking'?

- ☐ 
- ☐ 
- ☐ 
- ☐ 

This sign is placed on sections of road where overtaking would be dangerous. There'll often be a plate underneath the sign, telling you how far the restriction extends. You mustn't overtake until you pass another sign cancelling the restriction.

15.66 Mark one answer HC p102, KYTS p90

What does this motorway sign mean?

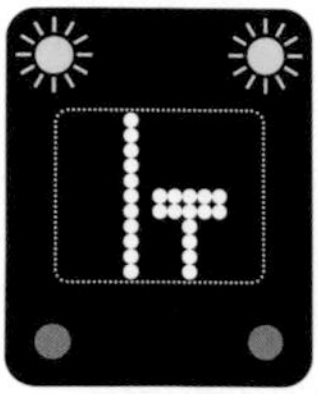

- ☐ Right-hand lane closed ahead
- ☐ 1 tonne weight limit ahead
- ☐ Left-hand lane closed ahead
- ☐ T-junction 1 mile ahead

Four amber lights flash in alternate horizontal pairs to provide a warning and to draw the driver's attention to the message displayed.

15.67 Mark one answer KYTS p18

What does this sign mean?

- ☐ No U-turns
- ☐ Two-way traffic
- ☐ One-way system
- ☐ End of one-way system

You mustn't make a U-turn on roads where this sign is displayed.

15.68 Mark one answer KYTS p12

What does this sign mean?

- ☐ Low bridge
- ☐ Tunnel ahead
- ☐ Accident blackspot
- ☐ Speed camera

A tunnel can present a number of dangers to large vehicles. Look out for height or width limits and always be aware of your vehicle's size. Reduce speed and turn on your headlights before entering the tunnel. Remove sunglasses if you're wearing them, and remember, your eyes may take a little time to adjust to the sudden darkness.

15.69 Mark one answer KYTS p12

What does this sign mean?

- ☐ Rumble strips
- ☐ Road humps
- ☐ Uneven road
- ☐ Double hump bridge

Slow down when you see a warning sign for an uneven road. Be especially careful if you're carrying passengers, livestock or fragile items.

15.70 Mark one answer KYTS p137

Why should you slow down when you see this sign?

- ☐ To reduce the risk of a blow-out
- ☐ To avoid splashing others with water
- ☐ To avoid throwing up loose chippings
- ☐ To prevent grit from blocking your air filter

This sign is usually accompanied by a reduced speed limit. When chippings fly out from your tyres, they can cause a lot of damage to other vehicles, or injury to pedestrians, cyclists and motorcyclists. Slowing down will reduce the risk.

15.71 Mark one answer KYTS p89–91

You're driving on a smart motorway. What's the speed limit when the hard shoulder is being used as a running lane?

- ☐ The speed limits displayed are advisory
- ☐ The national speed limit applies
- ☐ The set speed limit is displayed
- ☐ The speed limit is always 30 mph

When you're driving on a smart motorway, speed limits will show above each lane. These limits are shown inside a red circle and must be obeyed. A red cross will usually appear above the hard shoulder. When the hard shoulder is being used as a running lane, an appropriate speed limit will be displayed instead.

15.72 Mark one answer DBC s4, DGV s4, HC r108

In which part of the United Kingdom do traffic officers operate?

- ☐ England only
- ☐ England, Scotland and Wales
- ☐ England and Wales
- ☐ England, Wales and Northern Ireland

Traffic officers have powers to stop vehicles on safety grounds; for example, if you have an insecure load. They only operate in England and Wales. It's an offence not to comply with their directions.

15.73 Mark one answer DBC s4, DGV s4

What's the purpose of an emergency refuge area on a smart motorway?

- ☐ It's for use in case of emergency or breakdown
- ☐ It's for use if you think you'll be involved in a road-rage incident
- ☐ It's for a police patrol to park and watch traffic
- ☐ It's for construction and road workers to store emergency equipment

Emergency refuge areas may be found next to the hard shoulder on a smart motorway. If possible, you should use one of these areas, rather than the hard shoulder, when you break down. If the hard shoulder is being used as a running lane, take care when you rejoin the motorway.

15.74 Mark one answer DBC s4, DGV s4, HC r108

What are traffic officers authorised to do?

- ☐ Carry out enforcement duties
- ☐ Issue fixed penalty notices
- ☐ Test drivers suspected of drink-driving
- ☐ Stop and direct anyone on a motorway

Traffic officers don't have enforcement powers, but they're able to stop and direct people on motorways and some 'A' class roads. They operate in England and Wales and work in partnership with the police at incidents. Traffic officers can be recognised by their orange-and-yellow jackets and their vehicles, which have yellow-and-black markings.

15.75 Mark one answer DBC s4, DGV s4, HC p102, KYTS p89

You're on a motorway. What does it mean when a red cross is displayed above the hard shoulder?

- ☐ Stop in this lane to answer your mobile phone
- ☐ You may use this lane as a running lane
- ☐ This lane can be used if you need a rest
- ☐ You shouldn't use this lane as a running lane

On smart motorways, the hard shoulder can be used as a running lane when a speed limit is shown above it. However, a red cross above the hard shoulder means that this lane shouldn't be used, except for emergencies and breakdowns.

15.76 Mark one answer DBC s4, DGV s4

You're on a smart motorway. What does it mean when a mandatory speed limit is displayed above the hard shoulder?

- ☐ You shouldn't use the hard shoulder as a running lane
- ☐ The hard shoulder can be used as a running lane
- ☐ You can park on the hard shoulder if you feel tired
- ☐ Large vehicles must use the hard shoulder

If a mandatory speed-limit sign is shown above the hard shoulder, this lane can be used as a running lane. You must stay within the signed speed limit. Look out for any vehicles that have broken down and are blocking the hard shoulder.

15.77 Mark one answer DBC s4, DGV s4

What's the aim of a smart motorway?

- ☐ To prevent overtaking
- ☐ To reduce rest stops
- ☐ To prevent tailgating
- ☐ To reduce congestion

Smart motorways are intended to reduce congestion and make journey times more reliable. In these areas, the hard shoulder may be used as a running lane to ease congestion at peak times or when an incident occurs. The variable speed limits shown in red circles are mandatory and help to keep traffic flowing at a steady speed.

15.78 Mark one answer DBC s4, DGV s4

You're on a smart motorway. What do the speed-limit signs indicate?

- ☐ The speed limit on the sign is advisory
- ☐ The national speed limit will always apply
- ☐ The speed limit is always 30 mph
- ☐ The speed limit on the sign must be obeyed

When a smart motorway is in operation, you must follow the instructions shown on the gantries above each lane. There may also be instructions above the hard shoulder. The signals displayed are mandatory and may be enforced with camera technology.

15.79 Mark one answer DBC s4, DGV s4

You're travelling on a motorway. A red cross is displayed on the gantry signals above the hard shoulder and mandatory speed limits are shown above all other lanes. What does the red cross mean?

- ☐ The hard shoulder can be used as a rest area if you feel tired
- ☐ The hard shoulder is for emergency or breakdown use only
- ☐ The hard shoulder can be used as a normal running lane
- ☐ The hard shoulder has a speed limit of 50 mph

A red cross above the hard shoulder means that it's closed as a running lane and should only be used for emergencies or breakdowns. At busy times, the hard shoulder may be used as a running lane; the red cross will then be replaced by a mandatory speed-limit sign.

15.80 Mark one answer DBC s4, DGV s4

You see this sign on a smart motorway. Which lane can you use?

- ☐ Any lane except the hard shoulder
- ☐ The hard shoulder only
- ☐ The three right-hand lanes only
- ☐ Permitted lanes, including the hard shoulder

Mandatory speed-limit signs above the hard shoulder and all other lanes show that you're on a smart motorway. In this case, you can use the hard shoulder as a running lane. You must stay within the speed limit shown. Look well ahead for any vehicles that have broken down and are blocking the hard shoulder.

15.81 Mark one answer DBC s4, DGV s4

When can you travel on the hard shoulder of a motorway?

- ☐ When taking the next exit
- ☐ When traffic has stopped
- ☐ When signs show that you can
- ☐ When traffic is moving slowly

Normally, you should only use the hard shoulder for emergencies and breakdowns. On smart motorways, the hard shoulder may be used as a running lane, but only when speed-limit signs are shown directly above it. This technology has been introduced to ease traffic congestion.

15.82 Mark one answer KYTS p12

What does this sign mean?

- ☐ Soft verges
- ☐ Only cars may park here
- ☐ Parking restrictions apply
- ☐ Fuel and water tanks will leak

This sign tells you that the verge is soft. If you park here, there's a danger that your vehicle will become stuck. This can be inconvenient and could be costly if you need to be pulled out. In extreme cases, your vehicle may even tip over.

15.83 Mark one answer KYTS p32

What does this sign mean?

- ☐ Buses only
- ☐ Bus lane
- ☐ No buses
- ☐ Bus stop

No buses with more than eight passenger seats are permitted past this sign. Exceptions apply for scheduled services, school and works buses.

15.84 Mark one answer HC r141, KYTS p32

Which road user is sometimes allowed to share a bus lane?

- ☐ Learner driver
- ☐ Lorry driver
- ☐ Disabled driver
- ☐ Taxi driver

The bus-lane signs will indicate which road users are allowed to share the lane with buses and any time restrictions that may apply. Taxis are often allowed to use bus lanes, as these lanes provide a route that has less congestion and can therefore be quicker.

15.85 Mark one answer KYTS p17

You're driving a fully loaded coach that's 12 metres long. What should you do when you approach this sign?

- ☐ Don't proceed past the sign but find another route
- ☐ Set down all your passengers at a safe place before the sign
- ☐ Stop and check the legal lettering on the side panel
- ☐ Proceed as normal; the sign doesn't apply to you

The sign shows the maximum authorised mass for goods vehicles. It's used to restrict heavy lorries (for example, in residential areas), but it doesn't apply to buses and coaches.

Section sixteen
Answers

1. Vehicle weights and dimensions

1.1	☐ No goods vehicles over 7.5 tonnes maximum authorised mass
1.2	☐ Use only the left-hand lane
1.3	☐ 50 mph
1.4	☐ When the left-hand lane is closed
1.5	☐ On traffic-calming humps
1.6	☐ The ground clearance
1.7	☐ Stop before the crossing and phone the signal operator
1.8	☐ Chains
1.9	☐ 40 mph
1.10	☐ 35 mph
1.11	☐ Red/yellow
1.12	☐ Through a greater arc than the cab
1.13	☐ Through a greater arc than the cab
1.14	☐ When turning
1.15	☐ Unload only within a 'white box' area
1.16	☐ 60 mph
1.17	☐ Tri-axles with single wheels
1.18	☐ The driver has very little warning
1.19	☐ The pendulum effect
1.20	☐ To keep the load safe
1.21	☐ 
1.22	☐ 2 metres (6 feet 6 inches)
1.23	☐ When the load's height exceeds 5.25 metres (17 feet 6 inches)
1.24	☐ 2.9 metres (9 feet 5 inches)
1.25	☐ Red/white
1.26	☐ Using twist locks
1.27	☐ To make room for the trailer cutting in
1.28	☐ Vehicles over the weight shown are prohibited
1.29	☐ Overhead electrified cable
1.30	☐ No vehicles over 14 feet 6 inches (4.4 metres) high
1.31	☐ Tunnel ahead
1.32	☐ The distance to a tunnel
1.33	☐ When driving under overhead cables
1.34	☐ 5.0 metres (16 feet 6 inches)
1.35	☐ The police
1.36	☐ The bridge number
1.37	☐ Overhead electric cables ahead
1.38	☐ No vehicles over 6 feet 6 inches (2 metres) wide
1.39	☐ Lane ahead closed
1.40	☐ End of restriction

1.41	☐ In road tunnels
1.42	☐ Height
1.43	☐ Its length
1.44	☐ Icy
1.45	☐ 10 times
1.46	☐ 60 mph
1.47	☐ Your left-hand mirror
1.48	☐ During gentle acceleration
1.49	☐ No entry for vehicles over 32 feet 6 inches (9.9 metres) long
1.50	☐ Avoid going over the roundabout if possible
1.51	☐ Weight restrictions
1.52	☐ In the driver's cab
1.53	☐ Keep to the centre of the arch and give way to oncoming traffic
1.54	☐ 5.0 metres (16 feet 6 inches)
1.55	☐ In an off-road parking area
1.56	☐ When you're overtaking
1.57	☐ When overtaking
1.58	☐ Risk of grounding
1.59	☐ Your vehicle's length
1.60	☐ Straddle the lanes to make more room for the turn
1.61	☐ At an authorised speed-limiter centre
1.62	☐ On the vehicle plate
1.63	☐ 1 tonne
1.64	☐ 15
1.65	☐ By a plate fitted beneath the weight-limit sign
1.66	☐ At any time of the day
1.67	☐ 60 mph
1.68	☐ Buses with a maximum of 16 passenger seats
1.69	☐ On the side of the vehicle
1.70	☐ D
1.71	☐ You might hit something close to the road
1.72	☐ A plate in the cab showing the limited speed
1.73	☐ In the driver's cab
1.74	☐ The weight of your vehicle, with both luggage and passengers
1.75	☐ By reducing the fuel supplied to the engine
1.76	☐ The amount of space needed for the vehicle to turn

2. Drivers' hours and rest periods

2.1	☐ To help road safety
2.2	☐ Road safety
2.3	☐ Keep a written record of hours worked
2.4	☐ Every 24 hours
2.5	☐ 24 hours
2.6	☐ Use a spare chart and attach it to the damaged one
2.7	☐ Remove the chart and make a manual record of the break period
2.8	☐ 45 minutes
2.9	☐ Open the window and turn down the heating
2.10	☐ Stop in a safe place and seek help
2.11	☐ Make sure all doors and windows are locked
2.12	☐ Park with the rear doors close to another vehicle
2.13	☐ Make sure the doors and windows are secure
2.14	☐ Vary your routes and rest stops
2.15	☐ On secure premises
2.16	☐ When it's left unattended
2.17	☐ The starting point of your day's journey
2.18	☐ Take the chart with you and use it in the other vehicle
2.19	☐ Parking in a well-lit lorry park
2.20	☐ Keep plenty of cool, fresh air moving through the cab
2.21	☐ Fit an alarm and immobiliser
2.22	☐ Your local crime-prevention officer
2.23	☐ 00.00 hours Monday to 24.00 hours the following Sunday
2.24	☐ Continue your journey but make a manual record
2.25	☐ They can be heavily fined
2.26	☐ Carry enough approved charts
2.27	☐ One week
2.28	☐ The officer
2.29	☐ With their name
2.30	☐ To replace the original chart if it gets dirty
2.31	☐ 4.5 hours
2.32	☐ At least 45 minutes
2.33	☐ 9 hours, extended to 10 hours on two days of the week
2.34	☐ 9 hours
2.35	☐ Three days
2.36	☐ 11 hours
2.37	☐ 45 hours

2.38	☐ Use their own tachograph chart
2.39	☐ As a break in daily driving
2.40	☐ Driving time
2.41	☐ 
2.42	☐ Driving
2.43	☐ Other work
2.44	☐ 34
2.45	☐ Continue using the same tachograph chart and write an explanation on the back
2.46	☐ Having insufficient breaks from driving
2.47	☐ Stop as soon as it's safe to do so
2.48	☐ Stop as soon as it's safe to do so
2.49	☐ Leave by the next exit and find a place to stop
2.50	☐ At a service station
2.51	☐ Secure the vehicle and lock the doors
2.52	☐ 30 minutes
2.53	☐ 10 hours
2.54	☐ Two days
2.55	☐ 28
2.56	☐ One of 15 minutes, plus one of 30
2.57	☐ Within 7 days
2.58	☐ Five years
2.59	☐ The Driver and Vehicle Licensing Agency
2.60	☐ The Driver and Vehicle Licensing Agency
2.61	☐ Unlimited
2.62	☐ At no time
2.63	☐ Between 00.00 hours on Monday and 24.00 hours the following Sunday
2.64	☐ Workshop card
2.65	☐ Control card
2.66	☐ The Driver and Vehicle Licensing Agency
2.67	☐ The driver's employer
2.68	☐ From a plaque on or near the tachograph
2.69	☐ Every six years
2.70	☐ Take it with you, and use it in the new vehicle
2.71	☐ 90 hours
2.72	☐ Every two years
2.73	☐ EU only
2.74	☐ 45 minutes
2.75	☐ If it's fitted with a bunk
2.76	☐ Ask to see a warrant card
2.77	☐ Check that the parking brake is applied
2.78	☐ When the journey is more than 50 km

3. Braking systems

3.1	☐ While you're braking on a bend
3.2	☐ While driving on muddy construction sites
3.3	☐ Blue
3.4	☐ Red
3.5	☐ Auxiliary
3.6	☐ Drain the air tanks daily
3.7	☐ Select an appropriate gear before downhill gradients
3.8	☐ At high engine speed, while in low gears
3.9	☐ The service brake
3.10	☐ The vehicle could jack-knife
3.11	☐ While it's under heavy braking
3.12	☐ As soon as the vehicle is moving
3.13	☐ Blue
3.14	☐ It's the emergency line
3.15	☐ The trailer auxiliary line should be left unconnected
3.16	☐ Apply the footbrake firmly and continuously until the vehicle has stopped
3.17	☐ Brake fade
3.18	☐ To make the wheels less likely to spin
3.19	☐ They help you keep steering control while braking heavily
3.20	☐ Moisture drawn in with the air may freeze and cause a blockage
3.21	☐ Stop and get help without delay
3.22	☐ Keep the parking brake on
3.23	☐ Stop safely in an emergency
3.24	☐ Before every journey
3.25	☐ Reduced effectiveness
3.26	☐ Apply firm, continuous pressure to the brake pedal
3.27	☐ Reduction of braking effectiveness
3.28	☐ On a long downhill gradient
3.29	☐ Use the endurance brake
3.30	☐ Maintain steering control
3.31	☐ There's no need to alter the way you brake
3.32	☐ Down a steep hill
3.33	☐ When your brakes have failed
3.34	☐ While driving down long hills
3.35	☐ The endurance brake (retarder)

3.36	☐ An endurance brake (retarder)
3.37	☐ On long downhill slopes
3.38	☐ When driving in a straight line
3.39	☐ Be prepared to reduce your speed
3.40	☐ Drive in a low gear with the footbrake lightly applied
3.41	☐ The air brakes
3.42	☐ When the brakes get too hot
3.43	☐ Use the endurance brake
3.44	☐ Continuous use of the brakes
3.45	☐ On slippery roads
3.46	☐ Because the service brake may not stop you
3.47	☐ The vacuum pump isn't working
3.48	☐ The brake pedal feels hard when pressed
3.49	☐ There isn't enough fluid in the braking system
3.50	☐ The hydraulic-brake fluid level

3.51	☐ A buzzer or light
3.52	☐ Blue
3.53	☐ Low air pressure
3.54	☐ The brake system has a loss of vacuum
3.55	☐ The brake linings
3.56	☐ Stop and have the fault put right
3.57	☐ The drive wheels could lock
3.58	☐ The drive wheels locking
3.59	☐ Find out whether you need to drain the system manually
3.60	☐ Frozen moisture in the storage tanks
3.61	☐ When road speed is 6 mph (10 km/h) or more
3.62	☐ Service brake
3.63	☐ Apply the parking brake after stopping
3.64	☐ Put your foot on the footbrake
3.65	☐ Improved passenger safety and comfort

4. The driver

4.1	☐ Go forward if your vehicle will clear the junction
4.2	☐ It's required by law
4.3	☐ Check your right-hand mirror and blind spot

4.4	☐ Headlight dazzle
4.5	☐ Stay calm and accept the error

4.6	☐ When you're making deliveries that are less than 50 metres apart
4.7	☐ You, the driver
4.8	☐ Traffic further ahead may be stopping
4.9	☐ Drive on until you find a safe place to stop
4.10	☐ Shopping areas
4.11	☐ The noise of your vehicle
4.12	☐ Drive very slowly
4.13	☐ When driving close to the kerb
4.14	☐ Heart disorders
4.15	☐ Your distance from the controls
4.16	☐ To protect your neck in a collision
4.17	☐ When reversing
4.18	☐ Drop back to leave the correct separation distance
4.19	☐ Try to avoid splashing the pedestrians
4.20	☐ Slow down
4.21	☐ Stop in a safe place
4.22	☐ To warn others of your presence
4.23	☐ Be patient and stay well behind
4.24	☐ Stop on the left as soon as it's safe
4.25	☐ Signal with the left indicator
4.26	☐ Find a safe place to stop
4.27	☐ When you're stopped in a safe place
4.28	☐ As one continuous crossing
4.29	☐ Give way to pedestrians already on the crossing
4.30	☐ Always leave the crossing clear in traffic queues
4.31	☐ Do nothing and let the driver decide
4.32	☐ Stop to let it pass as soon as it's safe to do so
4.33	☐ Yes, because it's a doctor going to an emergency
4.34	☐ Doctor's car
4.35	☐ Be patient and wait
4.36	☐ There may be another vehicle coming
4.37	☐ The nearside mirror striking the heads of pedestrians
4.38	☐ Near the brow of a hill
4.39	☐ Your mirrors are clean
4.40	☐ Check your nearside mirror
4.41	☐ Allow the vehicle to overtake
4.42	☐ Slow down and be ready to stop
4.43	☐ You'll feel more confident
4.44	☐ Tiredness
4.45	☐ It reduces your concentration

4.46	☐ Insurance premiums
4.47	☐ Go home by public transport
4.48	☐ Avoid drinking alcohol completely
4.49	☐ You may still be over the legal drink-drive limit
4.50	☐ Some types of medicine can cause your reactions to slow down
4.51	☐ Only after checking with your doctor
4.52	☐ Ask your doctor for advice
4.53	☐ Check the label to see whether the medicine will affect your driving
4.54	☐ Leave the motorway at the next exit and take a rest
4.55	☐ Wait until you're fit and well before driving
4.56	☐ Ensure a supply of fresh air
4.57	☐ When driving while you're tired
4.58	☐ Seek medical advice
4.59	☐ Leave the motorway at the next exit
4.60	☐ Park in a suitable place and rest
4.61	☐ Withdrawal of a driving licence
4.62	☐ Hold the steering wheel firmly
4.63	☐ It allows easy location by the emergency services
4.64	☐ When you use a mobile phone
4.65	☐ Stop in a safe place before making the call
4.66	☐ Only when you're suitably parked
4.67	☐ Your attention will be diverted from the road
4.68	☐ Check your location from the marker posts on the left
4.69	☐ Stop safely before answering
4.70	☐ Move back to the left when you judge it's safe to do so
4.71	☐ Slow down and give way if it's safe to do so
4.72	☐ Stay where you are and allow them to cross in their own time
4.73	☐ Pull in when you can
4.74	☐ It will divert your attention
4.75	☐ Stop in a safe place when you can
4.76	☐ A false sense of confidence
4.77	☐ Concentrate on the road ahead
4.78	☐ It makes it easier for you to steer
4.79	☐ Keep calm and don't retaliate
4.80	☐ Stop and take a break
4.81	☐ Stay calm and don't retaliate

4.82	☐ Switch on your hazard warning lights
4.83	☐ Wait until it's clear in both directions
4.84	☐ The length of your vehicle
4.85	☐ Allow plenty of time for your journey
4.86	☐ Have regular meals and rest breaks
4.87	☐ Bottled water
4.88	☐ Eat a meal at the beginning of your shift
4.89	☐ Between 2 am and 7 am
4.90	☐ 40%
4.91	☐ Heart disease
4.92	☐ Stop and rest
4.93	☐ On motorways
4.94	☐ 30 years and under
4.95	☐ Hold the steering wheel firmly and slow down gradually
4.96	☐ When you hold a medical exemption certificate
4.97	☐ You're more likely to fall asleep at the wheel
4.98	☐ A head restraint
4.99	☐ When you're reversing the vehicle
4.100	☐ The emergency-exit location
4.101	☐ After overtaking, to see whether it's safe to return to the left
4.102	☐ Keep junctions clear

5. Carrying passengers

5.1	☐ The safety and comfort of your passengers
5.2	☐ By stopping close to the kerb
5.3	☐ To improve passenger comfort
5.4	☐ Plan ahead and take early action on all stops
5.5	☐ To the front of the bus
5.6	☐ Plan ahead and take early action
5.7	☐ When the vehicle is cornering
5.8	☐ Slow down to avoid the need to stop suddenly
5.9	☐ To keep the passengers safe
5.10	☐ Towards the outside of the bend
5.11	☐ Thinking and planning well ahead
5.12	☐ Passengers getting off the bus
5.13	☐ At all times
5.14	☐ A fire extinguisher

5.15	☐ The location of the first-aid equipment
5.16	☐ They have poor vision and hearing
5.17	☐ Be courteous and polite
5.18	☐ To be courteous and polite
5.19	☐ Look at them when speaking to them
5.20	☐ To allow the step height to be raised and lowered
5.21	☐ The bus may be carrying children
5.22	☐ When children are getting on or off the vehicle
5.23	☐ To limit injuries in the event of a rear-end collision
5.24	☐ Keep the interior lit
5.25	☐ Get all the passengers off the bus
5.26	☐ Access for passengers
5.27	☐ So that passengers can see to move around
5.28	☐ Those attempting to board the bus
5.29	☐ Wait until the passenger has sat down before moving away
5.30	☐ When children are boarding the bus
5.31	☐ Leave a responsible person on the coach
5.32	☐ Lampposts
5.33	☐ 5 tonnes
5.34	☐ 15
5.35	☐ School-bus signs
5.36	☐ Stop and have the fault put right
5.37	☐ Overhanging trees
5.38	☐ Helping passengers who need directions
5.39	☐ Fully trained people
5.40	☐ Access for disabled people
5.41	☐ The front of the bus
5.42	☐ Passengers leaving the bus
5.43	☐ Finish issuing tickets to passengers
5.44	☐ Move passengers to the front of the bus
5.45	☐ The safety and comfort of passengers
5.46	☐ Make sure you've cancelled any signal
5.47	☐ By making full use of the internal mirror system
5.48	☐ Passengers using the stairs could fall if the bus brakes or swerves
5.49	☐ Close to the kerb
5.50	☐ Close to the kerb
5.51	☐ Check for traffic on the left
5.52	☐ Check it's clear of traffic on the left
5.53	☐ Check mirrors before opening doors
5.54	☐ Keep the passenger doors closed until the bus has stopped

5.55	☐ When there are no passengers
5.56	☐ No, this isn't allowed
5.57	☐ Passenger comfort will be increased
5.58	☐ Smile and offer to help them
5.59	☐ When it can't be seen by the driver
5.60	☐ Priority seating
5.61	☐ Those with disabilities
5.62	☐ When the boarding device has failed to work
5.63	☐ When the designated space is occupied
5.64	☐ In a medical emergency
5.65	☐ High-visibility vest
5.66	☐ The boarding device
5.67	☐ When the lift has stopped working
5.68	☐ Passengers are standing in a wheelchair space and can't move elsewhere
5.69	☐ Wait until they're inside the bus

6. The road

6.1	☐ Leave plenty of room and check your nearside mirror before returning to the left
6.2	☐ High-level bridges may be closed to certain vehicles in windy weather
6.3	☐ Build up your speed on the slip road before joining the motorway
6.4	☐ Plan your route in advance
6.5	☐ Ease off and drop behind the vehicle you were trying to overtake
6.6	☐ Tie both curtains open
6.7	☐ When it's travelling empty
6.8	☐ To reduce the effect of side wind
6.9	☐ When travelling empty
6.10	☐ Lane closures
6.11	☐ An unladen lorry with box body
6.12	☐ A motorcycle
6.13	☐ High-sided lorry
6.14	☐ When there's a headwind
6.15	☐ Double your dry-weather separation distance
6.16	☐ Hose down the wheels
6.17	☐ It's an offence to deposit mud on a road
6.18	☐ Solid and level
6.19	☐ When the motorway has two lanes
6.20	☐ Stop as soon as it's safe to do so

6.21	☐ Four seconds
6.22	☐ When the escort vehicle of an oversized load signals you to pass
6.23	☐ Use dipped headlights
6.24	☐ The brakes will be less effective
6.25	☐ Fallen trees
6.26	☐ The effect of strong side wind
6.27	☐ There's ice or frost on the road
6.28	☐ Check ahead and then your nearside mirror
6.29	☐ Be prepared to stop in good time
6.30	☐ On exposed sections
6.31	☐ Across suspension bridges
6.32	☐ Ease off the accelerator
6.33	☐ They can park without lights
6.34	☐ Wait until the vehicle starts to turn in
6.35	☐ There's a long uphill gradient ahead
6.36	☐ The driver's door may suddenly open
6.37	☐ Use the parking lights
6.38	☐ When you're cornering
6.39	☐ More braking effort will be required
6.40	☐ When turning at tight junctions
6.41	☐ Anticipate how the weather may affect other road users
6.42	☐ Open roads
6.43	☐ On high-level roads
6.44	☐ 
6.45	☐ A safety lane for vehicles blown off course
6.46	☐ On high-level bridges
6.47	☐ When it's windy
6.48	☐ Drive slowly, in a higher gear than normal
6.49	☐ A motorcycle
6.50	☐ The buffeting effect
6.51	☐ They may be blown into your path
6.52	☐ Car drivers towing caravans
6.53	☐ Drivers towing caravans
6.54	☐ To see if the rider is in control of their motorcycle
6.55	☐ Curtain-sided vehicles
6.56	☐ Across viaducts
6.57	☐ The road may still be slippery
6.58	☐ To reduce the risk of skidding
6.59	☐ Twice the normal distance
6.60	☐ Switch on your dipped headlights
6.61	☐ Your tyres may lose grip

6.62	☐ Visibility will be reduced by spray from traffic
6.63	☐ More spray will be thrown up
6.64	☐ Move out earlier than normal
6.65	☐ Their vision will be reduced
6.66	☐ To reduce spray on wet roads
6.67	☐ Before setting out on every journey
6.68	☐ Spray reducers
6.69	☐ Rain
6.70	☐ Slowly, in a low gear, with engine speed high
6.71	☐ There may be deep snow ahead
6.72	☐ To allow for an increased stopping distance
6.73	☐ To reduce the amount of spray thrown up
6.74	☐ When visibility is reduced to 100 metres (328 feet) or less
6.75	☐ Water reducing the tyres' grip on the road
6.76	☐ It will take longer to stop
6.77	☐ To avoid splashing them
6.78	☐ Your view ahead is reduced
6.79	☐ When you're on a motorway and traffic ahead slows suddenly
6.80	☐ Children crossing the road
6.81	☐ Near a school
6.82	☐ You'll have reduced visibility
6.83	☐ Speed and distance are harder to judge
6.84	☐ The left-hand lane
6.85	☐ Traffic behind in the right-hand lane, returning to the middle lane
6.86	☐ Check your mirrors carefully
6.87	☐ The left-hand lane
6.88	☐ Be ready to adjust your speed
6.89	☐ Move to the next lane if it's safe
6.90	☐ Traffic in the centre lane may be travelling much faster than you
6.91	☐ When the motorway has two lanes
6.92	☐ Cutting in
6.93	☐ To enable slow-moving traffic to move further over to the left on uphill gradients
6.94	☐ You're approaching a long uphill slope
6.95	☐ When letting faster traffic overtake you
6.96	☐ Look ahead for road junctions
6.97	☐ When you would have to break the speed limit
6.98	☐ By checking your nearside mirror

6.99	☐ Stay behind the slower vehicle
6.100	☐ Move back to the left when it's safe to do so
6.101	☐ Signal right on approach
6.102	☐ When they need to avoid mounting the kerb
6.103	☐ Check your mirrors
6.104	☐ Allow extra room
6.105	☐ Go in any direction
6.106	☐ Look well ahead for uphill gradients
6.107	☐ They may move back to the middle lane as you move out
6.108	☐ It's facing towards you
6.109	☐ So that other road users can see the size of your vehicle
6.110	☐ When visibility falls below 100 metres (328 feet)
6.111	☐ When visibility is more than 100 metres (328 feet)
6.112	☐ When the roads are dry
6.113	☐ Increase your distance from the vehicle in front
6.114	☐ Slow down
6.115	☐ At least a two-second gap
6.116	☐ Drop further back
6.117	☐ You can see well ahead
6.118	☐ Ten times as long
6.119	☐ Brake gently and in good time
6.120	☐ Gently

6.121	☐ To keep a safe distance from the vehicle in front when conditions are good
6.122	☐ Your field of vision will be seriously reduced
6.123	☐ At least four seconds
6.124	☐ Brake in good time
6.125	☐ Make sure that your vehicle is travelling in a straight line
6.126	☐ You'll take longer to get past
6.127	☐ A motorcycle
6.128	☐ At least 53 metres (175 feet)
6.129	☐ Use dipped headlights
6.130	☐ Some roads may be closed to certain vehicles in high winds
6.131	☐ Spray could affect their control
6.132	☐ Steering will be more difficult
6.133	☐ Strong wind
6.134	☐ Pass wide in case they're blown off course
6.135	☐ Rain
6.136	☐ To avoid mounting the kerb
6.137	☐ They must leave their parking lights switched on
6.138	☐ There may be unseen dips or bends in the road
6.139	☐ To make them easier to see at junctions
6.140	☐ On a hump bridge

7. Incidents, accidents and emergencies

7.1	☐ Cut off the fuel supply
7.2	☐ How to use fire-fighting equipment
7.3	☐ 5 metres (16 feet 6 inches)
7.4	☐ Stop at the scene
7.5	☐ 24 hours
7.6	☐ It's carrying dangerous goods
7.7	☐ Douse the burns with cool water
7.8	☐ Apply firm pressure to the wound
7.9	☐ Warn other traffic
7.10	☐ Switch on your own hazard warning lights
7.11	☐ Check their airway is open
7.12	☐ Check that they're breathing
7.13	☐ Keep injured people warm and comfortable
7.14	☐ Deal with the danger from further collisions or fire
7.15	☐ Reassure them confidently
7.16	☐ Warn other traffic to prevent further collisions
7.17	☐ Reassure them confidently
7.18	☐ Warn other traffic
7.19	☐ Until medical help arrives and takes over
7.20	☐ When they can breathe without help
7.21	☐ Try to stem the flow of blood
7.22	☐ Apply firm pressure to the wound
7.23	☐ Because there's further danger
7.24	☐ Leave them where they are
7.25	☐ Keep them in the vehicle
7.26	☐ Your driving licence
7.27	☐ By talking to them quietly and firmly
7.28	☐ You could cause more injury
7.29	☐ A length of wood
7.30	☐ Facing the oncoming traffic
7.31	☐ Water (red)
7.32	☐ Call for help using the nearest emergency telephone
7.33	☐ Shut off the fuel supply
7.34	☐ Remove your sunglasses
7.35	☐ Use dipped headlights
7.36	☐ Tune your radio to a local channel
7.37	☐ Keep a safe distance from the vehicle in front

7.38	☐ Switch on hazard warning lights
7.39	☐ Drive it out of the tunnel if you can do so
7.40	☐ Follow the instructions given by variable message signs
7.41	☐ Douse the burns with cool water
7.42	☐ By opening their airway
7.43	☐ Stop, then report the incident to the police within 24 hours
7.44	☐ Switch off the engine and switch on hazard warning lights
7.45	☐ Stop at the side of the road and switch off the engine
7.46	☐ Rapid shallow breathing
7.47	☐ Drive out of the tunnel
7.48	☐ Talk to them quietly and firmly
7.49	☐ Rapid pulse and sweating
7.50	☐ Make sure the emergency services have been called
7.51	☐ Tell your passengers what's happening without upsetting them
7.52	☐ Switch off the fuel supply
7.53	☐ Report the incident to the police within 24 hours
7.54	☐ Keep them warm
7.55	☐ When they can breathe on their own
7.56	☐ Get medical help
7.57	☐ To reduce the risk of injury from a rear-end collision
7.58	☐ Immediately, to the railway authority
7.59	☐ Get the passengers off and keep them together
7.60	☐ Get everyone off the bus quickly and safely
7.61	☐ Move your passengers to the nearest tunnel exit

8. Vehicle condition

8.1	☐ To stop mud dropping onto the road
8.2	☐ To make sure bricks or debris aren't wedged between the wheels
8.3	☐ The diff-lock is disengaged
8.4	☐ Stop on the hard shoulder and phone for assistance
8.5	☐ 1 mm
8.6	☐ Release the trailer parking brake
8.7	☐ Regularly

8.8	☐ The power steering is faulty
8.9	☐ Soon after initial tightening
8.10	☐ The kingpin release handle
8.11	☐ Monthly
8.12	☐ Eyelet coupling
8.13	☐ To connect the tractor unit to the trailer
8.14	☐ Every 6000 miles (10 000 km)
8.15	☐ It expands
8.16	☐ Air pressure
8.17	☐ Park and phone for assistance
8.18	☐ Using a tyre with exposed ply or cord
8.19	☐ Drain the tanks daily
8.20	☐ Braking-system fault
8.21	☐ Stop on the hard shoulder as soon as possible
8.22	☐ On the left of the road
8.23	☐ Park and get help
8.24	☐ A fault with the power-assisted steering
8.25	☐ Park and phone for help
8.26	☐ To remove moisture drawn in from the atmosphere
8.27	☐ Leave the vehicle parked and report the fault immediately
8.28	☐ Low air pressure
8.29	☐ Three-quarters

8.30	☐ Immediately
8.31	☐ Anti-waxing additives
8.32	☐ Don't drive until they're repaired
8.33	☐ Look for a check switch on the dashboard
8.34	☐ They'll be more likely to overheat and disintegrate
8.35	☐ You'll break the law and risk prosecution
8.36	☐ To heat the combustion chambers
8.37	☐ Into the combustion chamber
8.38	☐ It could increase pressure in the engine and cause damage
8.39	☐ To protect the coolant from freezing
8.40	☐ Replace the valve
8.41	☐ When the engine is cold
8.42	☐ All year round
8.43	☐ Low oil pressure
8.44	☐ There's an electrical fault
8.45	☐ Maximum load
8.46	☐ It will reduce breakdowns
8.47	☐ Use wheel chocks if available
8.48	☐ Shortly afterwards
8.49	☐ By regular servicing
8.50	☐ To reduce friction and wear
8.51	☐ On flat ground

8.52	☐ Some gaskets might be damaged
8.53	☐ Apply the parking brake
8.54	☐ Apply the trailer parking brake
8.55	☐ Before any journey
8.56	☐ Stop in a safe place and get help
8.57	☐ Tyres
8.58	☐ Running tyres under-inflated
8.59	☐ To change either up or down in gear
8.60	☐ To give quicker acceleration
8.61	☐ Faulty power steering
8.62	☐ Scattered debris
8.63	☐ It will be prohibited from further use until the defects are rectified
8.64	☐ The Traffic Commissioner
8.65	☐ It may break down
8.66	☐ The engine may be damaged
8.67	☐ Speed capability
8.68	☐ Reduced rolling resistance
8.69	☐ Tighten the nuts evenly with a torque wrench
8.70	☐ They have a reduced rolling resistance
8.71	☐ The vehicle handbook
8.72	☐ Workshop manuals
8.73	☐ Loss of power
8.74	☐ To collect metal particles from the oil
8.75	☐ Torque wrench
8.76	☐ The fuel partly solidifying
8.77	☐ The diesel from solidifying
8.78	☐ To check for any defects
8.79	☐ It expands
8.80	☐ A system for reporting vehicle defects
8.81	☐ You may scrub the rear tyres
8.82	☐ Drive at a very low speed
8.83	☐ 1 mm
8.84	☐ Get it repaired before continuing
8.85	☐ Have the fault checked before setting off
8.86	☐ Punctures can be more difficult to detect
8.87	☐ The tyres on the smaller wheels are more likely to overheat
8.88	☐ Daily
8.89	☐ Allow the engine to idle
8.90	☐ Every day before starting out
8.91	☐ Allow the engine to idle, then switch it off
8.92	☐ Wait for the breakdown services
8.93	☐ At least once a week when they're cold
8.94	☐ When needing brisk acceleration

8.95	☐ Faulty power steering
8.96	☐ Punctures can be difficult to detect
8.97	☐ To top up the oil or water levels
8.98	☐ They're more difficult to detect
8.99	☐ The tyre pressures are all correct
8.100	☐ The tyres will wear more quickly
8.101	☐ Park on a firm, level surface

9. Leaving the vehicle

9.1	☐ Your nearside mirror doesn't strike the head of a pedestrian
9.2	☐ Check the rear-view mirrors
9.3	☐ At a service area
9.4	☐ In a lorry park
9.5	☐ Using the mirrors
9.6	☐ Slippery walkways
9.7	☐ Use a suitable set of steps
9.8	☐ There'll be a risk of theft or hijack
9.9	☐ Roof markings
9.10	☐ Before opening your door
9.11	☐ Apply the parking brake
9.12	☐ Switch off the engine
9.13	☐ Vehicles passing near the door
9.14	☐ When approaching queuing traffic on a motorway
9.15	☐ Whenever you leave the vehicle
9.16	☐ Stop the engine
9.17	☐ When on motorways or dual carriageways, to warn drivers behind of a hazard ahead
9.18	☐ Look ahead when the load is secure
9.19	☐ Have a stable position
9.20	☐ In a stable position
9.21	☐ When children are getting off a school bus
9.22	☐ When you're parked close to another vehicle
9.23	☐ Take everything with them
9.24	☐ Apply the parking brake
9.25	☐ Apply the parking brake
9.26	☐ High-visibility vest

10. Vehicle loading

10.1	☐ Spread the load evenly
10.2	☐ The container must be secured using the locking levers
10.3	☐ 
10.4	☐ To stop wind and rain from getting under the sheets
10.5	☐ A dolly knot
10.6	☐ Ropes can wear and snap
10.7	☐ To stop the load from blowing away
10.8	☐ Park and resecure the load before continuing
10.9	☐ Baffle plates
10.10	☐ It will try to push the vehicle forward
10.11	☐ Keep pressure on the brake pedal until you've stopped
10.12	☐ Glass
10.13	☐ 3.5 metres (11 feet 5 inches)
10.14	☐ When it's unloaded
10.15	☐ Try to move forward with the trailer parking brake on
10.16	☐ The trailer brake is applied
10.17	☐ Stow it away safely
10.18	☐ The surface is firm and level
10.19	☐ To give a view down each side of the trailer
10.20	☐ A short-wheelbase empty vehicle
10.21	☐ Steel plates
10.22	☐ Tubular metal
10.23	☐ Over the rear axles
10.24	☐ Tied down securely
10.25	☐ An unladen trailer
10.26	☐ When empty
10.27	☐ Driving too fast
10.28	☐ The lorry may overturn when cornering
10.29	☐ In high winds
10.30	☐ Restraining straps
10.31	☐ The load will push the lorry straight on
10.32	☐ You could lose control of your vehicle
10.33	☐ Use chains and tensioners
10.34	☐ Place chocks under the wheels
10.35	☐ When only the top deck is loaded
10.36	☐ Demount the body without a crane or lift
10.37	☐ Diamond
10.38	☐ To reduce the wave effect
10.39	☐ 

10.40	☐ Make sure the vehicle is on a firm, level surface
10.41	☐ The railway authority
10.42	☐ The maximum load the vehicle can carry
10.43	☐ Hygiene procedures
10.44	☐ Lower the stabilising legs
10.45	☐ Heavy items near the centre line, light items towards the sides
10.46	☐ Walk forward near the edges
10.47	☐ Frozen foods
10.48	☐ Practical experience in the care of animals
10.49	☐ Create compartments using moveable panels
10.50	☐ 25 tonnes
10.51	☐ You might overload an axle
10.52	☐ Reduced braking efficiency
10.53	☐ Damage to the road surface
10.54	☐ Covered
10.55	☐ Rope hook
10.56	☐ Split-link
10.57	☐ 10 mm
10.58	☐ Rear
10.59	☐ To prevent any spillage
10.60	☐ Perishable foods
10.61	☐ Chemicals
10.62	 ☐
10.63	☐ Oxidising agents
10.64	 ☐
10.65	 ☐
10.66	☐ Check the load by carefully entering through the rear door or opposite curtain
10.67	☐ Protection from the weather
10.68	☐ Vehicle stabilisers
10.69	☐ Between the shoulder and the waist
10.70	☐ 30 tonnes

11. Restricted view

11.1	☐ Parking
11.2	☐ The site should be level
11.3	☐ Check the blind spots before you start moving
11.4	☐ There's a motorcycle on the left
11.5	☐ Brake smoothly and allow it to emerge
11.6	☐ Get someone to guide you into the bay
11.7	☐ The motorcyclist from the right passing the parked van
11.8	☐ On the road
11.9	☐ When traffic is merging from the right or left
11.10	☐ It increases your blind spots
11.11	☐ The motorcyclist
11.12	☐ The nearside mirror
11.13	☐ The offside mirror
11.14	☐ Use your mirrors and look behind
11.15	☐ After passing cars parked on the left
11.16	☐ Mirrors, signal, manoeuvre
11.17	☐ With-flow bus and cycle lane
11.18	☐ Adjust your seat properly
11.19	☐ Between 7.00 am and 11.30 pm
11.20	☐ Your view ahead will be reduced
11.21	☐ They can be hidden in blind spots
11.22	☐ Someone may be overtaking you
11.23	☐ The left-hand mirror
11.24	☐ Cyclists on your left
11.25	☐ Use the left-hand of the two lanes
11.26	☐ You'll need to take all-round observation
11.27	☐ An area of road that can't be seen, even with mirrors
11.28	☐ It will have more blind spots than smaller vehicles
11.29	☐ Ease forward until you can see clearly in all directions
11.30	☐ Look all around
11.31	☐ Be cautious and keep looking all around
11.32	☐ When you're pulling in to park at the kerb
11.33	☐ The blind area close in front
11.34	☐ Open the window and look down and round to the right
11.35	☐ Cyclists and pedestrians
11.36	☐ The front wheel in relation to the kerb

12. Essential documents

12.1	☐ A valid ADR driver training certificate
12.2	☐ The driver
12.3	☐ Four
12.4	☐ A trailer up to 750 kg
12.5	☐ Hygiene procedures
12.6	☐ Your national driving licence
12.7	☐ Partial blindness
12.8	☐ Your entitlement to drive all motor vehicles
12.9	☐ The Driver and Vehicle Licensing Agency
12.10	☐ Until the vehicle is taxed, sold or scrapped
12.11	☐ A notification to tell DVLA that a vehicle isn't being used on the road
12.12	☐ Unlimited
12.13	☐ Up to £2000
12.14	☐ The vehicle hirer
12.15	☐ By using an effective vehicle security system
12.16	☐ Your operator documentation
12.17	☐ For each person they bring in
12.18	☐ 35 hours
12.19	☐ Five years
12.20	☐ 35 hours
12.21	☐ An eyesight test
12.22	☐ D
12.23	☐ For national security
12.24	☐ To prevent the carriage of illegal immigrants
12.25	☐ A list of passengers
12.26	☐ D
12.27	☐ 750 kg
12.28	☐ You can tow a trailer not exceeding 750 kg
12.29	☐ D
12.30	☐ D
12.31	☐ The Traffic Commissioner

13. Environmental issues

13.1	☐ They'll reduce the wind resistance
13.2	☐ Fuel consumption
13.3	☐ Fit wind deflectors
13.4	☐ Five years
13.5	☐ Braking in good time
13.6	☐ Drag effect
13.7	☐ Fit a cab-mounted wind deflector
13.8	☐ For authorised purposes only
13.9	☐ To prevent rubbish from falling out of it
13.10	☐ Make sure the lorry is always attended
13.11	☐ Police permission
13.12	☐ Changing filters regularly
13.13	☐ Save fuel
13.14	☐ Red diesel
13.15	☐ To reduce aerodynamic drag
13.16	☐ Road surfaces
13.17	☐ It uses electricity
13.18	☐ Lorries
13.19	☐ Underground services
13.20	☐ By braking in good time
13.21	☐ By using bypasses and avoiding town centres
13.22	☐ Plan routes to avoid busy times and congestion
13.23	☐ That the filler cap is properly closed and secure
13.24	☐ Close and secure the filler cap
13.25	☐ The road may be very slippery
13.26	☐ Vehicles with faulty suspension
13.27	☐ By planning routes to avoid congestion
13.28	☐ Low-sulphur diesel
13.29	☐ Low-sulphur diesel
13.30	☐ Securely close the filler caps
13.31	☐ Before reversing on a road that has a 30 mph speed limit
13.32	☐ Continuous use of the air conditioning
13.33	☐ To reduce wear to roads
13.34	☐ Between 11.30 pm and 7.00 am
13.35	☐ You'll be breaking the law
13.36	☐ Damage to the road surface

13.37	☐ Plan journeys to avoid busy times
13.38	☐ Switch off the engine
13.39	☐ Poorly maintained vehicles
13.40	☐ Motorcyclists
13.41	☐ To warn another road user who poses a danger
13.42	☐ Motorcycles
13.43	☐ Get a replacement before driving
13.44	☐ Look at a map
13.45	☐ A motoring organisation
13.46	☐ Use a route planner on the internet
13.47	☐ Print or write down the route
13.48	☐ It will help to ease congestion
13.49	☐ You're less likely to be delayed
13.50	☐ Your original route may be blocked
13.51	☐ You'll reduce fuel consumption by about 15%
13.52	☐ Missing out some gears
13.53	☐ Green
13.54	☐ You would damage the engine
13.55	☐ Reduced engine emissions
13.56	☐ They have a reduced rolling resistance
13.57	☐ An excessive amount of fuel will be used
13.58	☐ A reduction in fuel consumption
13.59	☐ Exhaust emissions
13.60	☐ Being aware of hazards
13.61	☐ Cruise control
13.62	☐ You'll use less fuel
13.63	☐ Using air conditioning
13.64	☐ Air conditioning
13.65	☐ By switching off your engine
13.66	☐ To reduce exhaust emissions
13.67	☐ Switch off your engine
13.68	☐ Increase it by about 15%
13.69	☐ With the engine ticking over
13.70	☐ Your filler caps are securely closed

14. Other road users

14.1	☐ Motorcyclists
14.2	☐ Motorcycles
14.3	☐ Keep calm and be patient
14.4	☐ Sounding your horn
14.5	☐ Give way to them
14.6	☐ 
14.7	☐ Wait and allow them to cross
14.8	☐ Give way to the pedestrians who are already crossing
14.9	☐ They might be overtaking on your right
14.10	☐ Cyclists can use it
14.11	☐ Pedestrians and cyclists may cross
14.12	☐ On the rear of a school bus or coach
14.13	☐ A route for pedestrians and cyclists
14.14	☐ Be patient and allow them to cross in their own time
14.15	☐ Be careful; they might misjudge your speed
14.16	☐ Give the cyclist plenty of room
14.17	☐ They may use the left-hand lane to turn right
14.18	☐ Motorcycles are small and hard to see
14.19	☐ To make themselves more easily seen
14.20	☐ To help other road users to see them
14.21	☐ They need to check for traffic in their blind area
14.22	☐ Give them plenty of room
14.23	☐ Drive very slowly
14.24	☐ You're approaching an organised walk
14.25	☐ Reduce speed until you're clear of the area
14.26	☐ Prepare to slow down and stop
14.27	☐ To allow an unobstructed view of the area
14.28	☐ It would obstruct everyone's view of the area
14.29	☐ Any direction
14.30	☐ Stay behind until the moped has passed the junction
14.31	☐ Stay well back
14.32	☐ Be patient and prepare for them to react more slowly
14.33	☐ Be patient, as you expect them to make mistakes
14.34	☐ Pedestrians

14.35	☐ Be aware that the driver's reactions may not be as fast as yours
14.36	☐ Hold back until the cyclist has passed the junction
14.37	☐ Go in any direction
14.38	☐ To check for overtaking vehicles
14.39	☐ Slow down and be prepared to stop for children
14.40	☐ Check for traffic passing on your right
14.41	☐ Pedestrians may come from behind the bus
14.42	☐ Drive slowly and leave plenty of room
14.43	☐ At a road junction
14.44	☐ You're approaching an organised walk
14.45	☐ Cyclists
14.46	☐ The rider intends to move to the right
14.47	☐ They may swerve to avoid potholes
14.48	☐ Give them extra room
14.49	☐ Motorcycles
14.50	☐ Drive past slowly
14.51	☐ Stay well back and allow them room
14.52	☐ Cyclists
14.53	☐ A mobility scooter
14.54	☐ Drive slowly until you're clear of the area

14.55	☐ Allow extra room in case the rider swerves to avoid potholes
14.56	☐ Wait until the cyclist has passed the junction
14.57	☐ Allow them to move in front of you
14.58	☐ It's a safer route for a cyclist
14.59	☐ Allow plenty of room
14.60	☐ Brake gently to a stop until they've passed
14.61	☐ Bicycle
14.62	☐ They're easily overlooked
14.63	☐ Other drivers
14.64	☐ Filtering motorcyclists
14.65	☐ Motorcyclists
14.66	☐ In gusting winds
14.67	☐ Be prepared to stop
14.68	☐ By displaying a 'stop' sign
14.69	☐ Stay behind
14.70	☐ Cyclists
14.71	☐ Check your mirrors and prepare to slow down
14.72	☐ Wait patiently and give them time to finish crossing
14.73	☐ Allow extra room, in case they swerve to avoid potholes
14.74	☐ They may be filtering between the lanes

14.75	☐ Slow down and be prepared to stop for a cyclist
14.76	☐ You mustn't wait or park on these lines

14.77	☐ Check for traffic overtaking on your right
14.78	☐ At junctions
14.79	☐ Look for people leaving buses

15. Road and traffic signs

15.1	☐ 50 mph
15.2	☐ 60 mph
15.3	☐ Drive past the sign because it doesn't apply to you
15.4	☐ 
15.5	☐ Find an alternative route to your destination
15.6	☐ Change to a lower gear
15.7	☐ You mustn't use the right-hand lane
15.8	☐ 60 mph (96 km/h)
15.9	☐ When you're carrying a load that overhangs the front or rear of the vehicle by more than 2 metres (6 feet 6 inches)
15.10	☐ Maximum speed 60 mph
15.11	☐ Maximum speed 60 mph
15.12	☐ Maximum speed 50 mph
15.13	☐ Stopping or parking

15.14	☐ End of restricted parking area
15.15	☐ 
15.16	☐ You have priority over vehicles from the opposite direction
15.17	☐ Stop
15.18	☐ Minimum speed 30 mph
15.19	☐ 
15.20	☐ Route for trams
15.21	☐ 
15.22	☐ Contraflow bus lane
15.23	☐ Road narrows

15.24	☐ 
15.25	☐ Side wind
15.26	☐ Danger ahead
15.27	☐ Hold back until you can see clearly ahead
15.28	☐ Water across the road
15.29	☐ Red alone
15.30	☐ At a level crossing
15.31	☐ The distance to the next exit
15.32	☐ No stopping at any time
15.33	☐ The left-hand lane
15.34	☐ 
15.35	☐ 
15.36	☐ To allow cyclists to position in front of other traffic
15.37	☐ The driver intends to turn left
15.38	☐ All lanes closed ahead
15.39	☐ Overhead electrified cable
15.40	☐ Oncoming traffic
15.41	☐ Queues likely

15.42	☐ Loose chippings
15.43	☐ Don't cross the bridge if your weight exceeds the limit
15.44	☐ A concealed level crossing
15.45	☐ No stopping
15.46	☐ On the approach to an arched bridge
15.47	☐ Red and amber, then green
15.48	☐ To overtake a pedal cycle travelling at 10 mph or less
15.49	☐ Slow down; the road may be narrow ahead
15.50	☐ Because children may be crossing
15.51	☐ Emergency diversion route for motorway traffic
15.52	☐ The left-hand lane
15.53	☐ Drivers of trams
15.54	☐ An area where special waiting restrictions apply
15.55	☐ Amber
15.56	☐ They mark the lanes in a contraflow system
15.57	☐ Tramway speed limit
15.58	☐ Flashing amber
15.59	☐ Until the pedestrians have reached a safe position
15.60	☐ The bus lane is in operation 24 hours a day
15.61	☐ Steady amber
15.62	☐ Give way to pedestrians on the crossing

15.63	☐ As a place to stop when an emergency arises
15.64	☐ When you're signalled to do so by flashing red lights
15.65	☐ 
15.66	☐ Right-hand lane closed ahead
15.67	☐ No U-turns
15.68	☐ Tunnel ahead
15.69	☐ Uneven road
15.70	☐ To avoid throwing up loose chippings
15.71	☐ The set speed limit is displayed
15.72	☐ England and Wales
15.73	☐ It's for use in case of emergency or breakdown
15.74	☐ Stop and direct anyone on a motorway
15.75	☐ You shouldn't use this lane as a running lane
15.76	☐ The hard shoulder can be used as a running lane
15.77	☐ To reduce congestion
15.78	☐ The speed limit on the sign must be obeyed
15.79	☐ The hard shoulder is for emergency or breakdown use only
15.80	☐ Permitted lanes, including the hard shoulder
15.81	☐ When signs show that you can
15.82	☐ Soft verges
15.83	☐ No buses
15.84	☐ Taxi driver
15.85	☐ Proceed as normal; the sign doesn't apply to you

15